Geometric Continuity of Curves and Surfaces

Synthesis Lectures on Visual Computing: Computer Graphics, Animation, Computational Photography, and Imaging

Editor
Brian A. Barsky, *University of California, Berkeley*

This series presents lectures on research and development in visual computing for an audience of professional developers, researchers, and advanced students. Topics of interest include computational photography, animation, visualization, special effects, game design, image techniques, computational geometry, modeling, rendering, and others of interest to the visual computing system developer or researcher.

Geometric Continuity of Curves and Surfaces
Przemysław Kiciak
2016

Heterogeneous Spatial Data: Fusion, Modeling, and Analysis for GIS Applications
Giuseppe Patanè and Michela Spagnuolo
2016

Geometric and Discrete Path Planning for Interactive Virtual Worlds
Marcelo Kallmann and Mubbasir Kapadia
2016

An Introduction to Verification of Visualization Techniques
Tiago Etiene, Robert M. Kirby, and Cláudio T. Silva
2015

Virtual Crowds: Steps Toward Behavioral Realism
Mubbasir Kapadia, Nuria Pelechano, Jan Allbeck, and Norm Badler
2015

Finite Element Method Simulation of 3D Deformable Solids
Eftychios Sifakis and Jernej Barbic
2015

Efficient Quadrature Rules for Illumination Integrals: From Quasi Monte Carlo to Bayesian Monte Carlo
Ricardo Marques, Christian Bouville, Luís Paulo Santos, and Kadi Bouatouch
2015

Numerical Methods for Linear Complementarity Problems in Physics-Based Animation
Sarah Niebe and Kenny Erleben
2015

Mathematical Basics of Motion and Deformation in Computer Graphics
Ken Anjyo and Hiroyuki Ochiai
2014

Mathematical Tools for Shape Analysis and Description
Silvia Biasotti, Bianca Falcidieno, Daniela Giorgi, and Michela Spagnuolo
2014

Information Theory Tools for Image Processing
Miquel Feixas, Anton Bardera, Jaume Rigau, Qing Xu, and Mateu Sbert
2014

Gazing at Games: An Introduction to Eye Tracking Control
Veronica Sundstedt
2012

Rethinking Quaternions
Ron Goldman
2010

Information Theory Tools for Computer Graphics
Mateu Sbert, Miquel Feixas, Jaume Rigau, Miguel Chover, and Ivan Viola
2009

Introductory Tiling Theory for Computer Graphics
Craig S.Kaplan
2009

Practical Global Illumination with Irradiance Caching
Jaroslav Krivanek and Pascal Gautron
2009

Wang Tiles in Computer Graphics
Ares Lagae
2009

Virtual Crowds: Methods, Simulation, and Control
Nuria Pelechano, Jan M. Allbeck, and Norman I. Badler
2008

Interactive Shape Design
Marie-Paule Cani, Takeo Igarashi, and Geoff Wyvill
2008

Real-Time Massive Model Rendering
Sung-eui Yoon, Enrico Gobbetti, David Kasik, and Dinesh Manocha
2008

High Dynamic Range Video
Karol Myszkowski, Rafal Mantiuk, and Grzegorz Krawczyk
2008

GPU-Based Techniques for Global Illumination Effects
László Szirmay-Kalos, László Szécsi, and Mateu Sbert
2008

High Dynamic Range Image Reconstruction
Asla M. Sá, Paulo Cezar Carvalho, and Luiz Velho
2008

High Fidelity Haptic Rendering
Miguel A. Otaduy and Ming C. Lin
2006

A Blossoming Development of Splines
Stephen Mann
2006

Geometric Continuity of Curves and Surfaces
Przemysław Kiciak

ISBN: 978-3-031-01462-8 paperback
ISBN: 978-3-031-02590-7 ebook

DOI 10.1007/978-3-031-02590-7

A Publication in the Springer series
SYNTHESIS LECTURES ON VISUAL COMPUTING: COMPUTER GRAPHICS, ANIMATION, COMPUTATIONAL PHOTOGRAPHY, AND IMAGING

Lecture #25
Series Editor: Brian A. Barsky, *University of California, Berkeley*
Series ISSN
Print 2469-4215 Electronic 2469-4223

Geometric Continuity of Curves and Surfaces

Przemysław Kiciak
University of Warsaw

SYNTHESIS LECTURES ON VISUAL COMPUTING: COMPUTER GRAPHICS, ANIMATION, COMPUTATIONAL PHOTOGRAPHY, AND IMAGING #25

ABSTRACT

This book is written for students, CAD system users and software developers who are interested in geometric continuity—a notion needed in everyday practice of Computer-Aided Design and also a hot subject of research. It contains a description of the classical geometric spline curves and a solid theoretical basis for various constructions of smooth surfaces. Textbooks on computer graphics usually cover the most basic and necessary information about spline curves and surfaces in order to explain simple algorithms. In textbooks on geometric design, one can find more details, more algorithms and more theory. This book teaches how various parts of the theory can be gathered together and turned into constructions of smooth curves and smooth surfaces of arbitrary topology.

The mathematical background needed to understand this book is similar to what is necessary to read other textbooks on geometric design; most of it is basic linear algebra and analysis. More advanced mathematical material is introduced using elementary explanations. Reading *Geometric Continuity of Curves and Surfaces* provides an excellent opportunity to recall and exercise necessary mathematical notions and it may be your next step towards better practice and higher understanding of design principles.

KEYWORDS

parametric curves and surfaces, Bézier curves and patches, B-spline curves and patches, Coons patches, cubic splines of interpolation, trigonometric spline functions, geometric splines, ν-splines, β-splines, γ-splines, geometric continuity, tangent line continuity, tangent plane continuity, curvature continuity, torsion continuity, Frenet frame continuity, modules, mesh refinement, compatibility conditions, filling polygonal holes, shape badness measures, shape optimisation, shape functions, shape visualisation, Fàa di Bruno's formula

Contents

Preface

Bézier and B-spline curves and surfaces developed years ago in the automotive industry as a design tool are now omnipresent in CAD systems and in computer graphics. Their most important feature is the ease of designing neatly shaped curve arcs and surface patches. There are many excellent textbooks from which a reader may find their properties and learn how to use them to make a design or how to implement an algorithm for processing curves or surfaces that may be embedded in a graphical program.

Usually textbooks on computer graphics cover only the most elementary information about curves and surfaces. For example, a definition of B-spline curves is followed by their basic properties and algorithms of computing points of the curve, computing derivatives and inserting knots. Readers of textbooks about Computer-Aided Design need to invest a little more in mathematics, but in return they get more advanced material, including e.g. algorithms of degree elevation and computing the Frenet frame and curvatures, or approximation properties of spline curves. Geometric continuity of curves and surfaces is one of the issues described in these books. It is essential when a sophisticated design requires obtaining a smooth surface made of many patches. As it is one of many issues present in the textbooks, geometric continuity is often only touched upon and not studied in detail. On the other hand, geometric continuity is needed in everyday practice of Computer-Aided Design and, therefore, it is a rapidly developing subject of research.

This book is addressed to readers who work or intend to work with spline curves and surfaces. In particular, it is aimed at students who have learned about spline curves and surfaces and want to know more about them, designers willing to understand the possibilities and limitations of their CAD software packages in order to enhance their designs, and authors of graphic design software who will find here theoretical bases of constructions to implement or develop; there are plenty of algorithms, known and yet unknown, whose elements are described in this book. The mathematics needed to understand its contents is elementary linear algebra (vectors, matrices, linear vector spaces, bases, linear and affine transformations, systems of linear equations) and analysis (vector functions, derivatives, Taylor expansions). On the other hand, reading this book is an opportunity to recall and exercise necessary notions, and the more advanced mathematical material needed in some places is given with elementary explanations.

What does a reader get in return? Knowledge. Perhaps the most important purpose of this book is to show a complete path from mathematical theory to the practical constructions of curves and surfaces. The book contains a description of the classical geometric spline curves: ν-splines, β-splines and γ-splines, and a solid theoretical basis for various constructions of smooth surfaces. It consists of an analysis of smooth junctions of pairs of parametric patches, which uses the theory of modules and translates to constructions of such pairs, and an analysis of compatibility conditions at a common corner of patches, using as a tool trigonometric splines, which allow one to find all degrees of freedom for possible constructions of smooth surfaces. An example of an application of this theory is a complete construction of surfaces filling polygonal holes in surfaces of arbitrary topology. This construction contains an optimisation algorithm producing surfaces of high quality. In addition, basic tools for the visualisation of shape of surfaces are discussed.

I would like to thank Prof. Brian Barsky for encouraging me to write this book, for piloting the project and for the amount of patience he found for it. I would also like to thank the reviewers, who spent their time reading an early version of this book and whose remarks helped me to improve it in many places. My special thanks are for my niece Anna Lycett who honed my prose to make it smooth.

Przemysław Kiciak
September 2016

Notation

$\mathbb{R}$ — set (field) of real numbers.

$\mathbb{R}^m$ — set (linear vector space) of real column matrices with m rows.

$\mathbb{R}^{m\times n}$ — set (linear vector space) of real matrices with m rows and n columns.

$\mathbb{R}[\cdot]$ — set (ring) of real polynomials of one variable.

$\mathbb{R}[\cdot]_n$ — set (linear vector space) of polynomials of degree not greater than n.

$\mathbb{R}[\cdot]^m$ — set (module) of real polynomial vector functions in $\mathbb{R}^m$.

$\mathbb{R}[\cdot]^{m\times n}$ — set of $m \times n$ matrices whose entries are real polynomials.

$\mathbb{Z}$ — set of integer numbers.

a, v, α, δ, f — scalars and functions taking scalar values.

$\boldsymbol{a}, \boldsymbol{v}, \boldsymbol{f}$ — points, vectors and functions whose values are points or vectors.

A, B, C — matrices; their rows and columns are numbered from 1.

A^+ — pseudoinverse of a matrix A.

A, B, Ω, Γ — planar areas and curves.

$\mathcal{M}, \boldsymbol{p}$ — surfaces and their parametrisations.

$f \circ g$ — composition of functions, $(f \circ g)(x) = f\big(g(x)\big)$.

$|a|$ — absolute value.

$|\boldsymbol{\alpha}|$ — sum of coordinates of a multi-index (integer vector) $\boldsymbol{\alpha} = (\alpha_1, \dots, \alpha_n)$.

$\Delta \boldsymbol{p}_i, \Delta_1 \boldsymbol{p}_{ij}, \Delta_2 \boldsymbol{p}_{ij}$ — differences, $\Delta \boldsymbol{p}_i = \boldsymbol{p}_{i+1} - \boldsymbol{p}_i$, $\Delta_1 \boldsymbol{p}_{ij} = \boldsymbol{p}_{i+1,j} - \boldsymbol{p}_{ij}$, $\Delta_2 \boldsymbol{p}_{ij} = \boldsymbol{p}_{i,j+1} - \boldsymbol{p}_{ij}$.

$f[u_i, \dots, u_{i+k}]$ — k-th order divided difference.

$f', \frac{\mathrm{d}f}{\mathrm{d}t}, f'', f^{(n)}$ — derivatives of a function of one variable.

$f_u, \frac{\partial f}{\partial u}, f_{uv}, \frac{\partial^2 f}{\partial u \partial v}, \frac{\partial^{|\alpha|} f(t_1,\dots,t_n)}{\partial^{\alpha_1} t_1 \dots \partial^{\alpha_n} t_n}$ — derivatives of a function of two or more variables.

∇f — gradient of a function f.

$\nabla_{\mathcal{M}} f$ — gradient of a function f on a surface $\mathcal{M}$.

Δf — Laplacian of a function f.

$\mathrm{D}\boldsymbol{p}$ — differential of a parametrisation $\boldsymbol{p}$.

$\langle \boldsymbol{u}, \boldsymbol{v} \rangle$ — scalar product of vectors. If $\boldsymbol{u}, \boldsymbol{v} \in \mathbb{R}^m$ then $\langle \boldsymbol{u}, \boldsymbol{v} \rangle = \boldsymbol{v}^T \boldsymbol{u}$.

$\|\boldsymbol{v}\|_2$ — Euclidean norm (or length) of a vector $\boldsymbol{v}$; $\|\boldsymbol{v}\|_2 = \sqrt{\langle \boldsymbol{v}, \boldsymbol{v} \rangle}$.

$\boldsymbol{v}_1 \wedge \dots \wedge \boldsymbol{v}_{m-1}$ — vector product in $\mathbb{R}^m$.

$\ell, \hbar$ — lines and halflines.

Δ — partition of the full angle.

$\mathcal{H}^{(l)}$ — space of homogeneous bivariate polynomials of degree l.

$\mathcal{H}_\Delta^{(l,n)}$ — space of homogeneous bivariate splines of degree l and of class $C^n(\mathbb{R}^2)$.

$\mathcal{T}_\Delta^{(l,n)}$ — space of trigonometric splines of degree l and of class C^n.

B_i^n, N_i^n — Bernstein basis polynomials and normalised B-spline functions of degree n.

$V \oplus W$ — direct algebraic sum of linear spaces V and W.

$f \otimes g$ — tensor product of functions f and g.

$V \otimes W$ — tensor product of linear spaces V and W.

$\mathcal{F}(\boldsymbol{c}_1, \dots, \boldsymbol{c}_k)$ — highest common factor of polynomial curves $\boldsymbol{c}_1, \dots, \boldsymbol{c}_k$.

$\mathcal{M}(\boldsymbol{c}_1, \dots, \boldsymbol{c}_k)$ — module of polynomial curves linearly dependent over $\mathbb{R}[\cdot]$ with $\boldsymbol{c}_1, \dots, \boldsymbol{c}_k$.

$\mathcal{L}(\boldsymbol{c}_1, \dots, \boldsymbol{c}_k)$ — module of curves being linear combinations of $\boldsymbol{c}_1, \dots, \boldsymbol{c}_k$ over $\mathbb{R}[\cdot]$.

CHAPTER 1
Introduction

Smoothness of curves and surfaces is one of the most important attributes of their shape; it is often the first thing noticed and judged by people seeing a new object and taking decisions: *to buy or not to buy*. But the connection between the smoothness and the aesthetic of an object is not the whole story. There are a variety of reasons why objects in technical applications need to be smooth. If a machine part is subject to vibrations, then a notch may initiate cracking caused by fatigue. The influence of shape on aero- and hydrodynamic properties of objects (airplane fuselages, propellers, ship hulls etc.) is obvious. Shape matters also for nonmaterial objects. In computer animation the smoothness of a trajectory of a point often helps to imitate natural motion. Moreover, the smooth motion of a robot part may decrease forces in its mechanism and thus improve its durability.

Shapes of curves and surfaces may be evaluated on two levels. On the macroscopic level one can notice defects such as self-intersections, undulations or flattenings. Possible blemishes on the microscopic level are singular points, ridges or curvature discontinuities. In practical design it is hard (if possible at all) to separate the two levels, but their theoretical issues are different. In this book we focus on the microscopic level, where the notion of geometric continuity is relevant. However, details of shape noticeable on the macroscopic level are touched upon in Chapter 5, where explicit measures of shape quality are used in constructions of smooth surfaces, and in Chapter 6, whose subject is the visualisation of surface shape.

Curves and surfaces may be described mathematically in a number of ways. One possibility is taking the graph of a scalar function of one or two variables. This method restricts the set of available shapes, in particular not allowing to obtain a closed curve.

Another method uses an implicit definition of a planar curve as the set of points, at which a given function of two variables is equal to zero, or a surface, defined as the set of zeros of a function of three variables. This method is popular (perhaps more so in computer graphics than in Computer-Aided Design), as it allows one to design complicated shapes relatively easily, whether by adding terms to the function (a popular example of this technique is known under the name blobs) or by applying Constructive Solid Geometry. On the other hand, the implicit approach has its restrictions. Apart from special (and simple, like a circle or a sphere) cases, points of the curve or surface must be found numerically, by solving nonlinear equations, which is computationally expensive. These costs, with today's computers, are not prohibitive, but they are also not insignificant. To define a curve in space one should specify two scalar functions of three variables; the curve is the intersection of two surfaces defined implicitly by the two functions, but this approach is awkward.

The third possibility is the one studied in this work: the parametric definition. Here we specify a vector function of one (to obtain a curve) or two (to obtain a surface) variables. The definition of the function by an explicit closed formula or by a simple algorithm makes it easy and cheap to find points, and the same formulae may be used both for planar curves and curves located in space. Derivatives of the vector function called parametrisation make it possible to find geometric attributes of curve or surface, like the tangent line or plane, curvatures, etc. The parametric definition also has some drawbacks. The parameters, i.e., arguments of the parametrisation, are often artificial in applications where only the shape matters; on the other hand, sometimes they are natural, especially when the parameter of a curve may be identified with time. Surfaces often have to be described piecewise, by a number of patches, and this is the case even for spheres. Making sure that junctions of the patches are smooth is not a trivial task. How to ensure the smoothness of junctions is the main thread of this work.

Any curve or surface has infinitely many parametrisations; some of them have many derivatives, while others are not even continuous. Geometric properties of a curve or surface, which are intuitively interpreted as "smoothness", occur when parametrisations with certain analytic properties exist. A number of notions related to shapes and their relations with parametrisations were gathered by Veltkamp [1992]. In this book we focus on the notion called geometric continuity, which in practice is the most important of them all.

Definition 1.1 *A curve is said to be of class G^n (has geometric continuity of order n, where $n \geq 1$) if there exists a local regular parametrisation of class C^n of this curve in a neighbourhood of each point of this curve.*

Definition 1.2 *A surface is said to be of class G^n (has geometric continuity of order n, where $n \geq 1$) if there exists a local regular parametrisation of class C^n of this surface in a neighbourhood of each point of this surface.*

Though the continuity of derivatives of a function is essential for the smoothness of its graph, the continuity of derivatives of any particular parametrisation (e.g. the one used as a representation) is neither a necessary nor a sufficient condition for the smoothness of the curve or the surface. This is why regularity is required by the definitions above. A vector function $\boldsymbol{f}$ of k variables is **regular** at a point if its partial derivatives at that point are linearly independent vectors; the derivative of a regular function of one variable is a non-zero vector. The role of regularity may be seen in Figure 1.1, showing two planar curves and their parametrisations. The *graph* of a parametrisation $\boldsymbol{p}(t)$ made of two scalar functions, $x(t)$ and $y(t)$, is a three-dimensional curve, whose parametrisation is obtained by attaching the identity function, $z(t) = t$, to describe the third coordinate. The graphs of the functions $x(t)$ and $y(t)$ are projections of this spatial curve on the xz and yz planes, while our planar curve is the projection on the xy plane.

Both functions, $x(t)$ and $y(t)$, making the parametrisation of the top curve in Figure 1.1, have continuous first- and second-order derivatives. However, the curve has a point of discontinuity of the tangent line. This defect corresponds to the parameter t, at which the derivatives

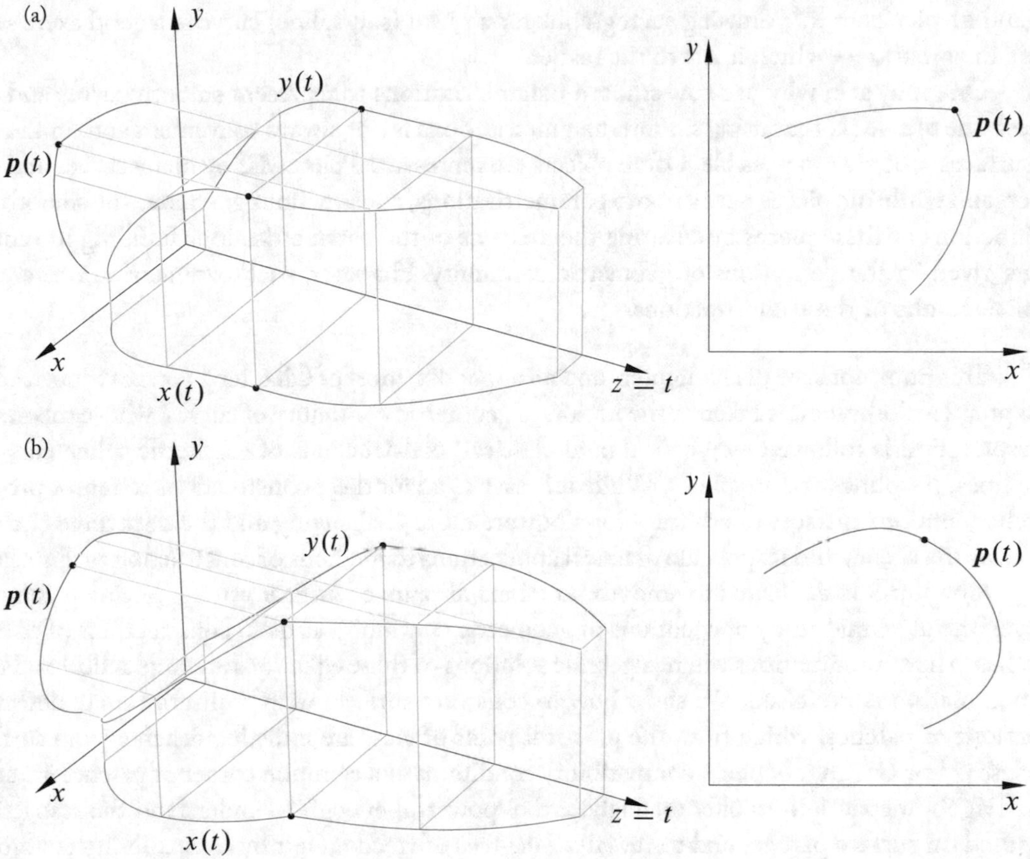

Figure 1.1: (a) A non-smooth curve with a parametrisation of class C^2, (b) A curve of class G^2, whose parametrisation has a discontinuous derivative

of both functions, $x(t)$ and $y(t)$, are zero, i.e., the derivative of the parametrisation, $\boldsymbol{p}'(t)$, is the zero vector. On the other hand, the curve shown below is represented by a parametrisation whose derivative is discontinuous at a certain point. This does not prevent the curve from being smooth—at each of its points the tangent line is defined; moreover, the curvature of this curve is continuous.

Analogous pictures could be drawn for parametric curves and surfaces in the three-dimensional space; however, they would be less obvious, because having never seen any four- or five-dimensional objects (as opposed to three-dimensional curves, like graphs of parametrisations of planar curves, and their projections on planes), one has to rely only on his or her own imagination to reconstruct the scene from a picture drawn on a planar sheet. Graphs of three scalar functions, $x(t)$, $y(t)$ and $z(t)$, are projections of the graph of a parametrisation on the xt,

yt and zt planes in $\mathbb{R}^4$; drawing such graphs for a given (say, spline) curve is a good exercise for one's imagination—which is left to the reader.

One may ask, why use non-smooth parametrisations to represent smooth curves and surfaces? The answer is that using smooth parametrisations is not always convenient and, in the case of surfaces, not always possible. Often objects are represented piecewise by many curves and surfaces, and while the pieces have smooth parametrisations, constructions guarantee the smoothness of junctions of these pieces by ensuring the *existence* of the parametrisations fulfilling the conditions given in the definitions of geometric continuity. However, such parametrisations are *not* explicit results of these constructions.

This book consists of six chapters and an appendix; most of Chapter 1 you have just read. In Chapter 2 a closer look is taken at the notion of geometric continuity of curves, whose theoretical interpretation is followed by practical (and classical) constructions of geometric spline curves—ν-splines, β-splines and γ-splines. While it is easy to adapt these constructions to tensor product patches, smooth surfaces of arbitrary topology are more challenging and the next three chapters describe the theory and its possible practical applications to methods of constructing such surfaces.

Chapter 3 is dedicated to smooth junctions of pairs of surface patches having a common curve. The algebraic study of equations of geometric continuity leads to constructions of pairs of patches. There are situations where algebraic solutions of these equations are impractical and then approximation is exercised. We show how to construct surfaces with arbitrarily small defects at junctions of patches, which from the practical point of view are indistinguishable from surfaces of class G^1 or G^2. In Chapter 4 compatibility conditions at a common corner of patches are studied. Trigonometric splines offer an analytic tool powerful enough to understand the restrictions imposed on surface patches and to reveal *all* degrees of freedom left by compatibility conditions for geometric continuity of any order. In Chapter 5 some methods of filling polygonal holes in surfaces are described as an example of applying theory in practice. A word of warning: such methods are never simple, and the ones in Chapter 5 are no exception. The surfaces filling the hole are obtained by a numerical algorithm whose goal is to minimise undulations and flattenings. The algorithm is a special case of the finite element method; its major part is the construction of a function space from which the algorithm chooses functions to describe the coordinates of the surface parametrisation. The construction of the function space applies most of the theory from the preceding chapters.

Chapter 6 describes basic tools used to visualise shapes of surfaces in order to evaluate their quality. These tools—shape functions—make it possible to show details at the microscopic level, revealing the actual class of geometric continuity of junctions of patches, as well as undulations and flattenings present at the macroscopic level.

Appendix A recalls basic information about interpolation problems, Bézier and B-spline curves and surfaces, surfaces of arbitrary topology represented by meshes, Coons patches, an absolutely minimal dose of differential geometry and Fàa di Bruno's formula, which is omnipresent in theoretical foundations and in practical constructions of smooth curves and surfaces. Though properties of curves and surfaces are known from textbooks and research papers, they are gathered here for the reader's convenience. A number of proofs for these are given in the appendix, either because of direct applications in constructions described in this book, or because many theorems (especially in Chapter 4) are proved using the same ideas, which is worth a special emphasis. Welcome to the lecture on geometric continuity.

CHAPTER 2

Geometric continuity of curves

The definition of geometric continuity has been given, and its connection with the analytical class of curve parametrisation has been explained in the Introduction. In this chapter we study the conditions which ensure the existence of a parametrisation with sufficiently many continuous derivatives of a curve defined piecewise. The basic idea is to make a substitution of the parameter of one arc; a new parametrisation of this arc is thus obtained. The second arc may be constructed to interpolate the derivatives of this new parametrisation. The reparametrisation of the first arc is a function defined with a number of arbitrary constants; their modifications are a means of changing the shape of the second arc.

It is worth recalling that to be successful we need a *regular* parametrisation of the first arc, i.e., its derivative must not be the zero vector; also the function used to reparametrise this arc must have a non-zero derivative. Geometric spline curves, whose practical constructions are described in Section 2.3, do not have to have smooth shapes if some of their control points or points to be interpolated coincide. Nevertheless, if these points are not positioned specifically to cause a singularity (i.e., *almost always*), they determine a smooth curve.

2.1 EQUATIONS OF GEOMETRIC CONTINUITY

Let $\boldsymbol{p}(t)$ be a parametrisation of class C^n whose domain is an interval $[a, t_0]$. Suppose that it is regular and it has at least n continuous derivatives. We can substitute $t = f(u)$, using a monotone function f of class C^n. If $\boldsymbol{q}(u) = \boldsymbol{p}(f(u))$, then by Fàa di Bruno's formula (A.54) we obtain

$$\frac{\mathrm{d}^j}{\mathrm{d}u^j}\boldsymbol{q}(u) = \sum_{k=1}^{j} a_{jk}(u)\frac{\mathrm{d}^k}{\mathrm{d}t^k}\boldsymbol{p}(t), \quad \text{for } j = 1, \dots, n, \tag{2.1}$$

where a_{jk} are functions determined by derivatives of the function f:

$$a_{jk}(u) = \sum_{\substack{m_1+\dots+m_k=j \\ m_1,\dots,m_k>0}} \frac{j!}{k!m_1!\dots m_k!}\frac{\mathrm{d}^{m_1}f(u)}{\mathrm{d}u^{m_1}}\cdots\frac{\mathrm{d}^{m_k}f(u)}{\mathrm{d}u^{m_k}}. \tag{2.2}$$

In particular,

$$\begin{aligned}
\boldsymbol{q}'(u) &= f'(u)\boldsymbol{p}'(t),\\
\boldsymbol{q}''(u) &= f''(u)\boldsymbol{p}'(t) + f'^2(u)\boldsymbol{p}''(t),\\
\boldsymbol{q}'''(u) &= f'''(u)\boldsymbol{p}'(t) + 3f'(u)f''(u)\boldsymbol{p}''(t) + f'^3(u)\boldsymbol{p}'''(t) \quad \text{etc.}
\end{aligned}$$

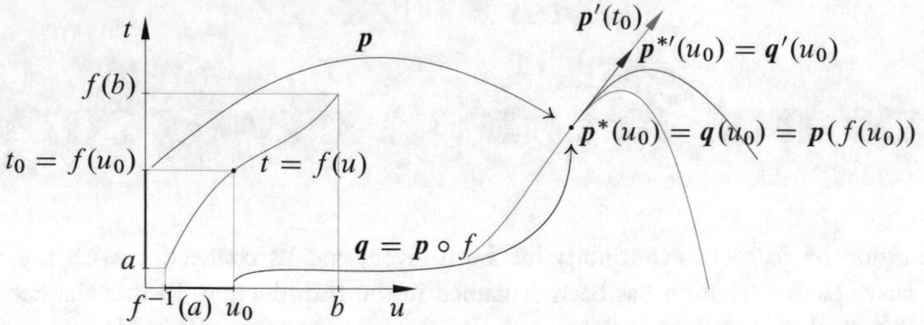

Figure 2.1: A smooth junction of two parametric curves

Suppose that $\boldsymbol{p}(t)$ is given and the goal is to construct a parametrisation $\boldsymbol{p}^*(u)$ of class C^n defined in an interval $[u_0, b]$ such that the arcs described by the two parametrisations form a curve of class G^n. To achieve this goal we impose some restrictions on the function f. To avoid singularities and cusps we need a function f increasing monotonically and regular; hence, the derivative of f ought to be positive. Such a function f has its inverse, f^{-1}. By taking f such that $f(u_0) = t_0$, the counterimage of the interval $[a, t_0]$ (i.e., the image of this interval under the mapping f^{-1}) will adhere to $[u_0, b]$ (see Figure 2.1), thus forming the domain $[f^{-1}(a), b]$ of the piecewise defined parametrisation

$$\boldsymbol{s}(u) = \begin{cases} \boldsymbol{p}\big(f(u)\big) & \text{if } u \in [f^{-1}(a), u_0), \\ \boldsymbol{p}^*(u) & \text{if } u \in [u_0, b]. \end{cases} \tag{2.3}$$

Then we specify the interpolation conditions for the parametrisation $\boldsymbol{p}^*$ at u_0: to obtain a continuous curve, we need

$$\boldsymbol{p}^*(u_0) = \boldsymbol{q}(u_0) = \boldsymbol{p}(t_0).$$

To obtain a curve of class G^1, we also need

$$\boldsymbol{p}^{*\prime}(u_0) = \boldsymbol{q}'(u_0) = f'(u_0)\boldsymbol{p}'(t_0).$$

To shorten the notation, below we use underlining and overlining to denote the evaluation of $\boldsymbol{p}^*$ at u_0 and $\boldsymbol{p}$ at t_0 respectively. The number t_k is the k-th order derivative of the function f at u_0. With this notation the last two equations take the form

$$\underline{\boldsymbol{p}^*} = \overline{\boldsymbol{p}}, \tag{2.4}$$
$$\underline{\boldsymbol{p}^{*\prime}} = t_1\overline{\boldsymbol{p}}'. \tag{2.5}$$

If the two equations are satisfied, then the arcs, whose parametrisations are $\boldsymbol{p}$ and $\boldsymbol{p}^*$, form a curve of class G^1. To obtain a curve of class G^2, we also need

$$\underline{\boldsymbol{p}^{*\prime\prime}} = t_2\overline{\boldsymbol{p}}' + t_1^2\overline{\boldsymbol{p}}'', \tag{2.6}$$

and geometric continuity of order 3 occurs if the following equation

$$\underline{p}^{*'''} = t_3\overline{p}' + 3t_1t_2\overline{p}'' + t_1^3\overline{p}''', \tag{2.7}$$

is satisfied together with the preceding three. Note that both parametrisations, $\boldsymbol{s}$ defined by (2.3) and the composition $\boldsymbol{s} \circ f^{-1}$ (whose domain is $[a, f(b)]$), are regular and have the required number of continuous derivatives.

To construct a parametrisation $\boldsymbol{p}^*$ such that the curve described by $\boldsymbol{p}$ and $\boldsymbol{p}^*$ is of class G^n we do not need a *function* f; what counts are the values of its derivatives of order $1, \dots, n$ at u_0. We can, therefore, choose arbitrary *numbers* $t_1, \dots, t_n$, which become **connection parameters**, use them to construct the *vectors*, being the derivatives of $\boldsymbol{q}$ at v_0, and then construct a parametrisation $\boldsymbol{p}^*$ satisfying the $n+1$ Hermite interpolation conditions at u_0. In constructions of curves made of a number of arcs, interpolation conditions will also be imposed at the point b, i.e., the other end of the domain of $\boldsymbol{p}^*$. But here we focus on junctions of just two arcs.

It is illustrative to write the relation between derivatives of the arcs $\boldsymbol{p}$ and $\boldsymbol{p}^*$ in matrix form. According to (2.4)–(2.6) and (2.4)–(2.7), for junctions of class G^2 and G^3 there is

$$\begin{bmatrix} \underline{p}^* \\ \underline{p}^{*\prime} \\ \underline{p}^{*\prime\prime} \end{bmatrix} = \begin{bmatrix} 1 & 0 & 0 \\ 0 & t_1 & 0 \\ 0 & t_2 & t_1^2 \end{bmatrix} \begin{bmatrix} \overline{p} \\ \overline{p}' \\ \overline{p}'' \end{bmatrix}, \quad \begin{bmatrix} \underline{p}^* \\ \underline{p}^{*\prime} \\ \underline{p}^{*\prime\prime} \\ \underline{p}^{*\prime\prime\prime} \end{bmatrix} = \begin{bmatrix} 1 & 0 & 0 & 0 \\ 0 & t_1 & 0 & 0 \\ 0 & t_2 & t_1^2 & 0 \\ 0 & t_3 & 3t_1t_2 & t_1^3 \end{bmatrix} \begin{bmatrix} \overline{p} \\ \overline{p}' \\ \overline{p}'' \\ \overline{p}''' \end{bmatrix}.$$

Matrices, which describe the relation between the derivatives of the arcs are called **connection matrices.** For geometric continuity of higher orders they may be obtained by appending more rows and columns, with the coefficients $a_{jk} = a_{jk}(u_0)$ from Formula (2.2). Such connection matrices are always lower triangular and non-singular, as their diagonal coefficients are consecutive powers of t_1, which is non-zero.

Remark. Connection matrices may be used to describe and define classes of continuity of junctions of curves other than G^n, e.g. the Frenet frame continuity considered in Section 2.2. Interested readers may find more information on this in the paper by Habib and Goldman [1996].

Now we consider junctions of arcs given in the homogeneous form (see Section A.7), representing rational curves in the space of dimension d, which is a plane if $d = 2$, and $d = 3$ in the majority of other practical situations. A homogeneous representation is a curve in the space of dimension $d + 1$. The coordinates of a parametrisation $\boldsymbol{p}$ in $\mathbb{R}^d$ are obtained by dividing the first d coordinates of the homogeneous parametrisation $\boldsymbol{P}$ by its last coordinate, called the **weight** coordinate and denoted by W; this is equivalent to the perspective projection of the homogeneous curve (with the centre at $\boldsymbol{0} \in \mathbb{R}^{d+1}$) on the plane $W = 1$.

An immediate observation is that by multiplying a homogeneous parametrisation by a non-zero function we obtain another homogeneous representation of the same rational curve. To cover this possibility by our analysis we consider a homogeneous parametrisation $\boldsymbol{P}$ over the interval $[a, t_0]$ and a parametrisation $\boldsymbol{Q}(u) = r(u)\boldsymbol{P}(f(u))$, obtained by a substitution of the parameter and by rescaling, with a non-zero function r of class C^n. Using the Leibniz formula for the derivatives of a product of functions, we obtain

$$\frac{\mathrm{d}^j}{\mathrm{d}u^j}\boldsymbol{Q}(u) = \sum_{i=0}^{j}\binom{j}{i}\frac{\mathrm{d}^{j-i}}{\mathrm{d}u^{j-i}}r(u)\frac{\mathrm{d}^i}{\mathrm{d}u^i}\boldsymbol{P}(f(u)), \quad \text{for } j = 0,\dots,n.$$

The i-th order derivative of $\boldsymbol{P}(f(u))$ with respect to u may be obtained with Fàa di Bruno's formula; after using it and doing some calculation, we obtain

$$\frac{\mathrm{d}^j}{\mathrm{d}u^j}\boldsymbol{Q}(u) = \frac{\mathrm{d}^j r}{\mathrm{d}u^j}\boldsymbol{P} + \sum_{k=1}^{j} A_{jk}\frac{\mathrm{d}^k}{\mathrm{d}t^k}\boldsymbol{P},$$

where

$$A_{jk} = \sum_{i=k}^{j}\binom{j}{i}\frac{\mathrm{d}^{j-i}r}{\mathrm{d}u^{j-i}} \sum_{\substack{m_1+\dots+m_k=i\\ m_1,\dots,m_k>0}} \frac{i!}{k!m_1!\dots m_k!}\frac{\mathrm{d}^{m_1}f}{\mathrm{d}u^{m_1}}\cdots\frac{\mathrm{d}^{m_k}f}{\mathrm{d}u^{m_k}}.$$

As before, we can use this formula to construct as many derivative vectors of the homogeneous parametrisation $\boldsymbol{P}^*$ of a rational curve at the parameter u_0 corresponding to the junction point as needed. In particular, to ensure geometric continuity of order 1, 2 or 3, we use the first 2, 3 or all 4 formulae of the following:

$$\underline{\boldsymbol{P}}^* = r_0\overline{\boldsymbol{P}}, \tag{2.8}$$
$$\underline{\boldsymbol{P}}^{*\prime} = r_1\overline{\boldsymbol{P}} + r_0t_1\overline{\boldsymbol{P}}', \tag{2.9}$$
$$\underline{\boldsymbol{P}}^{*\prime\prime} = r_2\overline{\boldsymbol{P}} + (r_0t_2 + 2r_1t_1)\overline{\boldsymbol{P}}' + r_0t_1^2\overline{\boldsymbol{P}}'', \tag{2.10}$$
$$\underline{\boldsymbol{P}}^{*\prime\prime\prime} = r_3\overline{\boldsymbol{P}} + (r_0t_3 + 3r_1t_2 + 3r_2t_1)\overline{\boldsymbol{P}}' + (3r_0t_1t_2 + 3r_1t_1^2)\overline{\boldsymbol{P}}'' + r_0t_1^3\overline{\boldsymbol{P}}''', \tag{2.11}$$

where $r_0,\dots,r_n$ and $t_1,\dots,t_n$ may be arbitrary numbers with the constraint that $r_0 \neq 0$ and $t_1 > 0$. Having obtained the derivative vectors, we can construct the homogeneous parametrisation $\boldsymbol{P}^*$ by solving a Hermite interpolation problem.

Junctions of rational arcs in the homogeneous representations may also be described using connection matrices. For example, the description of a junction of class G^2 is

$$\begin{bmatrix} \underline{\boldsymbol{P}}^* \\ \underline{\boldsymbol{P}}^{*\prime} \\ \underline{\boldsymbol{P}}^{*\prime\prime} \end{bmatrix} = \begin{bmatrix} r_0 & 0 & 0 \\ r_1 & r_0t_1 & 0 \\ r_2 & r_0t_2 + 2r_1t_1 & r_0t_1^2 \end{bmatrix} \begin{bmatrix} \overline{\boldsymbol{P}} \\ \overline{\boldsymbol{P}}' \\ \overline{\boldsymbol{P}}'' \end{bmatrix}.$$

Exercise. Extend the formula above to describe the case G^3.

2.2 INTERPRETATION

To better understand the notion of geometric continuity and the equations derived in the previous section, we introduce the following notion:

Definition 2.1 *A curve has the continuity of order n of the Frenet frame (is of class F^n) if it has a parametrisation such that the functions $\boldsymbol{e}_1, \dots, \boldsymbol{e}_n$, which describe the vectors of the curve's Frenet frame, and the curvatures $\kappa_1, \dots, \kappa_{n-1}$ are continuous.*

The Frenet frame in a space of dimension d is made of d vectors (see Section A.10.1), and only $d - 1$ curvatures may be defined, which means that a curve in such a space may be of class at most F^d. The notions of geometric continuity and Frenet frame continuity are, therefore, different. But it is illustrative to compare the two properties. So, let $\boldsymbol{p}$ and $\boldsymbol{p}^*$ be two regular parametrisations of class C^∞ of two arcs in $\mathbb{R}^2$ or $\mathbb{R}^3$ which form a curve. We consider a neighbourhood of the junction point of the two arcs.

Case $n = 1$. If the curve is of class G^1, then the arcs at the junction point have a common tangent line determined by the derivative vectors of both parametrisations. To avoid a cusp on the curve, these vectors must have the same orientation. This is the case if $t_1 > 0$ in Equation (2.5) or (2.9).

The F^1 continuity condition involves only the vector $\boldsymbol{e}_1$ of the Frenet frame—the tangent vector, which must be the same for both arcs at the junction point. Thus, the notions of F^1 and G^1 continuity are identical; they both mean the **continuity of direction of the tangent line of the curve.**

Case $n = 2$. If Equation (2.5) is satisfied with $t_1 > 0$, then the vectors $\overline{\boldsymbol{p}}'$ and $\underline{\boldsymbol{p}}^{*\prime}$ have the same direction and orientation. To satisfy Equation (2.6) the pairs of vectors $\overline{\boldsymbol{p}}', \overline{\boldsymbol{p}}''$ and $\underline{\boldsymbol{p}}^{*\prime}, \underline{\boldsymbol{p}}^{*\prime\prime}$ must span the same plane, and then both pairs of vectors determine the same vectors $\boldsymbol{e}_1$ and $\boldsymbol{e}_2$ of the Frenet frame at the junction point. Also, the first curvature $\kappa_1 = \kappa$ depends on the first two derivatives of the parametrisation in a continuous way. Therefore, a planar curve of class G^2 is also of class F^2.[1]

On the other hand, if a curve is of class F^2, then it is easy to prove that its arc length parametrisation is regular and of class C^2. Thus, the classes F^2 and G^2 of planar curves are identical and in a higher dimension the difference between the two classes is negligible. **Curvature continuity** is a property characteristic for curves of these classes.

Case $n = 3$. The F^3 continuity may be considered for three- or higher-dimensional curves with non-zero curvature. The derivatives of order 1 and 2 at the junction point of two arcs forming a curve of class G^2 determine the parameters t_1 and t_2; thus, the only degree of freedom in Equation (2.7) is in the choice of the parameter t_3 used to construct the vector $\underline{\boldsymbol{p}}^{*\prime\prime\prime}$.

[1]In a space of dimension 3 or higher, if the first curvature is zero, the vector $\boldsymbol{e}_2$ is undefined and the curve does not fulfill the definition of F^2 continuity.

The difference of vectors corresponding to different values of t_3 is parallel to $\overline{\boldsymbol{p}}'$, i.e., to the tangent line at the junction point. Taking into account the construction of the Frenet frame (via the Gram–Schmidt orthonormalisation, see Section A.10.1), we conclude that if a curve is of class G^3 and its curvature is non-zero, then it is also of class F^3.

The vectors $\boldsymbol{e}_1$ (tangent), $\boldsymbol{e}_2$ (normal) and $\boldsymbol{e}_3$ (binormal) of the Frenet frame and also the curvatures $\kappa_1 = \kappa$ (curvature) and $\kappa_2 = \tau$ (torsion) will not be changed if any vector in the osculating plane, spanned by $\boldsymbol{e}_1$ and $\boldsymbol{e}_2$, is added to $\underline{\boldsymbol{p}}^{*'''}$. To construct a curve of class F^3, instead of (2.7) we could use the formula

$$\underline{\boldsymbol{p}}^{*'''} = t_3\overline{\boldsymbol{p}}' + s_3\overline{\boldsymbol{p}}'' + t_1^3\overline{\boldsymbol{p}}''',$$

with arbitrary parameters t_3 and s_3. As this construction has two degrees of freedom, F^3 is a wider class of curves than G^3. In both cases the second curvature, i.e., torsion, is continuous.[2]

All equations of geometric continuity of junctions of two arcs have the form

$$\underline{\boldsymbol{p}}^{*(j)} = t_j\overline{\boldsymbol{p}}' + \boldsymbol{q}_j, \quad j = 1, \ldots, n, \tag{2.12}$$

where $\boldsymbol{q}_j$ is a vector determined by the parameters $t_1, \ldots, t_{j-1}$ and the vectors $\overline{\boldsymbol{p}}'', \ldots, \overline{\boldsymbol{p}}^{(j)}$ (in particular $\boldsymbol{q}_1 = \boldsymbol{0}$), i.e., derivatives of the parametrisation of the first arc. In the construction of the second arc, where G^n continuity of the junction is to be achieved, after constructing the derivatives of $\boldsymbol{p}^*$ of order less than n we can arbitrarily choose only one parameter, t_n. Equivalently, we can choose arbitrarily a vector $\overline{\boldsymbol{p}}^{*(n)}$, provided that the difference $\underline{\boldsymbol{p}}^{*(n)} - \boldsymbol{q}_n$ is parallel to the tangent line of the arcs at the junction point.

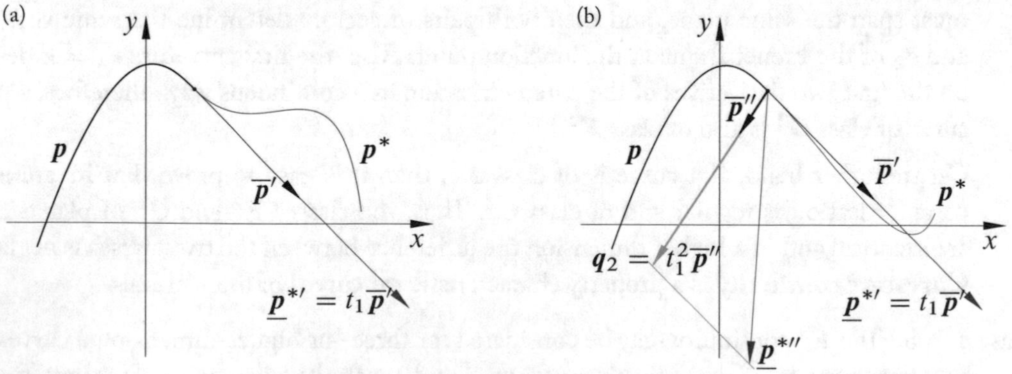

Figure 2.2: Geometric continuity at the junction point of two cubic curves

[2]The connection matrix, which describes a junction of class F^3 of two arcs, is lower-triangular, with consecutive powers of a number t_1 on the diagonal and, except for the first column, arbitrary coefficients below the diagonal. The matrices, which characterise the Frenet frame continuity in higher-dimensional spaces have the same structure.

Figure 2.2 shows two examples of junctions of two curves. In both cases we have two cubic curves with the parameter intervals of length 1. The junction in Figure 2.2a is of class G^1, where the parameter t_1 is equal to 2. The derivative vector of the curve $\boldsymbol{p}^*$ is twice longer than that of the curve $\boldsymbol{p}$ at the junction point. In Figure 2.2b there is a junction of class G^2, with the parameters $t_1 = 2$ and $t_2 = 0.9$; there is $\boldsymbol{q}_2 = t_1^2\overline{\boldsymbol{p}}''$ and the vector $\boldsymbol{q}_2 - \underline{\boldsymbol{p}}^{*\prime\prime}$ has the direction of the line tangent to both curves, which at their common point have the same curvature.

Equations for homogeneous arcs have a similar form:

$$\underline{\boldsymbol{P}}^{*(j)} = r_j\underline{\boldsymbol{P}} + r_0 t_j\underline{\boldsymbol{P}}' + \boldsymbol{Q}_j, \quad j = 1, \dots, n, \tag{2.13}$$

with the vector $\boldsymbol{Q}_j$ determined by the parameters $r_0, \dots, r_{j-1}$ and $t_1, \dots, t_{j-1}$ and the vectors $\overline{\boldsymbol{P}}'', \dots, \overline{\boldsymbol{P}}^{(j)}$. Choosing $\underline{\boldsymbol{P}}^{*(n)}$, we have two degrees of freedom, in the choice of r_n and t_n. The difference $\underline{\boldsymbol{P}}^{*(n)} - \boldsymbol{Q}_n$ must be parallel to the plane spanned by the vectors $\underline{\boldsymbol{P}}$ and $\underline{\boldsymbol{P}}'$. This is illustrated in Figure 2.3, where planar rational curves (in the plane $W = 1$) are shown together with their homogeneous representations. In Figure 2.3a there is a junction of class G^1; it is obtained by making sure that the vector $\underline{\boldsymbol{P}}^{*\prime}$ is a linear combination of $\overline{\boldsymbol{P}}$ and $\overline{\boldsymbol{P}}'$. Dashed lines are used to show a square in the plane $W = 1$ and a rectangle in the linear subspace spanned by the vectors $\overline{\boldsymbol{P}}$ and $\overline{\boldsymbol{P}}'$. The intersection of this subspace with the plane $W = 1$ is the line tangent to the rational arcs at the junction point.

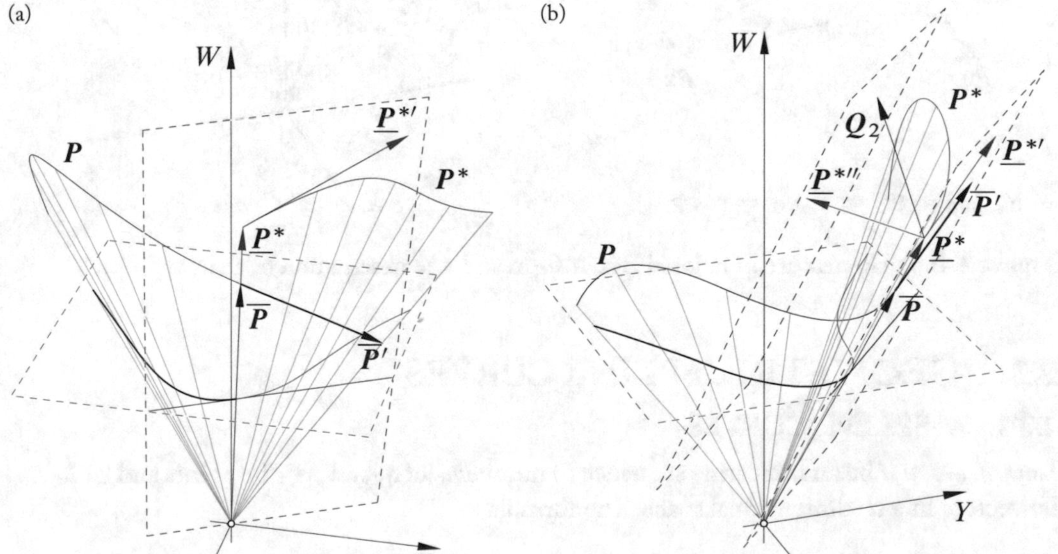

Figure 2.3: Geometric continuity at the junction point in homogeneous representation

The junction of class G^2 in Figure 2.3b is shown together with the vectors $\boldsymbol{Q}_2 = r_0 t_1^2 \overline{\boldsymbol{P}}''$ and $\underline{\boldsymbol{P}}^{*\prime\prime}$. With the first-order derivatives of the homogeneous curves fixed (in the way ensuring the G^1 continuity of the junction of the rational arcs), to obtain the G^2 continuity it suffices to choose the vector $\underline{\boldsymbol{P}}^{*\prime\prime}$ in such a way that the difference $\boldsymbol{Q}_2 - \underline{\boldsymbol{P}}^{*\prime\prime}$ be a linear combination of $\overline{\boldsymbol{P}}$ and $\overline{\boldsymbol{P}}'$. More information about smooth junctions of rational curves may be found in Geise and Jüttler [1993].

Exercise. A circular arc may be represented by a quadratic rational Bézier curve (see Section A.7), as shown in Figure 2.4. In particular, the full circle may be made of four such quarters, and then it is a quadratic rational spline curve of class G^∞. Considering a junction of two quarters whose domains are adjacent intervals of length 1, find the parameters $r_0, r_1, t_1, r_2, t_2, r_3, t_3$, which appear in Equations (2.8)–(2.11).

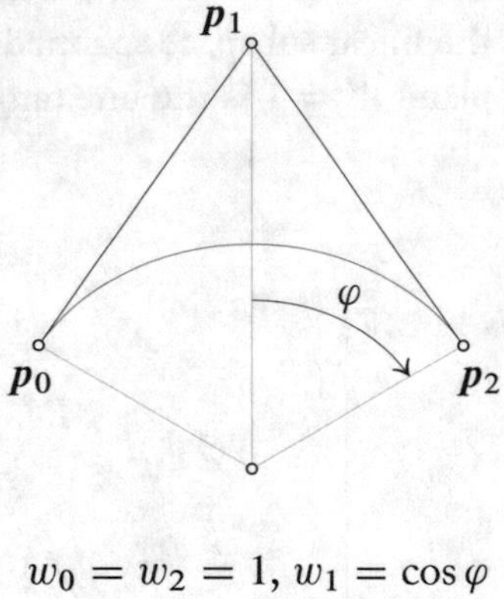

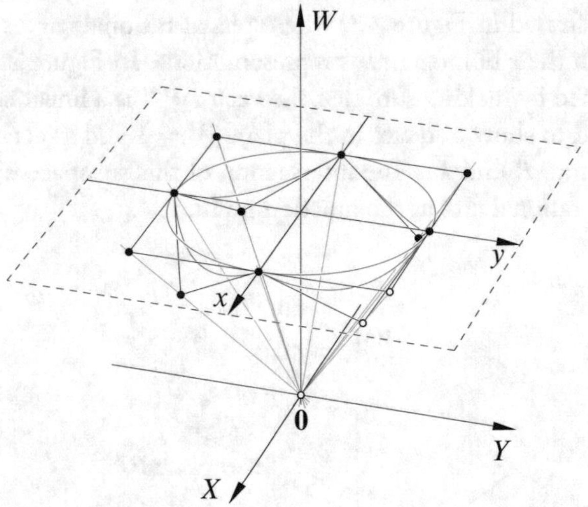

Figure 2.4: A circular arc in rational Bézier form and a representation of a circle

2.3 GEOMETRIC SPLINE CURVES

2.3.1 ν-SPLINE CURVES

Let $u_0, \ldots, u_N$ be an increasing sequence of numbers, let $\boldsymbol{a}_0, \ldots, \boldsymbol{a}_N$ be points and let $\boldsymbol{b}_0, \ldots, \boldsymbol{b}_N$ be vectors in a d-dimensional space. The formula

$$\boldsymbol{s}(t) = \boldsymbol{a}_i H_{i,00}(t) + \boldsymbol{b}_i H_{i,01}(t) + \boldsymbol{a}_{i+1} H_{i,10}(t) + \boldsymbol{b}_{i+1} H_{i,11}(t) \quad \text{if } t \in [u_i, u_{i+1}],$$

where $H_{i,00}, H_{i,01}, H_{i,10}, H_{i,11}$ are cubic polynomials defined by Formulae (A.39) and (A.41), defines a cubic spline parametrisation such that $\boldsymbol{s}(u_i) = \boldsymbol{a}_i$, $\boldsymbol{s}'(u_i) = \boldsymbol{b}_i$ for $i = 0, \ldots, N$. The numbers $u_0, \ldots, u_N$ are both knots of the spline and interpolation knots. From the considerations in Section A.8 it follows that for any choice of the points $\boldsymbol{a}_0, \dots, \boldsymbol{a}_N$ and vectors $\boldsymbol{b}_0, \dots, \boldsymbol{b}_N$ such a parametrisation is of class C^1.

The polynomial arcs of the curve $\boldsymbol{s}$ meeting at the knot u_i, denoted by $\boldsymbol{p}_{i-1}$ and $\boldsymbol{p}_i$, are determined by the points and vectors $\boldsymbol{a}_{i-1}, \boldsymbol{b}_{i-1}, \boldsymbol{a}_i, \boldsymbol{b}_i$ and $\boldsymbol{a}_i, \boldsymbol{b}_i, \boldsymbol{a}_{i+1}, \boldsymbol{b}_{i+1}$. There is

$$\begin{aligned} \boldsymbol{p}''_{i-1}(u_i) &= \frac{6}{h^2_{i-1}}(\boldsymbol{a}_{i-1} - \boldsymbol{a}_i) + \frac{2}{h_{i-1}}(\boldsymbol{b}_{i-1} + 2\boldsymbol{b}_i), \\ \boldsymbol{p}''_i(u_i) &= \frac{6}{h^2_i}(\boldsymbol{a}_{i+1} - \boldsymbol{a}_i) - \frac{2}{h_i}(2\boldsymbol{b}_i + \boldsymbol{b}_{i+1}), \end{aligned}$$

where $h_i = u_{i+1} - u_i$. Assuming that $\boldsymbol{p}''_i(u_i) = \boldsymbol{p}''_{i-1}(u_i)$, we can derive the equation[3]

$$\begin{aligned} h_i\boldsymbol{b}_{i-1} + 2(h_{i-1} + h_i)\boldsymbol{b}_i + h_{i-1}\boldsymbol{b}_{i+1} &= \\ 3\Big(\frac{h_i}{h_{i-1}}(\boldsymbol{a}_i - \boldsymbol{a}_{i-1}) &+ \frac{h_{i-1}}{h_i}(\boldsymbol{a}_{i+1} - \boldsymbol{a}_i)\Big). \end{aligned} \tag{2.14}$$

By solving the system of equations (2.14) for $i = 1, \dots, N-1$ (with unknown vectors $\boldsymbol{b}_0, \dots, \boldsymbol{b}_N$), complemented with suitable end conditions (see Section A.8), we can construct cubic spline curves of interpolation of class C^2.

A modification of this construction, proposed by Nielson [1974], allows us to obtain cubic arcs which make a parametrisation $\boldsymbol{s}$ of class C^1 of a curve of class G^2. The modified assumption is

$$\boldsymbol{p}''_i(u_i) = \boldsymbol{p}''_{i-1}(u_i) + \nu_i\boldsymbol{p}'_{i-1}(u_i).$$

This is in fact Equation (2.6) with the parameters $t_1 = 1$ and $t_2 = \nu_i$. After multiplying the sides by $h_{i-1}h_i/2$ and gathering the terms with the points $\boldsymbol{a}_{i-1}, \boldsymbol{a}_i, \boldsymbol{a}_{i+1}$ and vectors $\boldsymbol{b}_{i-1}, \boldsymbol{b}_i, \boldsymbol{b}_{i+1}$ on the opposite sides of the equality sign we obtain the following equation:

$$\begin{aligned} h_i\boldsymbol{b}_{i-1} + \big(2(h_{i-1} + h_i) + h_{i-1}h_i\nu_i/2\big)\boldsymbol{b}_i + h_{i-1}\boldsymbol{b}_{i+1} &= \\ 3\Big(\frac{h_i}{h_{i-1}}(\boldsymbol{a}_i - \boldsymbol{a}_{i-1}) &+ \frac{h_{i-1}}{h_i}(\boldsymbol{a}_{i+1} - \boldsymbol{a}_i)\Big). \end{aligned} \tag{2.15}$$

The curves obtained by solving Equations (2.15) instead of (2.14) are named **ν-spline curves.**

Each junction of polynomial arcs has its parameter ν_i, which may be chosen individually. In practice one can begin with $\nu_i = 0$ for all i and then modify the parameters interactively until a satisfactory result is obtained. Increasing ν_i increases the curvature of the curve in a vicinity of the point $\boldsymbol{a}_i$. Moreover, if all parameters ν_i tend to infinity, the vectors $\boldsymbol{b}_i$ tend to $\boldsymbol{0}$ and the curve tends to the polyline whose vertices are the points $\boldsymbol{a}_i$.

Figure 2.5 shows an example: two closed curves with equidistant knots ($u_i = i$ for all i). The sharpness of the teeth of the "comb" outlined by the ordinary cubic spline curve of interpolation (a)

[3]This is Equation (A.42), rewritten with scalars replaced by points and vectors.

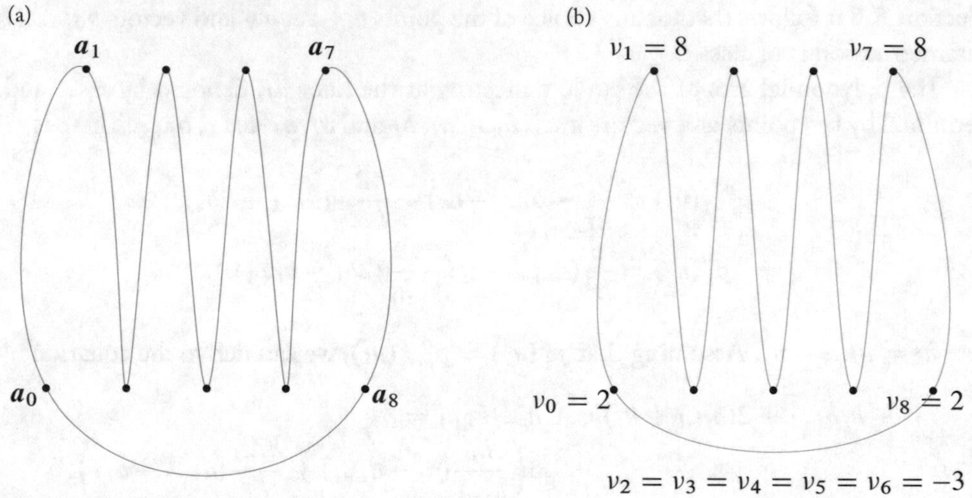

Figure 2.5: A cubic spline curve of interpolation of class C^2 (a) and a ν-spline curve (b)

varies, which may not be the designer's intention. Though it is possible to modify the shape by changing the knots, it is hard, if possible at all, to control the sharpness of the teeth in this way.[4] The curve (a) is a special case of a ν-spline curve with all parameters $\nu_i = 0$. The curve (b) is a result of the modification of the parameters. As the parametrisation is regular (i.e., the first-order derivative at each point is not the zero vector), the curvature of the entire curve is continuous.

2.3.2 β-SPLINE CURVES

A generalisation of B-spline curves called β-spline curves was invented by Barsky [1981], who (with DeRose in [1984]) coined the name "geometric continuity" for the subject of this book. The parametrisation of a β-spline curve is continuous, made of polynomial arcs of degree $n > 0$ and (if the arcs are regular) the curve is of class G^{n-1}. The representation of the curve has parameters, traditionally denoted by $\beta_{l,1}, \ldots, \beta_{l,n-1}$, associated with each junction of the arcs. In addition to changing knots and control points, modifications of the curve may be done by changing these parameters.

Definition

A β-**spline curve** of degree n in a d-dimensional space is described by the formula

$$s(t) = \sum_{i=0}^{N-n-1} d_i P_i^n(t), \quad t \in [u_n, u_{N-n}]. \tag{2.16}$$

[4]Often a cubic spline curve of interpolation with a good shape may be obtained after taking such a knot sequence that $h_i = u_{i+1} - u_i$ is proportional to the square root of the distance between the points a_i and a_{i+1} for all i. In the example discussed here this idea produces equidistant knots, and the result may be seen.

Here $N > 2n$ is the number of the last element of the increasing knot sequence $u_0, \dots, u_N$, and $\boldsymbol{d}_0, \dots, \boldsymbol{d}_{N-n-1} \in \mathbb{R}^d$ are the **control points**, often displayed as vertices of a polyline called the **control polygon**. The functions P_i^n, called **β-spline functions**, are defined by four properties:

(i) The function P_i^n takes non-zero values in the interval (u_i, u_{i+n+1}).

(ii) In each interval $[u_l, u_{l+1}] \subset [u_i, u_{i+n+1}]$ the function P_i^n is a polynomial of degree at most n, which we denote $p_{i,l}$.

(iii) The sum of the functions $P_0^n, \dots, P_{N-n-1}^n$ in the interval $[u_n, u_{N-n}]$ is equal to 1.

(iv) The polynomials $p_{i,l-1}$ and $p_{i,l}$, for all i and l, satisfy the following relation:

$$\begin{bmatrix} p_{i,l}(u_l) \\ p'_{i,l}(u_l) \\ \vdots \\ p_{i,l}^{(n-1)}(u_l) \end{bmatrix} = \begin{bmatrix} 1 & 0 & \dots & 0 \\ 0 & a_{l,1,1} & \ddots & \vdots \\ \vdots & \vdots & \ddots & 0 \\ 0 & a_{l,n-1,1} & \dots & a_{l,n-1,n-1} \end{bmatrix} \begin{bmatrix} p_{i,l-1}(u_l) \\ p'_{i,l-1}(u_l) \\ \vdots \\ p_{i,l-1}^{(n-1)}(u_l) \end{bmatrix}, \tag{2.17}$$

where

$$a_{l,j,k} = \sum_{\substack{m_1+\dots+m_k=j \\ m_1,\dots,m_k>0}} \frac{j!}{k!m_1!\dots m_k!} \beta_{l,m_1} \dots \beta_{l,m_k}. \tag{2.18}$$

In a short form (2.17) may be written as $\boldsymbol{q}_{i,l} = C_l \boldsymbol{p}_{i,l}$; a distinct connection matrix C_l is associated with each knot u_l and the same matrices are involved in the description of all functions P_i^n.

A proof of the existence and the uniqueness of functions having these properties (for almost all choices of the parameters β_{lj}) is given in Seroussi and Barsky [1991], together with an algorithm for constructing these functions based on a symbolic calculation. Below a numerical algorithm is described; a theoretical discussion follows the algorithm.

The first three of the properties above are shared with normalised B-spline functions; see Properties A.6, A.8 and A.10 in Section A.4.1. Each function P_i^n is described by $n+1$ non-zero polynomials, $p_{i,i}, \dots, p_{i,i+n}$. We can notice that $P_{l-n}^n(t) + \dots + P_l^n(t) = 1$ for $t \in [u_l, u_{l+1}]$. The connection matrices C_l are lower triangular and their coefficients are determined by the parameters $\beta_{l,1} > 0$ and $\beta_{l,2}, \dots, \beta_{l,n-1} \in \mathbb{R}$. Formula (2.18) is obtained from (2.2), and the parameter $\beta_{l,m}$ is the m-th order derivative of a reparametrisation function f considered in Section 2.1 (where symbols $t_1, t_2, \dots$ were used). If $\beta_{l,1} = 1$ and $\beta_{l,j} = 0$ for $j > 1$, then the matrix C_l is the $n \times n$ identity matrix, and if this is the case for all l, then the functions P_i^n are ordinary normalised B-spline functions N_i^n of class C^{n-1}. The relation between the polynomials $p_{i,l-1}$ and $p_{i,l}$ (for all l) assumed by the definition of β-spline functions implies the existence of a reparametrisation function f such that all compositions $P_i^n \circ f$ are functions of class $C^{n-1}(\mathbb{R})$ and the composition of a parametrisation $\boldsymbol{s}$ defined by Formula (2.16) with f is of class C^{n-1}. Hence, if the polynomial arcs of the parametrisation $\boldsymbol{s}$ are regular, the order of geometric continuity of the curve described by this parametrisation is $n-1$.

Remark. By taking connection matrices other than these described by (2.18) we can obtain geometric spline curves of other classes, e.g. of class F^{n-1} (see Dyn and Micchelli [1988]).

The diagonal coefficients of the matrix C_l, $a_{l,k,k} = \beta_{l,1}^k$ for $k = 0, \dots, n-1$, are non-zero, hence, this matrix is non-singular. From Formula (2.17) one can see that $p_{i,l-1}(u_l) = p_{i,l}(u_l)$ and, therefore, the functions P_i^n are continuous. If $n = 1$, then the connection matrices are 1×1; their only coefficient is 1 and there is no place for any connection parameters $\beta_{l,k}$; hence, the functions P_i^1 are just the normalised B-spline functions of degree 1.

Exercise. Prove that for a family of quadratic β-spline functions P_i^2 defined with a knot sequence $u_0, \dots, u_N$ and parameters $\beta_{0,1}, \dots, \beta_{N,1}$ there exists a knot sequence $\hat{u}_0, \dots, \hat{u}_N$ and a continuous spline function f of degree 1 such that $P_i^2 \circ f = N_i^2$, where N_i^2 are quadratic B-spline functions defined with the knots $\hat{u}_0, \dots, \hat{u}_N$. As a consequence, quadratic β-spline curves are quadratic B-spline curves subject to a piecewise linear reparametrisation.

Finding polynomial arcs of a β-spline curve

Before constructing β-spline functions, i.e., polynomials which describe the functions between the knots, it is necessary to choose their representation. A convenient representation, described below, uses "local" bases associated with the intervals $[u_k, u_{k+1}]$, obtained from the Bernstein polynomial basis of degree n (see Section A.2). Let $h_k = u_{k+1} - u_k$ for $k = 0, \dots, N-1$. After introducing local parameters s_k such that $t = u_k + h_k s_k$ (hence, $t \in [u_k, u_{k+1}] \Leftrightarrow s_k \in [0, 1]$), we take

$$B_{k,m}^n(t) \stackrel{\text{def}}{=} B_m^n(s_k), \quad k = 0, \dots, N-1, \; m = 0, \dots, n.$$

The polynomial $p_{i,k}(t)$ equal to $P_i^n(t)$ for $t \in [u_k, u_{k+1}]$ will be represented by the vector $\boldsymbol{b}_{i,k} = [b_{i,k,0}, \dots, b_{i,k,n}]^T$ such that

$$p_{i,k}(t) = b_{i,k,0} B_{k,0}^n(t) + \cdots + b_{i,k,n} B_{k,n}^n(t) \quad \text{for all } t \in \mathbb{R}.$$

With such a representation it is easy to find the Bézier control points of the polynomial arcs of a β-spline curve represented by the control points $\boldsymbol{d}_0, \dots, \boldsymbol{d}_{N-n-1}$. Indeed, since in the interval $[u_k, u_{k+1}] \subset [u_n, u_{N-n}]$ only the functions $P_{k-n}^n, \dots, P_k^n$ take non-zero values, the polynomial arc corresponding to this interval is determined by $n+1$ control points, $\boldsymbol{d}_{k-n}, \dots, \boldsymbol{d}_k$. The Bézier control points of this arc, $\boldsymbol{p}_{k,0}, \dots, \boldsymbol{p}_{k,n}$, may be obtained as follows: for $t \in [u_k, u_{k+1}]$ we have

$$\begin{aligned} \boldsymbol{s}(t) &= \sum_{i=k-n}^{k} \boldsymbol{d}_i p_{i,k}(t) = \sum_{i=k-n}^{k} \boldsymbol{d}_i \sum_{m=0}^{n} b_{i,k,m} B_{k,m}^n(t) = \sum_{m=0}^{n} \Bigl(\sum_{i=k-n}^{k} b_{i,k,m} \boldsymbol{d}_i \Bigr) B_{k,m}^n(t) \\ &= \sum_{m=0}^{n} \boldsymbol{p}_{k,m} B_{k,m}^n(t); \end{aligned}$$

hence, $\boldsymbol{p}_{k,m} = \sum_{i=k-n}^{k} b_{i,k,m}\boldsymbol{d}_i$ and

$$\begin{bmatrix} \boldsymbol{p}_{k,0} \\ \boldsymbol{p}_{k,1} \\ \vdots \\ \boldsymbol{p}_{k,n} \end{bmatrix} = \begin{bmatrix} b_{k-n,k,0} & b_{k-n+1,k,0} & \dots & b_{k,k,0} \\ b_{k-n,k,1} & b_{k-n+1,k,1} & \dots & b_{k,k,1} \\ \vdots & \vdots & & \vdots \\ b_{k-n,k,n} & b_{k-n+1,k,n} & \dots & b_{k,k,n} \end{bmatrix} \begin{bmatrix} \boldsymbol{d}_{k-n} \\ \boldsymbol{d}_{k-n+1} \\ \vdots \\ \boldsymbol{d}_k \end{bmatrix}. \tag{2.19}$$

The $(n+1) \times (n+1)$ matrix in this formula, whose columns are the vectors $\boldsymbol{b}_{k-n,k}, \dots, \boldsymbol{b}_{k,k}$, will be denoted by M_k. The representation of β-spline functions that we are going to find consists of $N - 2n$ matrices, $M_n, \dots, M_{N-n-1}$. Each of them allows us to find one of $N - 2n$ polynomial arcs of the β-spline curve.

It turns out that the first column of M_k consists of one non-zero coefficient followed by n zeros, the last column has only the last coefficient non-zero (in (2.19) the coefficients equal to 0 are blue) and, by Property (iii) in the definition of β-spline functions and properties of Bernstein basis polynomials, the sum of coefficients in each row is 1. As a consequence, if $n = 3$, the points $\boldsymbol{p}_{k,1}$ and $\boldsymbol{p}_{k,2}$ lie on the line $\boldsymbol{d}_{k-2}\boldsymbol{d}_{k-1}$ (see Figure 2.6).

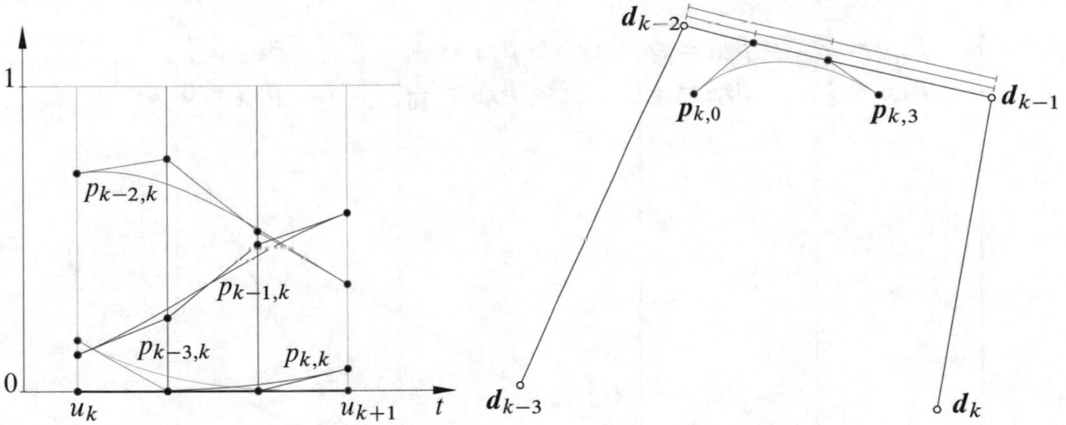

Figure 2.6: A polynomial arc of a cubic β-spline curve and its Bézier representation

A general construction of β-spline functions

Let $n > 1$. Given the degree n, the knots $u_0 < \cdots < u_N$ and the connection parameters $\beta_{j,k}$ for $j = 1, \dots, N$ and $k = 1, \dots, n-1$ we are going to find the polynomials $p_{i,l}$. It may be done by solving a number of systems of linear equations with unknown variables representing the polynomials. To derive the equations we define **auxiliary β-spline functions.**[5] The auxiliary function G_i^n in each interval $[u_k, u_{k+1}]$ is a polynomial $g_{i,k}$ such that

[5]The role of the auxiliary functions is similar to that of the truncated power function in the definition of B-spline functions (see Section A.4.1). The divided difference $f_t^n[u_i, \dots, u_{i+n+1}]$ is a linear combination of values of the function $f_t^n(u) =$

(i′) $g_{i,k} = 0$ for $k < i$ (hence $G_i^n(t) = 0$ for $t < u_i$),

(ii′) $g_{i,i}(t) = (t - u_i)^n$,

(iii′) if $k > i$ then the degree of the polynomial $g_{i,k}(t)$ is less than n,

(iv′) Formula (2.17) is satisfied by $g_{i,l-1}$ and $g_{i,l}$ substituted in place of $p_{i,l-1}$ and $p_{i,l}$.

Exercise. Prove that with a knot sequence and connection parameters fixed the conditions above uniquely define the functions G_i^n.

Note that the value and derivatives up to the order $n-1$ of the polynomial $g_{i,i}$ at u_i are zero, and the polynomials $g_{i,i-1} = 0$ and $g_{i,i}$ satisfy the relation described by Formula (2.17) in a trivial way. An example of a family of auxiliary functions is shown in Figure 2.7. The β-spline functions P_i^n may be obtained as linear combinations of the auxiliary functions G_i^n, which we can find first. In the construction of the functions P_i^n, for each function G_i^n it suffices to find explicitly the polynomials $g_{i,i}, \dots, g_{i,i+n}$.

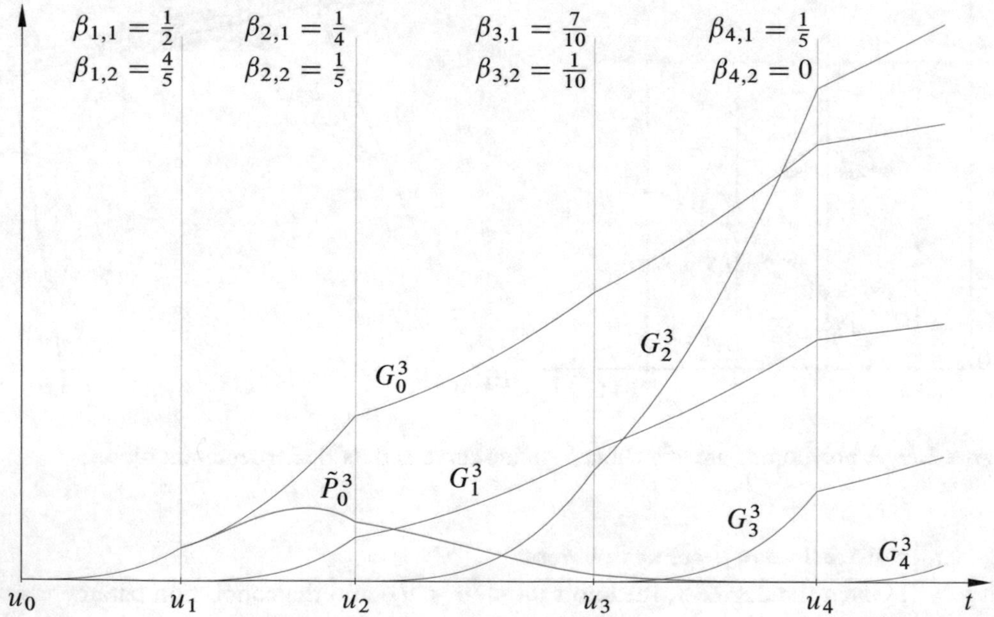

Figure 2.7: Auxiliary β-spline functions and their linear combination $\tilde{P}_0^3$ vanishing outside the interval $[u_0, u_4]$

$(t-u)_+^n$ at the knots $u_i, \dots, u_{i+n+1}$; hence, there exist numbers, say, $a_0, \dots, a_{n+1}$, such that the function $N_i^n(t)$ defined by Formula (A.24) is equal to $a_0(t-u_i)_+^n + \dots + a_{n+1}(t-u_{i+n+1})_+^n$. In Bartels, Beatty and Barsky [1985] the auxiliary β-spline functions are just called truncated power functions.

The functions $G_i^n, \dots, G_{i+n}^n$ in the interval $[u_{i+n+1}, u_{i+n+2}]$ are polynomials of degree less than n. Any set of $n+1$ polynomials (of one variable) of degree less than n is linearly dependent; hence, there exists a non-zero linear combination of the functions $G_i^n, \dots, G_{i+n}^n$ whose restriction to the interval $[u_{i+n+1}, u_{i+n+2}]$ is zero; in fact, this linear combination, denoted by $\tilde{P}_i^n$, is equal to 0 outside the interval (u_i, u_{i+n+1}) (an example is plotted in Figure 2.7). The function $\tilde{P}_i^n$ satisfies *almost all* properties that define the β-spline function P_i^n. The exception is the third property, that the sum of the functions P_i^n in the interval $[u_n, u_{N-n}]$ is equal to 1. This property may be satisfied by multiplying the functions $\tilde{P}_i^n$ by constant factors, which we deal with later.

In the calculation below we extensively use the linear mappings K_l, whose argument is a polynomial p and whose values are the vectors $[p(u_l), p'(u_l), \dots, p^{(n-1)}(u_l)]^T$. We already used such a mapping to write Formula (2.17), expressing a relation between the polynomials $p_{i,l-1}$ and $p_{i,l}$ via the vectors $\boldsymbol{p}_{i,l-1} = K_l(p_{i,l-1})$ and $\boldsymbol{q}_{i,l} = K_l(p_{i,l})$. In a similar way we define the vectors

$$\boldsymbol{g}_{i,l-1} \stackrel{\text{def}}{=} K_l(g_{i,l-1}) = \begin{bmatrix} g_{i,l-1}(u_l) \\ g'_{i,l-1}(u_l) \\ \vdots \\ g_{i,l-1}^{(n-1)}(u_l) \end{bmatrix} \quad \text{and} \quad \boldsymbol{h}_{i,l} \stackrel{\text{def}}{=} K_l(g_{i,l}) = \begin{bmatrix} g_{i,l}(u_l) \\ g'_{i,l}(u_l) \\ \vdots \\ g_{i,l}^{(n-1)}(u_l) \end{bmatrix}.$$

By definition of the functions G_i^n, $\boldsymbol{h}_{i,l} = C_l \boldsymbol{g}_{i,l-1}$. As $g_{i,i}(t) = (t-u_i)^n$, it follows that

$$\boldsymbol{g}_{i,i} = \begin{bmatrix} g_{i,i}(u_{i+1}) \\ g'_{i,i}(u_{i+1}) \\ \vdots \\ g_{i,i}^{(n-1)}(u_{i+1}) \end{bmatrix} = \begin{bmatrix} \frac{n!}{n!} h_i^n \\ \frac{n!}{(n-1)!} h_i^{n-1} \\ \vdots \\ \frac{n!}{1!} h_i \end{bmatrix}, \tag{2.20}$$

where $h_i = u_{i+1} - u_i$.

Let D_l be an $n \times n$ matrix whose coefficient in the j-th row and k-th column is

$$d_{l,j,k} = \begin{cases} 0 & \text{if } k < j, \\ \dfrac{h_l^{k-j}}{(k-j)!} & \text{if } k \geq j, \end{cases}$$

where $h_l = u_{l+1} - u_l$. The matrix D_l is upper triangular and its diagonal coefficients $d_{l,j,j}$ are equal to 1 for $j = 1, \dots, n$. The mappings K_l are isomorphisms (i.e., linear bijections) of the linear vector spaces $\mathbb{R}[\cdot]_{n-1}$ and $\mathbb{R}^n$; hence, for any polynomial p of degree less than n, the vectors $K_l(p)$ and $K_{l+1}(p)$ are complete representations of this polynomial which satisfy the following equality[6]: $K_{l+1}(p) = D_l K_l(p)$. In particular, $\boldsymbol{g}_{i,l} = D_l \boldsymbol{h}_{i,l}$. Using the matrices C_l and D_l, we

[6]The formula for the coefficients of D_l is obtained from the Taylor expansions of a polynomial of degree $n-1$ and its derivatives at the point u_l.

can compute recursively

$$h_{i,l} = C_l g_{i,l-1}, \quad g_{i,l} = D_l h_{i,l}. \tag{2.21}$$

As we noticed, the polynomials (of degree less than n) $g_{i,i+n+1}, \ldots, g_{i+n,i+n+1}$, which describe the functions $G_i^n, \ldots, G_{i+n}^n$ in the interval $[u_{i+n+1}, u_{i+n+2}]$, are linearly dependent. This is equivalent to the linear dependence of the vectors $\boldsymbol{h}_{i,i+n+1}, \ldots, \boldsymbol{h}_{i+n,i+n+1}$, and due to the fact that the matrix C_{i+n+1} is non-singular, the vectors $\boldsymbol{g}_{i,i+n}, \ldots, \boldsymbol{g}_{i+n,i+n}$ are also linearly dependent; moreover, a linear combination $\sum_{j=0}^{n} a_j g_{i+j,j+n+1}$ is the zero polynomial if and only if $\sum_{j=0}^{n} a_j \boldsymbol{g}_{i+j,i+n} = \boldsymbol{0}$. Therefore the function $\tilde{P}_i^n$ may be found by solving the system of linear equations

$$A_i \boldsymbol{x}_i = \boldsymbol{g}_{i,i+n} \tag{2.22}$$

with the $n \times n$ matrix $A_i = [\boldsymbol{g}_{i+1,i+n}, \ldots, \boldsymbol{g}_{i+n,i+n}]$. The solution $\boldsymbol{x}_i = [x_{i1}, \ldots, x_{in}]^T$ allows us to obtain the function

$$\tilde{P}_i^n = G_i^n - (x_{i1} G_{i+1}^n + \cdots + x_{in} G_{i+n}^n)$$

represented by the polynomials[7]

$$\begin{aligned} \tilde{p}_{i,i} &= g_{i,i}, \\ \tilde{p}_{i,i+k} &= g_{i,i+k} - (x_{i1} g_{i+1,i+k} + \cdots + x_{ik} g_{i+k,i+k}), \quad k = 1, \ldots, n, \end{aligned}$$

which describe $\tilde{P}_i^n$ in the intervals $[u_i, u_{i+1}], \ldots, [u_{i+n}, u_{i+n+1}]$.

Having the polynomials $\tilde{p}_{i,i}, \ldots, \tilde{p}_{i,i+n}$, we can find the normalisation factor c_i such that $P_i^n = c_i \tilde{P}_i^n$. By definition of β-spline functions the sum $S_i = P_i^n + \cdots + P_{i+n}^n$ in the interval $[u_{i+n}, u_{i+n+1}]$ is a constant function equal to 1. By summing the vectors associated (by the mapping K_{i+n}) with the polynomials, which describe the β-spline functions in the intervals $[u_{i+n-1}, u_{i+n}]$ and $[u_{i+n}, u_{i+n+1}]$, we obtain

$$\boldsymbol{t}_{i,i+n} \stackrel{\text{def}}{=} \sum_{j=i}^{i+n} \boldsymbol{p}_{j,i+n-1} \quad \text{and} \quad \boldsymbol{s}_{i,i+n} \stackrel{\text{def}}{=} \sum_{j=i}^{i+n} \boldsymbol{q}_{j,i+n}.$$

The polynomial $p_{i+n,i+n-1}$ is zero, hence $\boldsymbol{q}_{i+n,i+n} = C_{i+n} \boldsymbol{p}_{i+n,i+n-1} = \boldsymbol{0}$ and the last term of both expressions above may be dropped. The vectors defined above satisfy the equality $\boldsymbol{s}_{i,i+n} = C_{i+n} \boldsymbol{t}_{i,i+n}$. The coordinates of $\boldsymbol{s}_{i,i+n}$ are the value and derivatives at the point u_{i+n} of the polynomial $s_{i,i+n}$, which is equal to $1 = S_i(t)$ for $t \in [u_{i+n}, u_{i+n+1}]$; hence, $\boldsymbol{s}_{i,i+n}$ is equal to $\boldsymbol{e}_1 = [1, 0, \ldots, 0]^T$. Due to the structure of the matrix C_{i+n} (see Formula (2.17)) this is equivalent to $\boldsymbol{t}_{i,i+n} = \boldsymbol{e}_1$. As the polynomials $p_{i+1,k}, \ldots, p_{i+n,k}$ are linear combinations of

[7]Recall that $g_{i+j,i+k} = 0$ for $i + j > i + k$.

$g_{i+1,k}, \ldots, g_{i+n,k}$ and $p_{i,i+n} = c_i \tilde{p}_{i,i+n}$, there exist numbers $y_{i2}, \ldots, y_{in}$ such that

$$s_{i,i+n} = c_i \tilde{\boldsymbol{q}}_{i,i+n} + \sum_{j=i+1}^{i+n-1} y_{i,j-i+1} \boldsymbol{h}_{j,i+n},$$
$$\boldsymbol{t}_{i,i+n} = c_i \tilde{\boldsymbol{p}}_{i,i+n-1} + \sum_{j=i+1}^{i+n-1} y_{i,j-i+1} \boldsymbol{g}_{j,i+n-1}.$$

The last formula allows us to write the equality $\boldsymbol{t}_{i,i+n} = \boldsymbol{e}_1$ as a system of linear equations

$$B_i \boldsymbol{y}_i = \boldsymbol{e}_1, \tag{2.23}$$

with the $n \times n$ matrix $B_i = [\tilde{\boldsymbol{p}}_{i,i+n-1}, \boldsymbol{g}_{i+1,i+n-1}, \ldots, \boldsymbol{g}_{i+n,i+n-1}]$. The factor c_i is the first coordinate of the solution $\boldsymbol{y}_i = [y_{i1}, \ldots, y_{in}]^T$.

We can append the $n+1$-st coordinate, equal to the n-th order derivative of $g_{i,k}$ at u_{k+1}, to the vectors $\boldsymbol{g}_{i,k}$; for all $t \in \mathbb{R}$ there is $g_{i,i}^{(n)}(t) = n!$ and $g_{i,k}^{(n)}(t) = 0$ if $k \neq i$. The vectors $\hat{\boldsymbol{g}}_{i,k} \in \mathbb{R}^{n+1}$ obtained in this way consist of the value and derivatives up to the order n of the polynomials $g_{i,k}$ at the point u_{k+1}, which is a complete representation of any polynomial of degree $\leq n$. Therefore, by computing the vectors

$$\begin{aligned} \hat{\boldsymbol{p}}_{i,i} &= c_i \hat{\boldsymbol{g}}_{i,i}, \\ \hat{\boldsymbol{p}}_{i,i+k} &= c_i \big(\hat{\boldsymbol{g}}_{i,i+k} - (x_{i1} \hat{\boldsymbol{g}}_{i+1,i+k} + \cdots + x_{ik} \hat{\boldsymbol{g}}_{i+k,i+k}) \big), \quad k = 1, \ldots, n, \end{aligned} \tag{2.24}$$

we obtain a representation of the polynomials $p_{i,i}, \ldots, p_{i,i+n}$.

The last step of the construction is the conversion of representation of the polynomials representing β-spline functions in order to compute the matrices $M_n, \ldots, M_{N-n-1}$, which appear in Formula (2.19). The coefficients of the polynomial $p_{i,l}$ in the local Bernstein polynomial basis may be found by solving the Hermite interpolation problem with one knot, u_{l+1}, of multiplicity $n+1$. It may be done by solving the system of linear equations

$$E \boldsymbol{b}_{i,l} = F_l \hat{\boldsymbol{p}}_{i,l}, \tag{2.25}$$

with the $(n+1) \times (n+1)$ matrices E and F_l, whose coefficients are

$$e_{j,k} = \begin{cases} 0 & \text{for } j < k, \\ (-1)^{n-k} \binom{j-1}{n+1-j-k} & \text{for } j \geq k, \end{cases} \qquad f_{l,j,k} = \begin{cases} 0 & \text{for } j \neq k, \\ \frac{(n+1-j)!}{n!} h_l^{j-1} & \text{for } j = k. \end{cases}$$

For example, if $n = 3$ then

$$E = \begin{bmatrix} 0 & 0 & 0 & 1 \\ 0 & 0 & -1 & 1 \\ 0 & 1 & -2 & 1 \\ -1 & 3 & -3 & 1 \end{bmatrix}, \quad F_l = \begin{bmatrix} 1 & 0 & 0 & 0 \\ 0 & h_l/3 & 0 & 0 \\ 0 & 0 & h_l^2/6 & 0 \\ 0 & 0 & 0 & h_l^3/6 \end{bmatrix}.$$

Algorithm 2.1 Finding β-spline functions represented with local Bernstein bases

Input : degree n, knots $u_0 < \cdots < u_N$, parameters $\beta_{i,j}$, $i = 1, \ldots, N-1$, $j = 1, \ldots, n-1$

1. For $k = 1, \ldots, N-1$ compute the matrices C_k and D_k.
2. For $i = 0, \ldots, N-2$ construct the vectors $\boldsymbol{g}_{i,i}$ using (2.20).
3. For $i = 0, \ldots, N-3$, $j = 1, \ldots, n-1$, $i + j < N$ compute the vectors $\boldsymbol{g}_{i,i+j}$ using (2.21).
4. For $i = 0, \ldots, N-n-1$
   4.1. Assemble the matrix A_i and compute the vector $\boldsymbol{x}_i$ by solving System (2.22),
   4.2. Assemble the matrix B_i and compute c_i by solving System (2.23),
   4.3. For $k = \max\{i, n\}, \ldots, \min\{i+n, N-n-1\}$ compute the vector $\hat{\boldsymbol{p}}_{i,k}$ using (2.24), solve System (2.25) to find the vector $\boldsymbol{b}_{i,k}$, and store it as a column of M_k.

Output : coefficients $b_{i,k,m}$ of the polynomials $p_{i,k}$, stored in columns of the matrices M_k

Note that the matrix E is (lower right) triangular, which makes the system (2.25) particularly easy to solve. The entire construction of β-spline functions is done by Algorithm 2.1.

Exercise. Derive the formula for the coefficients of the matrix D_l in (2.21).

Exercise. Prove that the normalisation of β-spline functions, i.e., the choice of the factors c_i to multiply the functions $\tilde{P}_i^n$, is consistent by showing that if $\sum_{i=k-n}^{k} c_{i,k}\tilde{p}_{i,k} = 1$ and $\sum_{i=k+1-n}^{k+1} c_{i,k+1}\tilde{p}_{i,k+1} = 1$, then for $i = k+1-n, \ldots, k$ the numbers $c_{i,k}$ and $c_{i,k+1}$ are equal.

Exercise. Derive the system of equations (2.25) from Formula (A.15).

Existence and uniqueness of β-spline functions

Two steps of the algorithm described above may fail: depending on the knots and connection parameters, the systems of linear equations (2.22) and (2.23) may have unique solutions or not (the final conversion of the polynomials to the Bernstein bases is always feasible). Consider two examples, with $n = 3$ and the knot sequence $u_i = i$ for all i. By taking $\beta_{l,1} = 1$ and $\beta_{l,2} = 0$ for all l we obtain the matrices C_l equal to the 3×3 identity matrix, and the matrices A_i, B_i (the same for all i) are non-singular:

$$A_i = \begin{bmatrix} 19 & 7 & 1 \\ 15 & 9 & 3 \\ 6 & 6 & 6 \end{bmatrix}, \quad B_i = \begin{bmatrix} 1 & 7 & 1 \\ -3 & 9 & 3 \\ 6 & 6 & 6 \end{bmatrix}.$$

The functions P_i^3 obtained in this case are unique; they are the normalised cubic B-spline functions with uniform knots. On the other hand, if $\beta_{l,1} = 1$ and $\beta_{l,2} = -4$ for all l, then the ma-

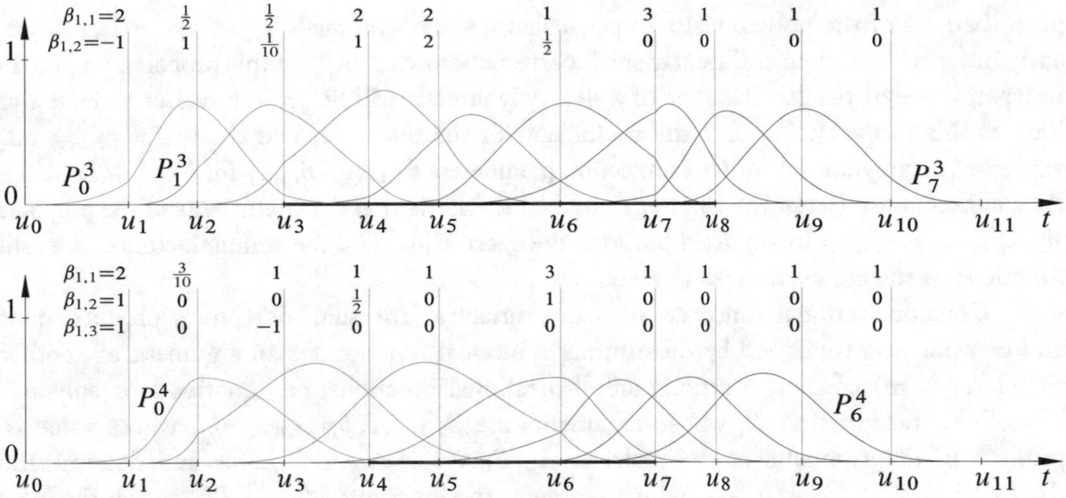

Figure 2.8: β-spline functions of degree 3 and 4

trix A_i, for all i, is singular; there is

$$A_i = \begin{bmatrix} 1 & 1 & 1 \\ 3 & -3 & 3 \\ 6 & -6 & 6 \end{bmatrix}, \quad g_{i,i+n} = \begin{bmatrix} 1 \\ -3 \\ -6 \end{bmatrix},$$

and though System (2.22) in this case is consistent, its solution is all but unique.

Below we outline a proof that such a situation is rare and one can expect that a random choice of the connection parameters will *almost always* result in a unique family of β-spline functions; the idea of this proof is similar to that in Seroussi and Barsky [1991]. We fix an increasing knot sequence $u_0, \ldots, u_N$. Let $\boldsymbol{\beta}$ be a vector in $\mathbb{R}^{(N-1)(n-1)}$ whose coordinates are the parameters $\beta_{1,1}, \ldots, \beta_{N-1,n-1}$. We are going to describe the set $S \subset \mathbb{R}^{(N-1)(n-1)}$ such that if $\boldsymbol{\beta} \in S$, then at least one of the systems of equations, (2.22) or (2.23), for some i, does not have a unique solution.

The key observation is that the function $\tilde{P}_i^n$ exists and is unique if the polynomials $g_{i+1,i+n+1}, \ldots, g_{i+n,i+n+1}$ are linearly independent; then, the columns of A_i are linearly independent and this matrix is non-singular. Similarly, the normalisation, i.e., rescaling the functions $\tilde{P}_i^n, \ldots, \tilde{P}_{i+n}^n$ so as to obtain functions $P_i^n, \ldots, P_{i+n}^n$ whose sum in the interval $[u_{i+n}, u_{i+n+1}]$ is equal to 1, is possible if the polynomials $\tilde{p}_{i,i+n}, \ldots, \tilde{p}_{i+n,i+n}$, which describe $\tilde{P}_i^n, \ldots, \tilde{P}_{i+n}^n$ in this interval, are linearly independent. In that case they form a basis of the space $\mathbb{R}[\cdot]_n$ (of polynomials of degree $\leq n$); any element of this space, in particular the polynomial 1, is a unique linear combination of $\tilde{p}_{i,i+n}, \ldots, \tilde{p}_{i+n,i+n}$.

All polynomials $g_{i,l}$ which describe the auxiliary functions G_i^n depend on the connection parameters in a continuous way; sufficiently small perturbations of the parameters cause small perturbations of the polynomials. In particular, if the polynomials $g_{i,i+n}, \dots, g_{i+n,i+n}$ are linearly independent, then sufficiently small perturbations cannot spoil this property, i.e., the linear independence of the resulting set of $n+1$ polynomials will be preserved. Let us take a closer look at this property. For all i the coefficients of the matrix A_i and the vector $g_{i,i+n}$ may be expressed as polynomials of the connection parameters $\beta_{l,1}, \dots, \beta_{l,n-1}$ for $l = i+1, \dots, i+n$. As a consequence (from the Cramer formulae) it follows that the coefficients of the polynomials $\tilde{p}_{i,i+n}, \dots, \tilde{p}_{i+n,i+n}$ in any fixed basis (of the space $\mathbb{R}[\cdot]_n$) and the scaling factors c_i are rational functions of the connection parameters.

Consider rational functions of many variables. The sum, difference, product, quotient and composition (obtained by substituting a rational function for an argument of another rational function) of such functions are also rational functions, i.e., quotients of polynomials. Therefore, the function R, whose arguments are $\beta_{1,1}, \dots, \beta_{N-1,n-1}$ and whose value is the product of determinants of the matrices $A_0, B_0, \dots, A_{N-n-1}, B_{N-n-1}$ and determinants of the matrices $M_n, \dots, M_{N-n-1}$, which represent the functions $P_0^n, \dots, P_{N-n-1}^n$ in the intervals $[u_n, u_{n+1}], \dots, [u_{N-n-1}, u_{N-n}]$, is a rational function of the connection parameters.

The set S, whose elements do not determine a unique family of β-spline functions $P_0^n, \dots, P_{N-n-1}^n$, is the union of the sets of zeros of the numerator and denominator of the function R, i.e., of two polynomials, and, therefore, it is an algebraic manifold in the space $\mathbb{R}^{(N-1)(n-1)}$. We shall prove that the interior of the set S is empty. Let $\boldsymbol{\beta}_0$ be the vector whose coordinates are $\beta_{l,1} = 1$ and $\beta_{l,j} = 0$ for $j > 1$ and $l = 1, \dots, N-1$. The β-spline functions corresponding to $\boldsymbol{\beta}_0$ exist and are unique; these are normalised B-spline functions with the knots $u_0, \dots, u_N$. Hence, $\boldsymbol{\beta}_0 \notin S$.

Let $\boldsymbol{\beta}_1 \in S$. The function $r(t) \stackrel{\text{def}}{=} R\big((1-t)\boldsymbol{\beta}_0 + t\boldsymbol{\beta}_1\big)$ is a non-zero rational function of the variable t; $r(0)$ is non-zero and finite, while $r(1)$ is either zero or undefined. The number of points of the line $\boldsymbol{\beta}_0\boldsymbol{\beta}_1$ being the elements of S is finite; these points correspond to zeros of the numerator or the denominator of the function r. Hence, no ball with the centre at $\boldsymbol{\beta}_1$ and a positive radius is contained in S, which completes the proof of the following:

Theorem 2.2 *Let an increasing knot sequence $u_0, \dots, u_N$ be fixed. The function, which maps a vector of connection parameters $\boldsymbol{\beta} = (\beta_{l,j})_{l=1,\dots,N-1,\, j=1,\dots,n-1}$ to a unique family of β-spline functions of degree n with these knots, is defined in $\mathbb{R}^{(N-1)(n-1)} \setminus S$; the set S, where this function is undefined, is an algebraic manifold whose interior is empty.*

As a side result of the considerations above we obtain the following property of β-spline functions: if the connection parameters do not determine a point of the set S, the functions $P_{k-n}^n, \dots, P_k^n$ restricted to the interval $[u_k, u_{k+1}]$ are linearly independent polynomials—which is analogous to Property A.13 of B-spline functions.

Remark. In the definition of β-spline functions in Seroussi and Barsky [1991] local linear independence and minimal support were assumed instead of the property (i).

Exercise. Analyse Systems (2.22) and (2.23) to prove that the function P_i^n, if it exists, does not depend on knots other than $u_i, \dots, u_{i+n+1}$ and connection parameters other than $\beta_{l,1}, \dots, \beta_{l,n-1}$ for $l = i+1, \dots, i+n$.

***β*-splines with equidistant knots and global shape parameters**
If the knots $u_0, \dots, u_N$ are equidistant and the connection parameters $\beta_{l,j}$ are the same for all l (i.e., there exist numbers $h > 0$ such that $u_{k+1} - u_k = h$ for $k = 0, \dots, N-1$ and $\beta_1, \dots, \beta_{n-1}$ such that $\beta_{l,j} = \beta_j$ for $j = 1, \dots, n-1$), then the β-spline functions are identical up to translations of the argument:

$$P_0^n(t) = P_i^n(t + ih) \quad \text{for } i = 1, \dots, N-n-1 \text{ and } t \in \mathbb{R}.$$

Such a restriction reduces the flexibility of β-spline curves as a design tool, as in this case we have only $n-1$ *global* shape parameters. On the other hand, processing such β-spline curves may be much easier than in the general case. To represent the entire set of β-spline functions $\{P_0^n, \dots, P_{N-n-1}^n\}$ we need only $n+1$ non-zero polynomials, $p_0, \dots, p_n$, such that

$$p_{i,i+k}(t + ih) = p_k(t) \quad \text{for } i = 1, \dots, N-n-1 \text{ and } t \in \mathbb{R}. \tag{2.26}$$

Below we derive a system of equations whose solution represents four cubic polynomials representing the cubic β-spline functions with equidistant knots and with two global parameters, β_1 and β_2. The unknown variables in this system of equations are the coefficients of the polynomials in the Bernstein polynomial basis of degree 3.

Without loss of generality we assume that $u_i = i$ for $i = 0, \dots, N$, i.e., $h = 1$. We need to find cubic polynomials $p_0, \dots, p_3$ such that

$$p_{i,i+k}(t + i) = p_k(t).$$

From Formula (2.17) it follows that the equations

$$p_k(0) = p_{k-1}(1), \tag{2.27}$$
$$p'_k(0) = \beta_1 p'_{k-1}(1), \tag{2.28}$$
$$p''_k(0) = \beta_2 p'_{k-1}(1) + \beta_1^2 p''_{k-1}(1) \tag{2.29}$$

must be satisfied for $k = 0, \dots, 4$. The polynomials p_{-1} and p_4, which appear above for $k = 0$ and $k = 4$ are zero. In addition, the sum of $p_0, \dots, p_3$ must be equal to 1.

Let $p_k(t) = \sum_{j=0}^{3} b_{kj} B_j^3(t)$ for $k = 0, \dots, 3$. As the polynomials p_{-1} and p_4 are zero, Formulae (A.15) and Equations (2.27)–(2.29) give us immediately $b_{00} = b_{01} = b_{02} = 0$, and $b_{31} = b_{32} = b_{33} = 0$. From (2.27) for $k = 1, 2, 3$ we obtain also $b_{10} = b_{03}$, $b_{20} = b_{13}$ and $b_{30} = b_{23}$. Thus, the number of unknown coefficients is reduced to 7: b_{03}, b_{11}, b_{12}, b_{13}, b_{21}, b_{22} and b_{23}. After substituting expressions obtained from Formulae (A.15) for the derivatives of the polynomials to Equations (2.28) and (2.29) for $k = 1, 2, 3$, we obtain the following equations:

$$\begin{aligned}
b_{11} - b_{03} &= \beta_1 b_{03}, & 2b_{12} - 4b_{11} + 2b_{03} &= \beta_2 b_{03} + 2\beta_1^2 b_{03}, \\
b_{21} - b_{13} &= \beta_1(b_{13} - b_{12}), & 2b_{22} - 4b_{21} + 2b_{13} &= \beta_2(b_{13} - b_{12}) + 2\beta_1^2(b_{13} - 2b_{12} + b_{11}), \\
-b_{23} &= \beta_1(b_{23} - b_{22}), & 2b_{23} &= \beta_2(b_{23} - b_{22}) + 2\beta_1^2(b_{23} - 2b_{22} + b_{21}).
\end{aligned}$$

A polynomial represented in a Bernstein basis is a constant if and only if all its coefficients are equal to this constant. The condition $p_0 + p_1 + p_2 + p_3 = 1$ is, thus, equivalent to four equalities, $b_{0j} + b_{1j} + b_{2j} + b_{3j} = 1$ for $j = 0, \dots, 3$. As we need just one normalisation equation, we can take any of these four. The simplest normalisation equations correspond to $j = 1$ or $j = 2$, as there are only two non-zero unknown variables in these equations: for $j = 1$ the equation is $b_{11} + b_{21} = 1$. In matrix form the entire system (with the equations reordered so as to facilitate solving it numerically) is:

$$\begin{bmatrix}
-1-\beta_1 & 1 & 0 & 0 & 0 & 0 & 0 \\
2-\beta_2-2\beta_1^2 & -4 & 2 & 0 & 0 & 0 & 0 \\
0 & 1 & 0 & 0 & 1 & 0 & 0 \\
0 & -2\beta_1^2 & \beta_2+4\beta_1^2 & 2-\beta_2-2\beta_1^2 & -4 & 2 & 0 \\
0 & 0 & \beta_1 & -1-\beta_1 & 1 & 0 & 0 \\
0 & 0 & 0 & 0 & -2\beta_1^2 & \beta_2+4\beta_1^2 & 2-\beta_2-2\beta_1^2 \\
0 & 0 & 0 & 0 & 0 & \beta_1 & -1-\beta_1
\end{bmatrix}
\begin{bmatrix} b_{03} \\ b_{11} \\ b_{12} \\ b_{13} \\ b_{21} \\ b_{22} \\ b_{23} \end{bmatrix}
=
\begin{bmatrix} 0 \\ 0 \\ 1 \\ 0 \\ 0 \\ 0 \\ 0 \end{bmatrix}.$$

To find the polynomial arcs of the curve we can use Formula (2.19). For all k the matrix M_k is the same, and

$$\begin{bmatrix} \boldsymbol{p}_{k,0} \\ \boldsymbol{p}_{k,1} \\ \boldsymbol{p}_{k,2} \\ \boldsymbol{p}_{k,3} \end{bmatrix}
=
\begin{bmatrix}
b_{23} & b_{13} & b_{03} & 0 \\
0 & b_{21} & b_{11} & 0 \\
0 & b_{22} & b_{12} & 0 \\
0 & b_{23} & b_{13} & b_{03}
\end{bmatrix}
\begin{bmatrix} \boldsymbol{d}_{k-3} \\ \boldsymbol{d}_{k-2} \\ \boldsymbol{d}_{k-1} \\ \boldsymbol{d}_k \end{bmatrix}.$$

Exercise. Derive a system of equations for quartic β-spline curves of class G^3 with equidistant (integer) knots and global shape parameters $\beta_1, \beta_2, \beta_3$. The solution of this system consists of coefficients of polynomials $p_0, \dots, p_4$ in Bernstein basis of degree 4.

Further properties and examples

A consequence of Property (iii) in the definition of β-spline functions is the **affine invariance** of the representation: if A is an affine transformation, then the image $A(\boldsymbol{s})$ of a curve given by Formula (2.16) is represented by the control points $A(\boldsymbol{d}_i)$.

The **convex hull property** occurs if all β-spline functions are nonnegative, which depends on the connection parameters. In particular, if $\beta_{l,1} = 1$ and $\beta_{l,j} = 0$ for $j > 1$, then the β-spline functions are B-spline functions, which are nonnegative. Small enough perturbations of these particular connection parameters do not destroy this property, but it *does not* hold in general. *If* the functions $P^n_{k-n}, \ldots, P^n_k$ are nonnegative then the arc $\{\, \boldsymbol{s}(t) : t \in [u_k, u_{k+1}]\,\}$ is contained in the convex hull of the control points $\boldsymbol{d}_{k-n}, \ldots, \boldsymbol{d}_k$.

Figure 2.9 shows two planar β-spline curves obtained with the knots and parameters that were used to define the cubic and quartic β-spline functions shown in Figure 2.8.

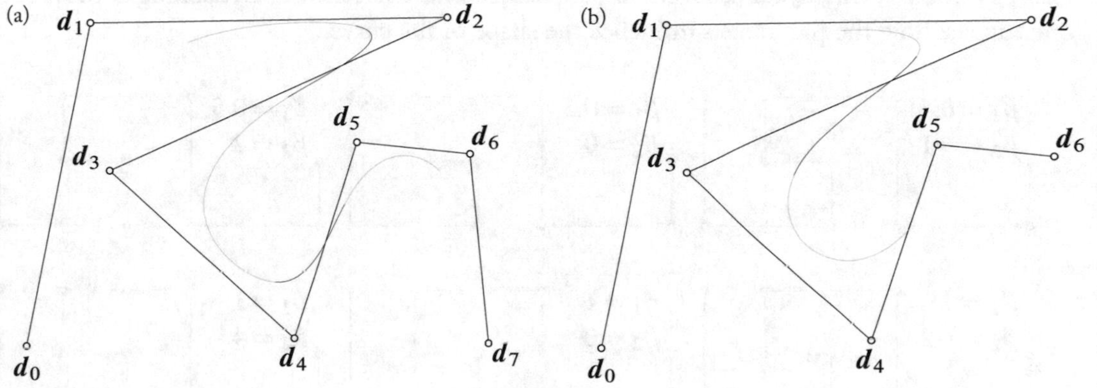

Figure 2.9: Planar β-spline curves and their control polygons: (a) cubic, (b) quartic

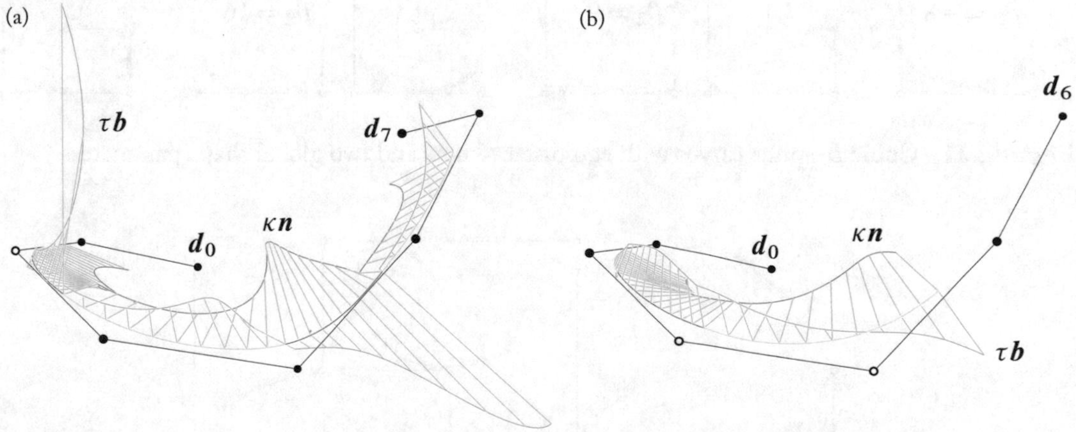

Figure 2.10: β-spline curves in $\mathbb{R}^3$, with graphs of curvature and torsion: (a) cubic, (b) quartic

Figure 2.10 shows two curves in the three-dimensional space: a cubic curve of class G^2 and a quartic curve of class G^3, with the same knots and connection parameters as the planar curves in the previous examples. The picture shows their curvature and torsion. To draw the picture, the Frenet frame has been found at a number of points of the curves and then the unit normal vector $\boldsymbol{n}$ was multiplied by the curvature κ and a constant factor, and the binormal vector $\boldsymbol{b}$ was multiplied by the torsion τ and a constant factor. The curvature of both the curves is continuous. We can see points of discontinuity of the torsion of the cubic curve, while the torsion of the quartic curve is continuous.

The cubic curves in Figure 2.11 have the same equidistant knots and the same (up to translations) control polygons; each of them was obtained with a different pair of global shape parameters, β_1 and β_2, which, for cubic curves, are named **bias** and **tension**. By looking at these curves one can see how the parameters influence the shape of the curve.

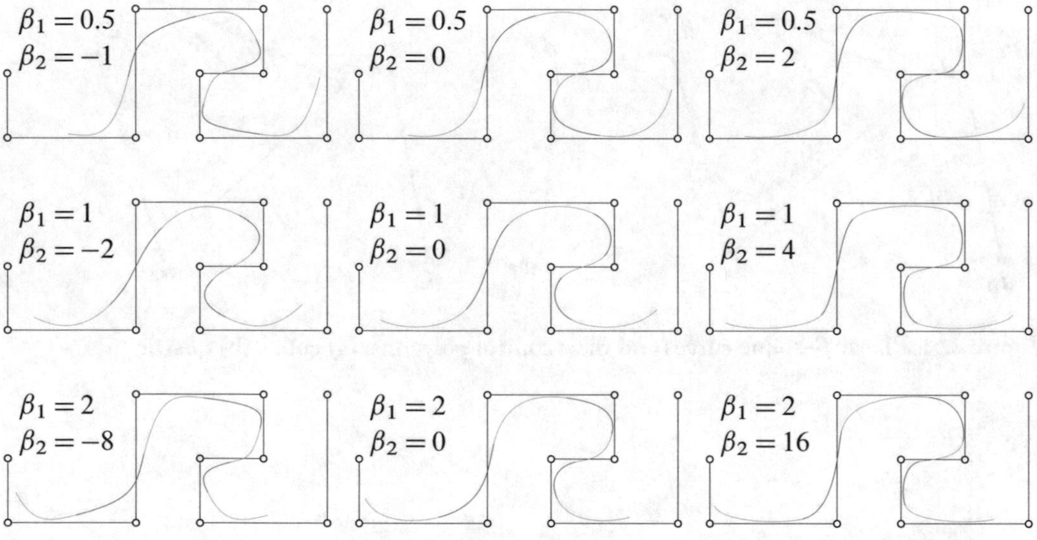

Figure 2.11: Cubic β-spline curves with equidistant knots and two global shape parameters

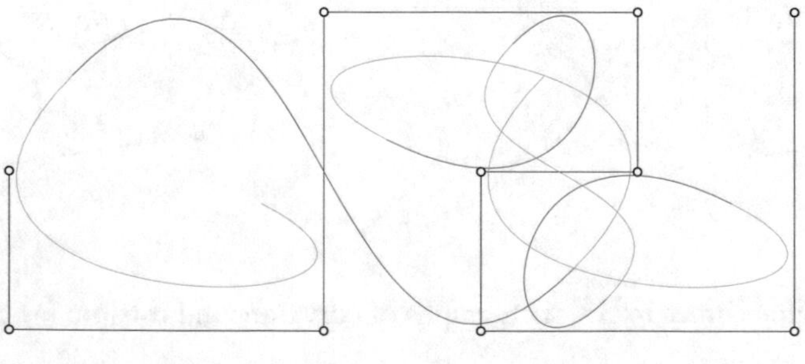

Figure 2.12: A cubic β-spline curve with no convex hull property

The last example (Fig. 2.12) shows a cubic curve with equidistant knots and global shape parameters $\beta_1 = 1$, $\beta_2 = -9.5$; the β-spline functions with these parameters take negative values and the representation does not have the convex hull property.

Knot insertion

Given a β-spline curve $\boldsymbol{s}$ of degree n with a knot sequence $u_0, \ldots, u_N$, connection parameters $\beta_{1,1}, \ldots, \beta_{N-1,n-1}$ and control points $\boldsymbol{d}_0, \ldots, \boldsymbol{d}_{N-n-1}$, we may want to find a representation of this curve with an additional knot, $\hat{u}_{k+1} \in (u_k, u_{k+1}) \subset [u_n, u_{N-n}]$. We assume that this knot does not coincide with any element of the original knot sequence. The goal is to find the control points $\hat{\boldsymbol{d}}_0, \ldots, \hat{\boldsymbol{d}}_{N-1}$ of the new representation of the curve.

By inserting the number $\hat{u}_{k+1}$ to the knot sequence we obtain an increasing sequence $\hat{u}_0, \ldots, \hat{u}_{N+1}$. It is obvious how to associate the connection parameters with these knots: $\hat{\beta}_{l,j} = \beta_{l,j}$ if $l \leq k$, $\hat{\beta}_{l,j} = \beta_{l-1,j}$ if $l > k+1$, and $\hat{\beta}_{k+1,1} = 1$, $\hat{\beta}_{k+1,j} = 0$ for $j = 2, \ldots, n-1$. The parametrisation must remain unchanged and in particular its derivatives at $\hat{u}_{k+1}$ have to remain continuous.

Consider a function $\tilde{P}_i^n(t)$, whose restriction to the interval $[u_i, u_{i+1}]$ is $(t-u_i)^n$; there is $P_i^n = c_i \tilde{P}_i^n$, where the normalisation factor c_i may be found by solving System (2.23). Similarly, for the β-spline functions, which form a basis for the new representation, we have $\hat{P}_i^n = \hat{c}_i \tilde{\hat{P}}_i^n$. If $i \in \{k-n, \ldots, k\}$, then $\tilde{P}_i^n$ is a linear combination

$$\tilde{P}_i^n = \tilde{\hat{P}}_i^n + b_i \tilde{\hat{P}}_{i+1}^n,$$

for some $b_i \in \mathbb{R}$. The factor multiplying $\tilde{\hat{P}}_i^n$ above is 1 because the restriction of both functions, $\tilde{P}_i^n$ and $\tilde{\hat{P}}_i^n$, to $[\hat{u}_i, \hat{u}_{i+1}]$ is the same polynomial $(t-u_i)^n$. Hence,

$$P_i^n = \frac{c_i}{\hat{c}_i}\hat{P}_i^n + \frac{c_{i+1}b_i}{\hat{c}_{i+1}}\hat{P}_{i+1}^n \quad \text{and} \quad P_{i-1}^n = \frac{c_{i-1}}{\hat{c}_{i-1}}\hat{P}_{i-1}^n + \frac{c_i b_{i-1}}{\hat{c}_i}\hat{P}_i^n.$$

From the property $\sum_{i=k-n}^{k} P_i^n(t) = 1 = \sum_{i=k-n}^{k+1} \hat{P}_i^n(t)$ for all $t \in [u_k, u_{k+1}]$ it follows that

$$\frac{c_i b_{i-1}}{\hat{c}_i} + \frac{c_i}{\hat{c}_i} = 1; \quad \text{hence,} \quad P_i^n = \alpha_i \hat{P}_i^n + (1-\alpha_{i+1})\hat{P}_{i+1}^n,$$

where $\alpha_i = c_i/\hat{c}_i$ and $\alpha_{i+1} = c_{i+1}/\hat{c}_{i+1}$. For $i \leq k-n$ we have $\alpha_i = 1$ and if $i > k$ then $\alpha_i = 0$. The final calculation is

$$\begin{aligned} \boldsymbol{s} = \sum_{i=0}^{N-n-1} \boldsymbol{d}_i P_i^n &= \sum_{i=0}^{N-n-1} \boldsymbol{d}_i\big(\alpha_i \hat{P}_i^n + (1-\alpha_{i+1})\hat{P}_{i+1}^n\big) \\ &= \sum_{i=0}^{N-n-1} \alpha_i \boldsymbol{d}_i \hat{P}_i^n + \sum_{i=1}^{N-n}(1-\alpha_i)\boldsymbol{d}_{i-1}\hat{P}_i^n = \sum_{i=0}^{N-n}\big((1-\alpha_i)\boldsymbol{d}_{i-1} + \alpha_i \boldsymbol{d}_i\big)\hat{P}_i^n. \end{aligned}$$

Hence,

$$\begin{aligned} \hat{\boldsymbol{d}}_i &= \boldsymbol{d}_i && \text{for } i \leq k-n, \\ \hat{\boldsymbol{d}}_i &= \Big(1 - \frac{c_i}{\hat{c}_i}\Big)\boldsymbol{d}_{i-1} + \frac{c_i}{\hat{c}_i}\boldsymbol{d}_i && \text{for } i = k-n+1, \ldots, k, \\ \hat{\boldsymbol{d}}_i &= \boldsymbol{d}_{i-1} && \text{for } i > k. \end{aligned} \tag{2.30}$$

An example result of using Formula (2.30) is shown in Figure 2.13. This formula is a generalisation of Formula (A.32), which describes knot insertion for B-spline curves.

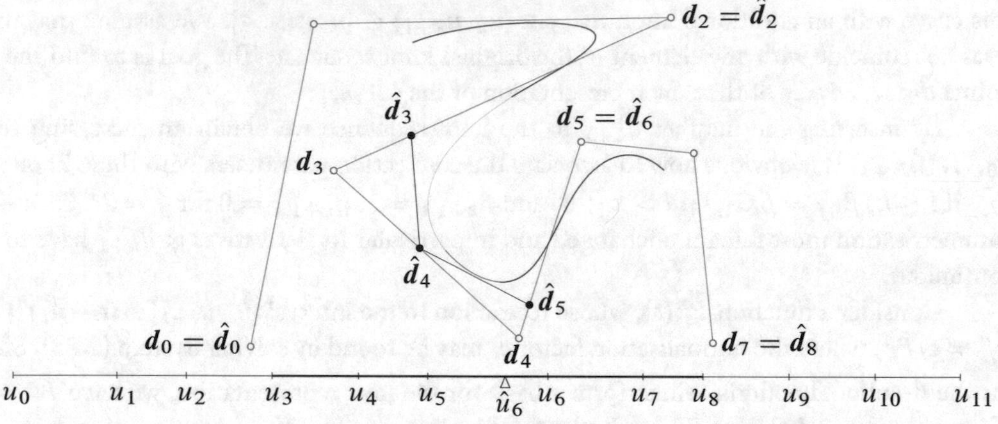

Figure 2.13: Knot insertion for the cubic β-spline curve from Figure 2.9a

Knot insertion may be considered also if the new knot coincides with an old one; however, that would require for the definition of β-spline functions to be extended by admitting knots of multiplicity greater than 1. Interested readers may refer to literature, e.g. Joe [1990], Seidel [1990].

2.3.3 γ-SPLINE CURVES

A cubic B-spline curve

$$s(t) = \sum_{i=0}^{N-4} d_i N_i^3(t), \quad t \in [u_3, u_{N-3})$$

with knots $u_4, \ldots, u_{N-4}$ of multiplicity 1 has continuous derivatives of order 1 and 2. If the first-order derivative is non-zero, then the curve is of class G^2, i.e., its curvature is continuous. By knot insertion (see Section A.4.2) we can obtain the Bézier control polygons of the polynomial arcs of the curve. Below we consider a modification of this construction; it will produce cubic polynomial arcs (in Bézier form) connected with each other in a way preserving the continuity of the first-order derivative. The second-order derivative will be discontinuous, but the curvature of any two arcs at their junction point will be the same.

To understand the construction, we take a look at the knot insertion for an ordinary cubic B-spline curve. Suppose that the knots $u_3, \ldots, u_{N-3}$, initially of multiplicity 1, are inserted twice (e.g. with Algorithm A.9) so as to raise their multiplicity to 3. The effect is shown in Figure 2.14. Let $h_i = u_{i+1} - u_i$ for $i = 1, \ldots, N-2$. On each line segment of the original control polygon between the points d_i and d_{i+1} there are two new points, e_i and f_i, dividing the line segment

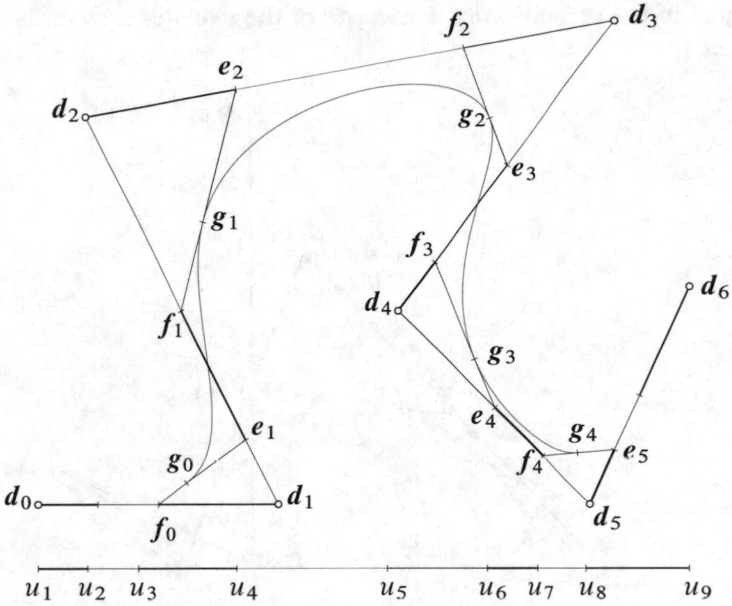

Figure 2.14: Dividing a cubic B-spline curve into polynomial arcs

in the proportion $h_{i+1} : h_{i+2} : h_{i+3}$. Then, the line segment $\overline{f_i e_{i+1}}$ is divided in the proportion $h_{i+2} : h_{i+3}$ by the point g_i, which is the junction point of the polynomial arcs of our spline curve. Using properties of B-spline functions in Section A.4 (in particular Formula (A.31)) we can calculate

$$
\begin{aligned}
s(u_{i+3}) &= g_i,\\
s'(u_{i+3}) &= \frac{3}{h_{i+2}+h_{i+3}}(e_{i+1}-f_i),\\
s''(u_{i+3}) &= \frac{6}{h_{i+2}+h_{i+3}}\left(\frac{1}{h_{i+2}}(e_{i+1}-d_{i+1}) - \frac{1}{h_{i+3}}(d_{i+1}-f_i)\right).
\end{aligned}
$$

Consider two cubic Bézier arcs, whose control points are g_{i-1}, e_i, f_i, g_i and g_i, e_{i+1}, f_{i+1} and g_{i+1} respectively. Their derivatives of the first and second order at the junction point are the same.[8] With the other control points fixed, we investigate the possibility of modifying the points e_i and f_{i+1} in a way that would preserve the curvature of the two arcs. Moving the two points changes the second-order derivatives of the arcs at u_3, leaving the junction point and the first-order derivative intact, i.e., preserving the continuity of the first-order derivative. If the points are moved in the direction of the vector $s'(u_{i+3})$, i.e., parallel to the line segment $\overline{f_i e_{i+1}}$ (see Figure 2.15), the increments of the second-order derivatives of the two arcs are vectors parallel to $s'(u_{i+3})$. In that case Equation (2.6), with $s'(u_{i+3})$ substituted for $\overline{p}'$ and with $\overline{p}''$ and $\underline{p}''$

[8]Here we consider derivatives with respect to the parameter of the spline curve, not to the local parameters of the Bézier arcs.

respectively equal to the second-order derivatives of the two arcs at u_{i+3}, is satisfied with some t_2 and with $t_1 = 1$.

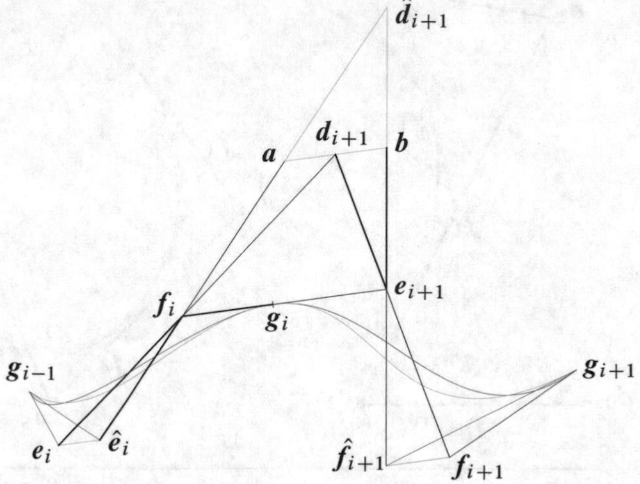

Figure 2.15: Modification of cubic Bézier arcs preserving curvature continuity

If the points f_i, d_{i+1} and e_{i+1} are not collinear, then points $\hat{e}_i$ and $\hat{f}_{i+1}$ such that the line segments $\overline{\hat{e}_i e_i}$ and $\overline{\hat{f}_{i+1} f_{i+1}}$ are parallel to $\overline{f_i e_{i+1}}$ may be obtained as follows: we choose (arbitrarily) a point $\hat{d}_{i+1}$, on the same side of the line $\overline{f_i e_{i+1}}$ as d_{i+1}. There is a positive constant γ such that the distances of d_{i+1} and $\hat{d}_{i+1}$ from this line are respectively d and γd. The triples of points (e_i, f_i, d_{i+1}) and $(d_{i+1}, e_{i+1}, f_{i+1})$ are collinear; the line segments $\overline{e_i d_{i+1}}$ and $\overline{d_{i+1} f_{i+1}}$ are divided by the points f_i and e_{i+1} in the proportion $h_{i+2} : h_{i+3}$. So, the points $\hat{e}_i$ and $\hat{f}_{i+1}$ we choose are such that the triples of points $(\hat{e}_i, f_i, \hat{d}_{i+1})$ and $(\hat{d}_{i+1}, e_{i+1}, \hat{f}_{i+1})$ are collinear, and the line segments $\overline{\hat{e}_i \hat{d}_{i+1}}$ and $\overline{\hat{d}_{i+1} \hat{f}_{i+1}}$ are divided by the points f_i and e_{i+1}, respectively in the proportions $h_{i+2} : \gamma h_{i+3}$ and $\gamma h_{i+2} : h_{i+3}$.

The considerations above justify Algorithm 2.2; the points d_i in it play the role of $\hat{d}_i$ from the construction above; a parameter γ_i is associated with each point d_i. If $\gamma_i = 1$ for $i = 0, \dots, N-4$, then the Bézier arcs produced by Algorithm 2.2 are the arcs of the ordinary cubic B-spline curve represented by the given knots and control points.

An example of a γ-spline curve is in Figure 2.16. Here $\gamma_0 = \gamma_6 = 1$ and $\gamma_1, \dots, \gamma_5 > 1$. Increasing the parameter γ_i pushes the junction point g_{i-1} away from the control point d_i. On the other hand, if γ_i tends to zero, then the point g_{i-1} is pulled towards d_i and the curvature of the curve at that point increases to infinity.

Though the β-spline representation of piecewise cubic curves is more flexible than the γ-spline (the γ-spline parametrisation has the first-order derivative continuous, while β-spline does not have to), both approaches may be used to obtain exactly the same curves. The γ-spline curves

Algorithm 2.2 Constructing cubic Bézier arcs making a γ-spline curve

Input : knots $u_1, \ldots, u_{N-1}$, control points $\boldsymbol{d}_1, \ldots, \boldsymbol{d}_{N-4}$, positive constants $\gamma_0, \ldots, \gamma_{N-4}$

Compute $h_i = u_{i+1} - u_i$ for $i = 1, \ldots, N-2$;
For $i = 0, \ldots, N-5$ find the points $\boldsymbol{e}_i$ and $\boldsymbol{f}_i$ which divide
 the line segments $\overline{\boldsymbol{d}_i \boldsymbol{d}_{i+1}}$ in the proportion $\gamma_i h_{i+1} : h_{i+2} : \gamma_{i+1} h_{i+3}$.
For $i = 0, \ldots, N-6$ find the points $\boldsymbol{g}_i$ which divide
 the line segments $\overline{\boldsymbol{f}_i \boldsymbol{e}_{i+1}}$ in the proportion $h_{i+2} : h_{i+3}$.

Output : control points $\boldsymbol{g}_i, \boldsymbol{e}_{i+1}, \boldsymbol{f}_{i+1}, \boldsymbol{g}_{i+1}$ of cubic Bézier arcs for $i = 0, \ldots, N-6$

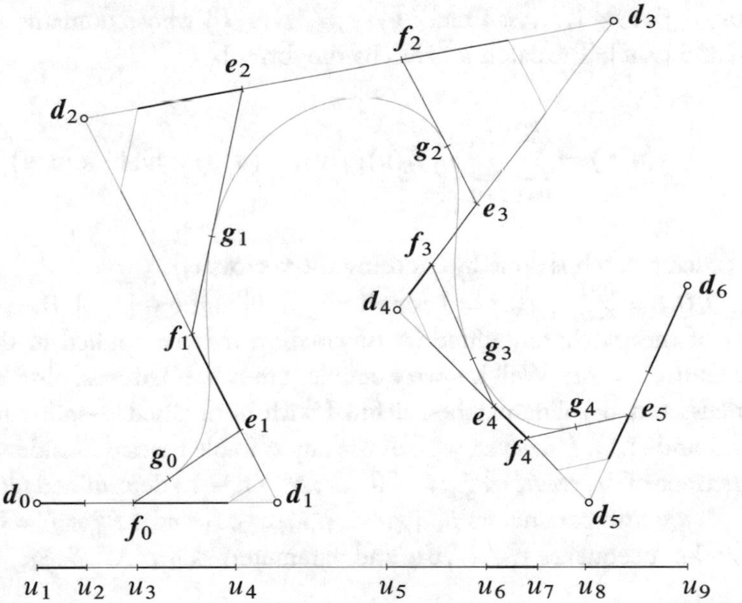

Figure 2.16: A γ-spline curve

seem easier to implement, but the decision which parameters, γ_i or $\beta_{i,1}$ and $\beta_{i,2}$, are more intuitive and convenient has to be taken individually by users of graphics software.

Exercise. Prove that cubic β-spline curves and γ-spline curves are the same, i.e., any cubic β-spline curve with knots $u_0, \ldots, u_N$, control points $\boldsymbol{d}_0, \ldots, \boldsymbol{d}_{N-4}$ and junction parameters $\beta_{1,1}, \beta_{1,2}, \ldots, \beta_{N-1,1}, \beta_{N-1,2}$ may be represented with a knot sequence $\hat{u}_0, \ldots, \hat{u}_N$, the same control points $\boldsymbol{d}_0, \ldots, \boldsymbol{d}_{N-4}$ and some junction parameters $\gamma_1, \ldots, \gamma_N$. Hints: begin with the special case of $\beta_{1,1} = \cdots = \beta_{N-1,1} = 1$ and $u_k = \hat{u}_k$ for all k. Use the fact that the Bézier control points $\boldsymbol{e}_k$, $\boldsymbol{f}_k$ and $\boldsymbol{p}_{k+2,1}, \boldsymbol{p}_{k+2,2}$ of cubic polynomial arcs in both constructions lie on the line $\boldsymbol{d}_k \boldsymbol{d}_{k+1}$ of the control polygon and the triples $(\boldsymbol{f}_k, \boldsymbol{g}_k, \boldsymbol{e}_{k+1})$ and $(\boldsymbol{p}_{k,2}, \boldsymbol{p}_{k,3} = \boldsymbol{p}_{k+1,0}, \boldsymbol{p}_{k+1,1})$

must be collinear. Then generalise, using a reparametrisation function f being a spline of degree 1 such that $f(\hat{u}_k) = u_k$ for all k.

The idea of γ-spline curves was given by Farin [1982] and further developed by Boehm [1985, 1987], who described a knot insertion algorithm for these curves and extended this idea to rational curves with continuous curvature and torsion.

2.4 TENSOR PRODUCT GEOMETRIC SPLINE PATCHES

Tensor product patches are defined using two bases of function spaces, i.e., linearly independent sets of functions $\{ f_i : i = 1, \ldots, k \}$ and $\{ g_j : j = 1, \ldots, l \}$ whose domains are respectively the intervals $[a, b]$ and $[c, d]$. The patch is given by the formula

$$s(u, v) = \sum_{i=1}^{k} \sum_{j=1}^{l} \boldsymbol{v}_{ij} f_i(u) g_j(v), \quad (u, v) \in [a, b] \times [c, d]$$

and modelling such a patch is done by choosing the vectors $\boldsymbol{v}_{ij}$.

If $\sum_{i=1}^{k} f_i(u) = \sum_{j=1}^{l} g(v) = 1$ for all $u \in [a, b]$ and $v \in [c, d]$, then the vectors $\boldsymbol{v}_{ij}$ are **control points** of the patch; any affine transformation may be applied to the patch by transforming all its control points. Well known examples are Bézier patches, obtained with Bernstein basis polynomials, and B-spline patches, defined with normalised B-spline functions (see Sections A.3, A.4.1 and A.5). However, we can use any suitable bases. Consider using two families of β-spline functions of degree n, $\{ P^n_{u:i} : i = 0, \ldots, N - n - 1 \}$ determined respectively by a knot sequence $u_0, \ldots, u_N$ and parameters $\beta_{u:1,1}, \ldots, \beta_{u:N-1,n-1}$ and $\{ P^n_{v:j} : j = 0, \ldots, M - n - 1 \}$ determined by a knot sequence $v_0, \ldots, v_M$ and parameters $\beta_{v:1,1}, \ldots, \beta_{v:M-1,n-1}$. The formula

$$s(u, v) = \sum_{i=0}^{N-n-1} \sum_{j=0}^{M-n-1} \boldsymbol{d}_{ij} P^n_{u:i}(u) P^n_{v:j}(v), \quad (u, v) \in [u_n, u_{N-n}] \times [v_n, v_{M-n}]$$

defines a surface patch with the control points $\boldsymbol{d}_{ij}$; if the parametrisation is piecewise regular (which depends on the control points), then this β-spline patch is of class G^{n-1} (it is also possible to use functions of different degrees—the order of geometric continuity is determined by the lower degree).

An example of a tensor product patch of class G^2 is shown in Figure 2.17. This patch is obtained as the **spherical product** of two planar curves called the **equator** and the **meridian**.[9] The

[9] A sphere may be obtained as the spherical product of a circle—an equator and a halfcircle—a meridian, hence the name of this operation.

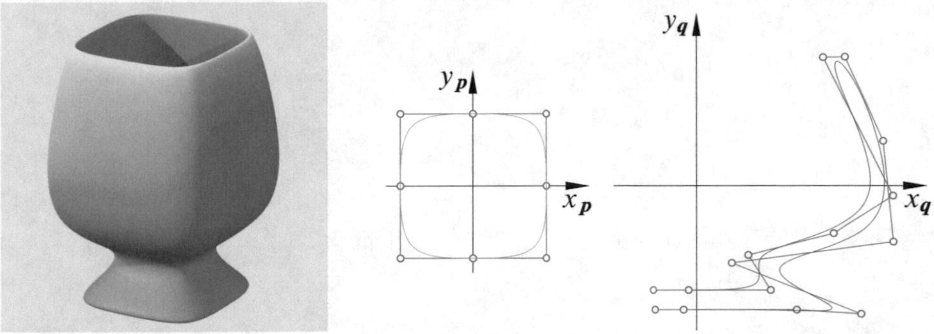

Figure 2.17: A tensor product patch obtained as the spherical product of two β-spline curves

two curves havc paramctrisations

$$p(u) = \begin{bmatrix} x_p(u) \\ y_p(u) \end{bmatrix}, \quad u \in [u_n, n_{N-n}],$$
$$q(v) = \begin{bmatrix} x_q(v) \\ y_q(v) \end{bmatrix}, \quad v \in [v_n, v_{M-n}],$$

and their spherical product is

$$s(u,v) = \begin{bmatrix} x_p(u)x_q(v) \\ y_p(u)x_q(v) \\ y_q(v) \end{bmatrix}, \quad (u,v) \in [u_n, u_{N-n}] \times [v_n, v_{M-n}].$$

If the control points of the curves are $p_i = [x_{p,i}, y_{p,i}]^T$ and $q_j = [x_{q,j}, y_{q,j}]^T$, then the control points of the spherical product are

$$d_{ij} = \begin{bmatrix} x_{p,i}x_{q,j} \\ y_{p,i}x_{q,j} \\ y_{q,j} \end{bmatrix}, \quad i = 0, \dots, N-n-1, \ j = 0, \dots, M-n-1.$$

In our example the two curves are cubic β-spline curves of class G^2.

The tensor product approach may also be used to define γ-spline patches (which describe the same class of surfaces as bicubic β-spline patches) and ν-spline patches of interpolation to a set of points given in a rectangular array. A drawback of this approach is that the influence of the connection parameters on the surface shape is not local. For example, changing the parameter $\beta_{u:i}$ of a tensor product β-spline patch s causes modifications of *all* points $s(u,v)$ for $(u,v) \in ([u_{i-n}, u_{i+n}] \cap [u_n, u_{N-n}]) \times [v_n, v_{M-n}]$. Also, the topology of tensor product geometric spline surfaces is restricted, which causes the necessity of extending the repertoire. In the chapters that follow we discuss some methods of overcoming these restrictions.

CHAPTER 3

Pairs of surface patches

In this chapter we study smooth junctions of parametric patches having a common boundary curve. We begin with deriving the equations which describe such junctions. Algebraic properties of these equations prompt some methods of constructing pairs of polynomial or rational patches whose junctions are smooth. Finally, we show examples of practical constructions of spline patches forming non-smooth surfaces which approximate smooth surfaces with good accuracy.

3.1 GEOMETRIC CONTINUITY AT A COMMON BOUNDARY

Consider two patches represented by smooth regular parametrisations, $\boldsymbol{p}(s,t)$ and $\boldsymbol{p}^*(u,v)$. We assume that the patches have a common boundary curve being a constant parameter curve of the first patch, and corresponding to $t = t_0$, and a constant parameter curve of the second patch, corresponding to $v = v_0$. Let I denote the line segment $v = v_0$, bounding the domain of the parametrisation $\boldsymbol{p}^*$. We assume that the two parametrisations of the common curve, obtained by restricting the parametrisations $\boldsymbol{p}$ and $\boldsymbol{p}^*$ and denoted by $\overline{\boldsymbol{p}}$ and $\underline{\boldsymbol{p}}^*$, are identical: $\overline{\boldsymbol{p}}(s) = \underline{\boldsymbol{p}}^*(u)$ for $s = u$. This assumption guarantees the **positional continuity** of the junction of the two patches:

$$\overline{\boldsymbol{p}} = \underline{\boldsymbol{p}}^*. \tag{3.1}$$

The derivation of equations of geometric continuity for a junction of two patches is similar to that of curve arcs. Using a function $\boldsymbol{f}\colon \mathbb{R}^2 \to \mathbb{R}^2$, whose coordinates are described by scalar functions s and t (see Figure 3.1), we obtain a new parametrisaton of the patch $\boldsymbol{p}$:

$$\boldsymbol{q}(u,v) = \boldsymbol{p}\big(s(u,v), t(u,v)\big).$$

We assume that the partial derivatives of $\boldsymbol{q}$ up to the order n at each point of the line segment I are equal to the corresponding derivatives of the parametrisation $\boldsymbol{p}^*$.

We can obtain the derivatives of the parametrisation $\boldsymbol{q}$ using the generalised Fàa di Bruno's formula (A.55) for functions of two variables. Then, we restrict them to the line segment I. We can notice that if the derivatives with respect to v of $\boldsymbol{q}$ and $\boldsymbol{p}^*$ of any order k are equal at each point of the line segment I, then also

$$\frac{\partial^{m+k}}{\partial u^m \partial v^k}\boldsymbol{q}\bigg|_{v=v_0} = \frac{\partial^{m+k}}{\partial u^m \partial v^k}\boldsymbol{p}^*\bigg|_{v=v_0}$$

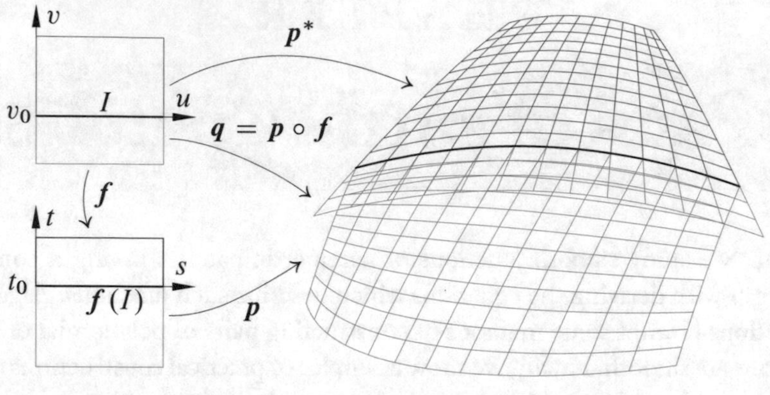

Figure 3.1: A smooth junction of two parametric patches

for all m such that these derivatives exist. Therefore, we can focus our attention on the partial derivatives with respect to v—the parameter changing across the boundary—which is why they are called the **cross-boundary derivatives** of the patches.

The cross-boundary derivatives of the parametrisation $\boldsymbol{q}$ are related to those of $\boldsymbol{p}$ in the following way:

$$\begin{aligned}\overline{\boldsymbol{q}}_v &= \overline{s}_v\overline{\boldsymbol{p}}_s + \overline{t}_v\overline{\boldsymbol{p}}_t,\\ \overline{\boldsymbol{q}}_{vv} &= \overline{s}_{vv}\overline{\boldsymbol{p}}_s + \overline{t}_{vv}\overline{\boldsymbol{p}}_t + \overline{s}_v^2\overline{\boldsymbol{p}}_{ss} + 2\overline{s}_v\overline{t}_v\overline{\boldsymbol{p}}_{st} + \overline{t}_v^2\overline{\boldsymbol{p}}_{tt}\end{aligned}$$

etc. The general (and rather impractical) formula, which is a special case of (A.55), is

$$\begin{aligned}\frac{\partial^j}{\partial v^j}\overline{\boldsymbol{q}} &= \sum_{k=1}^{j}\sum_{h=0}^{k} a_{jkh}\frac{\partial^k}{\partial s^h\partial t^{k-h}}\overline{\boldsymbol{p}}, \qquad (3.2)\\ a_{jkh} &= \binom{k}{h}\sum_{\substack{m_1+\dots+m_k=j\\ m_1,\dots,m_k>0}}\frac{j!}{k!m_1!\dots m_k!}\overline{s}_{vm_1}\dots\overline{s}_{vm_h}\overline{t}_{vm_{h+1}}\dots\overline{t}_{vm_k}.\end{aligned}$$

Here $\overline{s}_{vm_i}$ and $\overline{t}_{vm_i}$ are restrictions of partial derivatives of order m_i of the functions $s(u,v)$ and $t(u,v)$ with respect to v to the line segment I. Let $s_1,\dots,s_n$ and $t_1,\dots,t_n$ be functions of class C^n of one variable. If we define the functions

$$s(u,v) = u + \sum_{k=1}^{n}\frac{1}{k!}s_k(u)(v-v_0)^k \quad \text{and} \quad t(u,v) = t_0 + \sum_{k=1}^{n}\frac{1}{k!}t_k(u)(v-v_0)^k,$$

then we can see that their partial derivatives with respect to v up to the order n, restricted to the line segment I (where $v = v_0$) are the functions $s_1,\dots,s_n$ and $t_1,\dots,t_n$. Thus, we can use arbitrary functions $s_1, \ldots, s_n$ and $t_1, \ldots, t_n$, called the **junction functions,** to construct the cross-boundary derivatives of a patch $\boldsymbol{p}^*$ at the line segment I:

$$\underline{\boldsymbol{p}}_v^* = s_1\overline{\boldsymbol{p}}_s + t_1\overline{\boldsymbol{p}}_t, \tag{3.3}$$

$$\underline{\boldsymbol{p}}_{vv}^* = s_2\overline{\boldsymbol{p}}_s + t_2\overline{\boldsymbol{p}}_t + s_1^2\overline{\boldsymbol{p}}_{ss} + 2s_1t_1\overline{\boldsymbol{p}}_{st} + t_1^2\overline{\boldsymbol{p}}_{tt} \tag{3.4}$$

etc., using Formula (3.2) rewritten in the form

$$\frac{\partial^j}{\partial v^j}\underline{\boldsymbol{p}}^* = \sum_{k=1}^{j}\sum_{h=0}^{k} a_{jkh}\frac{\partial^k}{\partial s^h \partial t^{k-h}}\overline{\boldsymbol{p}}, \tag{3.5}$$

$$a_{jkh} = \binom{k}{h} \sum_{\substack{m_1+\cdots+m_k=j \\ m_1,\ldots,m_k>0}} \frac{j!}{k!m_1!\ldots m_k!} s_{m_1}\cdots s_{m_h} t_{m_{h+1}}\cdots t_{m_k}.$$

If we construct (by solving an interpolation problem) a parametrisation $\boldsymbol{p}^*$ whose cross-boundary derivatives for $j = 1, \ldots, n$ are as above, then the junction of the patches described by $\boldsymbol{p}$ and $\boldsymbol{p}^*$ is of class G^n. There is one more condition, though. The function t_1 must be non-zero at all points of I to prevent a singularity (i.e., linear dependence of the first-order partial derivatives) of $\boldsymbol{p}^*$. Moreover, the sign of the function t_1 must be chosen so as to avoid the creation of a cusp-like junction. If the domain of the parametrisation $\boldsymbol{q}$ and the domain of $\boldsymbol{p}^*$ are located on the opposite sides of the line segment I (as in Figure 3.1), then the function t_1 must be positive.

Equations governing smooth junctions of rational patches represented by homogeneous patches may be obtained in a similar way. Apart from changing the parameters of the homogeneous patch $\boldsymbol{P}$, we can also multiply it by any non-zero function. In this way we obtain the homogeneous patch

$$\boldsymbol{Q}(u,v) = r(u,v)\boldsymbol{P}\big(s(u,v), t(u,v)\big).$$

In the equations considered below we need functions r, s, t of class C^n. Using (3.2) and the Leibniz formula for derivatives of the product of functions, one can derive the formula

$$\frac{\partial^j}{\partial v^j}\underline{\boldsymbol{Q}} = \overline{r}_{v^j}\overline{\boldsymbol{P}} + \sum_{k=1}^{j}\sum_{h=0}^{k} A_{jkh}\frac{\partial^k}{\partial s^h \partial t^{k-h}}\overline{\boldsymbol{P}}, \tag{3.6}$$

$$A_{jkh} = \binom{k}{h}\sum_{i=k}^{j}\binom{j}{i}\overline{r}_{v^{j-i}} \sum_{\substack{m_1+\cdots+m_k=i \\ m_1,\ldots,m_k>0}} \frac{i!}{k!m_1!\ldots m_k!} \times$$
$$\times\, \overline{s}_{v^{m_1}}\ldots\overline{s}_{v^{m_h}}\overline{t}_{v^{m_{h+1}}}\ldots\overline{t}_{v^{m_k}}$$

for $j = 0, \ldots, n$.

The junction of rational patches $\boldsymbol{p}$ and $\boldsymbol{p}^*$ is of class G^n if the parametrisations $\boldsymbol{P}^*$ and $\boldsymbol{Q}$ have the same derivatives up to the order n along the line segment I. This is the case if there exist junction functions $r_0, \ldots, r_n, s_1, \ldots, s_n$ and $t_1, \ldots, t_n$ such that the cross-boundary derivatives of the homogeneous patches $\boldsymbol{P}$ and $\boldsymbol{P}^*$ satisfy the equations obtained by substituting r_j, s_j and t_j in place of $\overline{r}_{v^j}$, $\overline{s}_{v^j}$ and $\overline{t}_{v^j}$ respectively in (3.6). Thus, from (3.6) we obtain the formula

$$\frac{\partial^j}{\partial v^j}\underline{\boldsymbol{P}}^* = r_j\overline{\boldsymbol{P}} + \sum_{k=1}^{j}\sum_{h=0}^{k} A_{jkh}\frac{\partial^k}{\partial s^h \partial t^{k-h}}\overline{\boldsymbol{P}}, \tag{3.7}$$

$$A_{jkh} = \binom{k}{h}\sum_{i=k}^{j}\binom{j}{i} r_{j-i} \sum_{\substack{m_1+\cdots+m_k=i\\ m_1,\ldots,m_k>0}} \frac{i!}{k!m_1!\ldots m_k!} \times \\ \times s_{m_1}\ldots s_{m_h} t_{m_{h+1}}\ldots t_{m_k}.$$

To avoid creating a cusp-like junction, t_1 has to be greater than 0 when the domains of $\boldsymbol{Q}$ and $\boldsymbol{P}^*$ are on the opposite sides of I. In practice we are most often interested in junctions of class G^1 and G^2, where we need to ensure that

$$\begin{aligned}
\underline{\boldsymbol{P}}^* &= r_0\overline{\boldsymbol{P}},\\
\underline{\boldsymbol{P}}^*_u &= r_1\overline{\boldsymbol{P}} + r_0 s_1\overline{\boldsymbol{P}}_s + r_0 t_1\overline{\boldsymbol{P}}_t,\\
\underline{\boldsymbol{P}}^*_{uu} &= r_2\overline{\boldsymbol{P}} + (r_0 s_2 + 2r_1 s_1)\overline{\boldsymbol{P}}_s + (r_0 t_2 + 2r_1 t_1)\overline{\boldsymbol{P}}_t + {}\\
&\quad r_0 s_1^2\overline{\boldsymbol{P}}_{ss} + 2r_0 s_1 t_1\overline{\boldsymbol{P}}_{st} + r_0 t_1^2\overline{\boldsymbol{P}}_{tt}.
\end{aligned}$$

If we are not interested in junctions of class G^3 or higher, then the last equation may be rewritten in the simpler form

$$\underline{\boldsymbol{P}}^*_{uu} = r_2\overline{\boldsymbol{P}} + \tilde{s}_2\overline{\boldsymbol{P}}_s + \tilde{t}_2\overline{\boldsymbol{P}}_t + r_0 s_1^2\overline{\boldsymbol{P}}_{ss} + 2r_0 s_1 t_1\overline{\boldsymbol{P}}_{st} + r_0 t_1^2\overline{\boldsymbol{P}}_{tt},$$

with arbitrary functions $\tilde{s}_2$ and $\tilde{t}_2$ instead of s_2 and t_2.

3.2 INTERPRETATION

We take a closer look at the junctions of patches of class G^n for $n = 1$ and $n = 2$, bearing in mind that the patches have a common curve $\overline{\boldsymbol{p}} = \underline{\boldsymbol{p}}^*$. From the assumption that $\overline{\boldsymbol{p}}(s) = \underline{\boldsymbol{p}}^*(u)$ if $s = u$ it follows that all partial derivatives of the parametrisations $\boldsymbol{p}$ and $\boldsymbol{p}^*$ with respect to s and u at the junction points agree.

Case $n = 1$. At any point of the common curve the cross-boundary derivative of $\boldsymbol{p}^*$ is a linear combination of the first-order partial derivatives of $\boldsymbol{p}$. The partial derivatives of both patches at any point of their common curve determine the same plane (see Figure 3.2). Geometrically G^1 continuity is the **continuity of tangent plane** of the surface made of the two patches. One can also talk about the continuity of the normal vector, which is equivalent.

The same geometric interpretation applies to the equations for the homogeneous representations of the patches. The homogeneous patches reside in the four-dimensional space. At

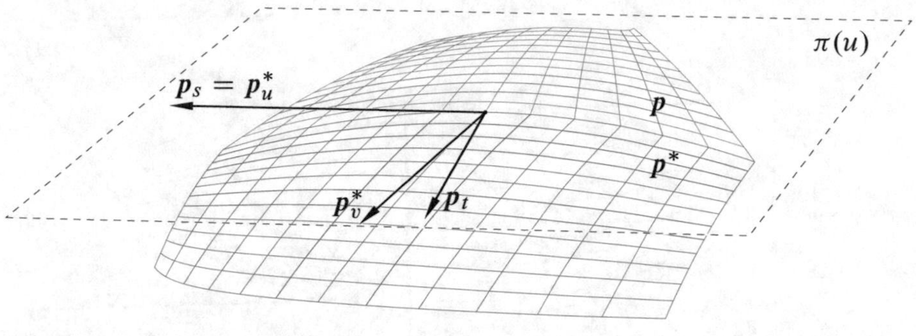

Figure 3.2: Junction of patches of class G^1

any junction point the triples of vectors, $\overline{P}(s)$, $\overline{P}_s(s)$, $\overline{P}_t(s)$ and $\underline{P}^*(u)$, $\underline{P}^*_u(u)$, $\underline{P}^*_v(u)$ span the same three-dimensional linear subspace (i.e., hyperplane) $\Pi(u)$ of $\mathbb{R}^4$. The common tangent plane $\pi(u)$ of the rational patches is represented by this hyperplane.[1]

Case $n = 2$**.** The first and second fundamental forms (see Section A.10.2) are expressed by the derivatives of the first and second order of the surface's parametrisation. The forms may be used to find the curvature of curves obtained by intersecting the surface with planes. If there exists a regular parametrisation of class C^2 of the surface, e.g. described piecewise by $\boldsymbol{q}$ and $\boldsymbol{p}^*$, then the curvature of the intersection of the surface with *any* plane not tangent to the surface is continuous.

On the other hand, having a surface whose all planar sections (with non-tangent planes) are curves with the curvature continuous, it is possible to find local regular parametrisations of class C^2 of this surface.[2] Thus geometric continuity of the second order is equivalent to the **curvature continuity** of the surface.

The equations of geometric continuity of junctions of patches have the same characteristic feature as the equations for curves. Consider the set of patches $\boldsymbol{p}^*$, whose junctions with a fixed patch $\boldsymbol{p}$ are of class G^{n-1}. The equation for the n-th order cross-boundary derivative of $\boldsymbol{p}^*$ has the form

$$\underline{\boldsymbol{p}}^*_{v^n} = s_n \overline{\boldsymbol{p}}_s + t_n \overline{\boldsymbol{p}}_t + \boldsymbol{q}_n,$$

where $\boldsymbol{q}_1 = \boldsymbol{0}$ and the curve $\boldsymbol{q}_n$ is fixed after choosing the functions $s_1, \dots, s_{n-1}$ and $t_1, \dots, t_{n-1}$. The functions s_n and t_n may be chosen arbitrarily in the construction of the n-th order cross-boundary derivative of the patch $\boldsymbol{p}^*$. Choosing the two junction functions, for a given u, we

[1]This representation is analogous to the representation of the common tangent line of planar rational arcs at their junction points by the plane spanned by the vectors $\overline{\boldsymbol{P}}, \overline{\boldsymbol{P}}'$ and $\underline{\boldsymbol{P}}^*, \underline{\boldsymbol{P}}^{*\prime}$, see Figure 2.3.

[2]One possibility is choosing the coordinate system whose xy plane is the plane $\pi(u)$ tangent to the patches at $\overline{\boldsymbol{p}}(u) = \underline{\boldsymbol{p}}^*(u)$. A part of the surface in a neighbourhood of this point is the graph of a function $f(x, y)$, and the parametrisation $\boldsymbol{s}(x, y) = [x, y, f(x, y)]^T$ is regular and of class C^2.

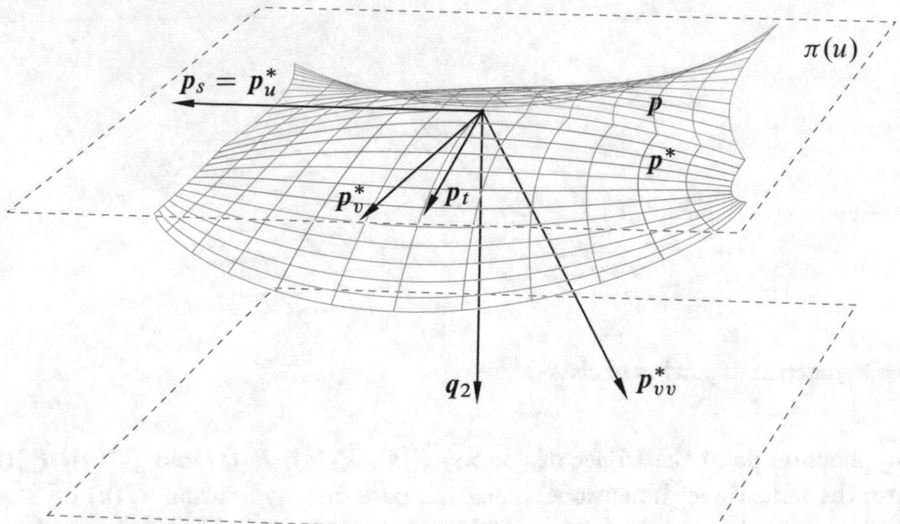

Figure 3.3: Junction of patches of class G^2

manipulate two numbers—the values of the functions—and the vector $\underline{p}_{vn}^*(u) - q_n$ remains parallel to the plane $\pi(u)$ tangent to the patches at the common point; see Figure 3.3. The same observation applies to Equation (3.6), which may be rewritten in the form

$$\underline{P}_{vn}^* = r_n\overline{P} + r_0 s_n\overline{P}_s + r_0 t_n\overline{P}_t + Q_n.$$

Changing the values of the functions r_n, s_n and t_n, for any fixed u, causes changes of the vector $\underline{P}_{vn}^*(u)$ by adding to it a linear combination of the vectors $\overline{P}(u)$, $\overline{P}_s(u)$ and $\overline{P}_t(u)$. Any such linear combination is an element of the hyperplane $\Pi(u) \subset \mathbb{R}^4$ which represents the plane $\pi(u)$ tangent to the patch p at the point $\overline{p}(u)$. Using the language of algebra, we can say that for a given u the vectors $\underline{p}_{vn}^*$ (or $\underline{P}_{vn}^*$) satisfying the equation of G^n continuity form a coset of the space $\mathbb{R}^3$, parallel to the plane $\pi(u)$, whose element is the vector $q_n(u)$ (or a coset of $\mathbb{R}^4$, parallel to the hyperplane $\Pi(u)$, whose element is the vector $Q_n(u)$).

3.3 A LITTLE BIT OF ALGEBRA

So far, we have not assumed any particular form of the parametrisations. Now we take a closer look at parametrisations described by polynomials and rational functions; in the latter case the homogeneous patches used to represent our surfaces will be made of polynomials. We are going to study the algebra hidden behind smooth junctions of two polynomial or rational patches. The key to solve the equations of geometric continuity is the theory of modules; the first to use this approach was Degen [1990]. The theoretical results form a solid ground for practical constructions of smooth surfaces discussed later.

A **module** is an algebraic structure similar to a linear vector space. The difference between the two notions is that the role of the field of scalars is played by a ring. Consider an example: let $\boldsymbol{e}_1 = [1,0]^T$ and $\boldsymbol{e}_2 = [0,1]^T$. Any vector $\boldsymbol{p} \in \mathbb{R}^2$ may be obtained as a linear combination: $\boldsymbol{p} = [x,y]^T = x\boldsymbol{e}_1 + y\boldsymbol{e}_2$. Moreover, we can take any pair of linearly independent vectors, say, $\boldsymbol{v}_1$ and $\boldsymbol{v}_2$, and find two real numbers, a_1 and a_2 such that $\boldsymbol{p} = a_1\boldsymbol{v}_1 + a_2\boldsymbol{v}_2$. Both pairs of vectors, $\{\boldsymbol{e}_1, \boldsymbol{e}_2\}$ and $\{\boldsymbol{v}_1, \boldsymbol{v}_2\}$, are bases of the linear vector space $\mathbb{R}^2$. If real polynomials of one variable are used *instead* of numbers to obtain linear combinations of vectors, we obtain polynomial vector functions, i.e., parametrisations of curves. For example, the parametrisation $\boldsymbol{p}(t) = [x(t), y(t)]^T = x(t)\boldsymbol{e}_1 + y(t)\boldsymbol{e}_2$, where $x(t)$ and $y(t)$ are polynomials, describes a curve in $\mathbb{R}^2$ and any polynomial parametrisation of a planar curve may be represented in this form. The set of all real polynomials of one variable (with the operations of addition and multiplication) is a ring, $\mathbb{R}[\cdot]$, and the set of all parametrisations made of two polynomials (with the operations of addition and multiplication by polynomials) is a module over the ring $\mathbb{R}[\cdot]$, which we denote by $\mathbb{R}[\cdot]^2$.

Elements of a module M over a ring, multiplied by elements of this ring are also elements of M. For example, if $a(t)$ is a polynomial and $\boldsymbol{p}(t) = [x(t), y(t)]^T \in \mathbb{R}[\cdot]^2$, then the product $a(t)\boldsymbol{p}(t) = [a(t)x(t), a(t)y(t)]^T$ is also an element of the module $\mathbb{R}[\cdot]^2$.

A basis of a linear vector space V is a set of vectors, say, $\{\boldsymbol{v}_1, \dots, \boldsymbol{v}_n\}$, such that any vector $\boldsymbol{p} \in V$ is a linear combination of the elements of the basis, $\boldsymbol{p} = \sum_{i=1}^{n} a_i\boldsymbol{v}_i$, and this representation is unique, i.e., the coefficients $a_1, \dots, a_n$ are determined by $\boldsymbol{p}$. Note that any real linear vector space having a non-empty basis has infinitely many bases; each basis has the same number of elements, which is (by definition) the dimension of the space V. In a similar way a **basis of a module** is defined. For example, the pair of vectors $\{\boldsymbol{e}_1, \boldsymbol{e}_2\}$ is a basis of the module $\mathbb{R}[\cdot]^2$, as each of its elements is a linear combination of the two vectors (with polynomial coefficients) and such a representation is unique. In contrast to linear spaces, there exist modules having no basis. A module having a basis is called a **free module**. The number of elements of a basis of a free module is called the **rank** of the module; it corresponds to the dimension of a linear vector space.

A module L being a subset of a module M is called a **submodule**, which corresponds to the notion of a linear vector subspace. Note the difference between modules and vector spaces: if the dimension of a subspace X of a linear vector space V is equal to the dimension of V, then the two spaces are identical. This is *not* the case for modules. If $a(t)$ is a polynomial of degree greater than 0, then the vectors $a(t)\boldsymbol{e}_1$ and $a(t)\boldsymbol{e}_2$ form a basis of a submodule L of rank 2 of the module $\mathbb{R}[\cdot]^2$, whose rank is also 2, but $L \neq \mathbb{R}[\cdot]^2$. The reason is that any element of L must be possible to obtain by multiplying the elements of its basis by some polynomials and adding the products. Thus, all elements of the submodule L are made of polynomials divisible by $a(t)$. Note, however, that any triple $\{a(t)\boldsymbol{e}_1, a(t)\boldsymbol{e}_2, \boldsymbol{p}(t)\}$, where $\boldsymbol{p}(t) \in \mathbb{R}[\cdot]^2$, is **linearly dependent over the ring** $\mathbb{R}[\cdot]$, i.e., there exist polynomials $c_1(t), c_2(t), c_3(t)$, not all equal to 0, such that the linear combination $c_1(t)a(t)\boldsymbol{e}_1 + c_2(t)a(t)\boldsymbol{e}_2 + c_3(t)\boldsymbol{p}(t)$ is the zero vector.

Exercise. Write down explicit formulae for non-zero polynomials $c_1(t), c_2(t), c_3(t)$ such that if $\boldsymbol{p}(t) = [x(t), y(t)]^T$, then $c_1(t)a(t)\boldsymbol{e}_1 + c_2(t)a(t)\boldsymbol{e}_2 + c_3(t)\boldsymbol{p}(t) = \boldsymbol{0}$.

Equations of geometric continuity of junctions of patches considered in this chapter have the general form

$$a_1\boldsymbol{c}_1 + \cdots + a_k\boldsymbol{c}_k + a_{k+1}\boldsymbol{c}_{k+1} = \boldsymbol{0}. \tag{3.8}$$

Hereafter, the argument of the functions is omitted for brevity. We assume that $\boldsymbol{c}_1, \ldots, \boldsymbol{c}_k$ are polynomial vector functions taking values in $\mathbb{R}^d$ (where $d > k$), determined by a patch $\boldsymbol{p}$ (or by a homogeneous patch $\boldsymbol{P}$). We seek the set of solutions of such equations; each solution consists of scalar functions $a_1, \ldots, a_{k+1}$ and a polynomial vector function $\boldsymbol{c}_{k+1}$, which may be used (in the way described later) to construct a cross-boundary derivative of the second patch, $\boldsymbol{p}^*$ (or $\boldsymbol{P}^*$).

The left-hand side of (3.8) is a linear combination of polynomial curves in $\mathbb{R}^d$, i.e., elements of the module $\mathbb{R}[\cdot]^d$, with coefficients $a_1, \ldots, a_{k+1}$ being functions. The equation expresses the linear dependency of the vectors $\boldsymbol{c}_1(t), \ldots, \boldsymbol{c}_{k+1}(t) \in \mathbb{R}^d$ for all $t \in \mathbb{R}$, which is equivalent to the linear dependency of the curves $\boldsymbol{c}_1, \ldots, \boldsymbol{c}_{k+1}$. It turns out that if continuous functions $a_1, \ldots, a_{k+1}$ satisfying Equation (3.8) exist, then one can find polynomials doing the job and our equation describes the linear dependency of the curves over the ring $\mathbb{R}[\cdot]$.

Consider the matrix $A \in \mathbb{R}[\cdot]^{d\times k}$ (having polynomial entries in its d rows and k columns) whose columns are the curves $\boldsymbol{c}_1, \ldots, \boldsymbol{c}_k$ from Equation (3.8). With $d > k$ we can cross out any $d - k$ rows and obtain a square matrix (with k rows and columns) made of polynomials. The determinant of such a matrix is a polynomial, which we call a **minor** of the matrix A. The highest common factor of all $\binom{d}{k}$ minors is a polynomial; it will be called the **highest factor** of the curves $\boldsymbol{c}_1, \ldots, \boldsymbol{c}_k$ and denoted by the symbol $\mathcal{F}(\boldsymbol{c}_1, \ldots, \boldsymbol{c}_k)$. If the curves are linearly independent (over the ring $\mathbb{R}[\cdot]$), then their highest factor is non-zero and it is determined up to a constant factor. We may assume that its leading coefficient in the power basis is 1. The theorems given below were proved in Kiciak [1995, 1996].

Theorem 3.1 *If there exist continuous functions $a_1, \ldots, a_{k+1}$, not all equal to zero, which satisfy Equation (3.8) with curves $\boldsymbol{c}_1, \ldots, \boldsymbol{c}_{k+1} \in \mathbb{R}[\cdot]^d$, then there exist polynomials $a_1, \ldots, a_k$, which satisfy this equation with these curves and with the polynomial $a_{k+1} = \mathcal{F}(\boldsymbol{c}_1, \ldots, \boldsymbol{c}_k)$.*

Theorem 3.2 *For fixed curves $\boldsymbol{c}_1, \ldots, \boldsymbol{c}_k \in \mathbb{R}[\cdot]^d$, linearly independent over the ring $\mathbb{R}[\cdot]$, there exist curves $\hat{\boldsymbol{c}}_1, \ldots \hat{\boldsymbol{c}}_k \in \mathbb{R}[\cdot]^d$ such that any curve $\boldsymbol{c}_{k+1}$ satisfying (3.8) together with some polynomials $a_1, \ldots, a_{k+1}$ is a linear combination*

$$\boldsymbol{c}_{k+1} = \hat{a}_1\hat{\boldsymbol{c}}_1 + \cdots + \hat{a}_k\hat{\boldsymbol{c}}_k$$

whose coefficients $\hat{a}_1, \ldots, \hat{a}_k$ are polynomials.

Theorem 3.3 *There exist polynomial curves* $\hat{\boldsymbol{c}}_1, \dots, \hat{\boldsymbol{c}}_k$ *such that for any polynomials* $\hat{a}_1, \dots, \hat{a}_k$ *the curve* $\boldsymbol{c}_{k+1} = \hat{a}_1\hat{\boldsymbol{c}}_1 + \cdots + \hat{a}_k\hat{\boldsymbol{c}}_k$ *satisfies* (3.8) *with some polynomials* $a_1, \dots, a_k$ *and the degree of this curve is equal to the maximal degree of the terms of this sum. The leading coefficients (in the power basis) of the curves* $\hat{\boldsymbol{c}}_1, \dots, \hat{\boldsymbol{c}}_k$ *are linearly independent vectors in* $\mathbb{R}^d$.

In other words, by assuming that a_{k+1} is the highest factor of the curves $\boldsymbol{c}_1, \dots, \boldsymbol{c}_k$ we do not reject any curve $\boldsymbol{c}_{k+1}$ satisfying Equation (3.8) together with some continuous functions. Theorem 3.2[3] states that the set of curves $\boldsymbol{c}_{k+1}$ linearly dependent on $\boldsymbol{c}_1, \dots, \boldsymbol{c}_k$ over the ring $\mathbb{R}[\cdot]$ is a free submodule of rank k of the module $\mathbb{R}[\cdot]^d$. Theorem 3.3 claims the existence of a specific basis of this module, which will be called a **basis of minimal degree**.

The module of curves $\boldsymbol{c}_{k+1}$ linearly dependent on $\boldsymbol{c}_1, \dots, \boldsymbol{c}_k$ over the ring $\mathbb{R}[\cdot]$, i.e., the set of all solutions of (3.8), will be denoted by the symbol $\mathcal{M}(\boldsymbol{c}_1, \dots, \boldsymbol{c}_k)$. To denote the module whose elements are linear combinations of those curves, we shall use the symbol $\mathcal{L}(\boldsymbol{c}_1, \dots, \boldsymbol{c}_k)$; the latter module is a submodule of the former one. The two modules are equal if and only if the highest factor of the curves $\boldsymbol{c}_1, \dots, \boldsymbol{c}_k$ is a constant (i.e., $\deg \mathcal{F}(\boldsymbol{c}_1, \dots, \boldsymbol{c}_k) = 0$). There exist curves $\hat{\boldsymbol{c}}_1, \dots, \hat{\boldsymbol{c}}_k$ which form a basis of the module $\mathcal{M}(\boldsymbol{c}_1, \dots, \boldsymbol{c}_k)$, and for such curves we can write $\mathcal{M}(\boldsymbol{c}_1, \dots, \boldsymbol{c}_k) = \mathcal{L}(\hat{\boldsymbol{c}}_1, \dots, \hat{\boldsymbol{c}}_k)$. Moreover, for any such a module there exists a basis of minimal degree.

Equations of geometric continuity of order higher than 1 have the form

$$a_1\boldsymbol{c}_1 + \cdots + a_k\boldsymbol{c}_k + a_{k+1}(\boldsymbol{k} - e\boldsymbol{s}) = \boldsymbol{0}, \tag{3.9}$$

with the curves $\boldsymbol{c}_1, \dots, \boldsymbol{c}_k$ and $\boldsymbol{k}$ fixed and also with a polynomial e fixed. Solutions of such equations are described by the following theorem:

Theorem 3.4 *If the set of polynomial curves* $\boldsymbol{s}$ *satisfying Equation* (3.9) *is non-empty, then there exists a curve* $\hat{\boldsymbol{k}}$*, such that each solution has the form*

$$\boldsymbol{s} = \hat{\boldsymbol{k}} + \hat{a}_1\hat{\boldsymbol{c}}_1 + \cdots + \hat{a}_k\hat{\boldsymbol{c}}_k, \tag{3.10}$$

where the curves $\hat{\boldsymbol{c}}_1, \dots, \hat{\boldsymbol{c}}_k$ *are elements of a basis of the module* $\mathcal{M}(\boldsymbol{c}_1, \dots, \boldsymbol{c}_k)$ *and* $\hat{a}_1, \dots, \hat{a}_k$ *are polynomials.*

According to this theorem the set of solutions $\boldsymbol{s}$ of (3.9) is either empty, or it is a coset of the module $\mathbb{R}[\cdot]^d$ parallel to the submodule $\mathcal{M}(\boldsymbol{c}_1, \dots, \boldsymbol{c}_k)$. This set is empty when no curve of the form $\boldsymbol{k} - b_1\hat{\boldsymbol{c}}_1 - \cdots - b_k\hat{\boldsymbol{c}}_k$, where $b_1, \dots, b_k$ are polynomials, is divisible by the polynomial e.

3.4 POLYNOMIAL SOLUTIONS OF EQUATIONS OF GEOMETRIC CONTINUITY

Now we return to the equations of geometric continuity derived in Section 3.1, assuming that the parametrisations of patches, whose junctions we are studying, are polynomial (e.g. in Bézier

[3]This is a special case of a more general theorem, that any submodule of a free module over a principal ideal domain is free (the ring $\mathbb{R}[\cdot]$ is a principal ideal domain); see Hungerford [1974].

representation). In that case the common boundary $\overline{\boldsymbol{p}} = \underline{\boldsymbol{p}}^*$ and the cross-boundary derivatives are polynomial curves in $\mathbb{R}^3$. We assume that the patch $\boldsymbol{p}$ is regular, which implies that the polynomial curves $\overline{\boldsymbol{p}}_s$ and $\overline{\boldsymbol{p}}_t$, which describe its partial derivatives, are linearly independent over the ring $\mathbb{R}[\cdot]$.

Theorem 3.5 *The existence of junction functions $s_1, \ldots, s_n$ and $t_1, \ldots, t_n$ which satisfy Equations (3.5) for $j = 1, \ldots, n$ is equivalent to the existence of polynomials $b_1, \ldots, b_n$ and $c_1, \ldots, c_n$ and d, which satisfy the equations*

$$b_j \overline{\boldsymbol{p}}_s + c_j \overline{\boldsymbol{p}}_t + d\big(\boldsymbol{k}_j - d^{2j-2}\underline{\boldsymbol{p}}^*_{v^j}\big) = \boldsymbol{0}, \tag{3.11}$$

where

$$\boldsymbol{k}_j = \sum_{k=2}^{j}\sum_{h=0}^{k} \hat{a}_{jkh}\overline{\boldsymbol{p}}_{s^h t^{k-h}},$$

$$\hat{a}_{jkh} = \binom{k}{h} \sum_{\substack{m_1+\cdots+m_k=j \\ m_1,\ldots,m_k>0}} \frac{j!}{k!m_1!\ldots m_k!} b_{m_1}\ldots b_{m_h} c_{m_{h+1}}\ldots c_{m_k}.$$

The proof may be found in Kiciak [1996]. It may be done by applying Theorems 3.1–3.4 to Equations (3.2), and as it is done by a rather tedious calculation, we omit it here. The two cases of Equation (3.11) that are the most important in practice[4] are

$$b_1 \overline{\boldsymbol{p}}_s + c_1 \overline{\boldsymbol{p}}_t - d\,\underline{\boldsymbol{p}}^*_v = \boldsymbol{0},$$
$$b_2 \overline{\boldsymbol{p}}_s + c_2 \overline{\boldsymbol{p}}_t + d\big(\underbrace{b_1^2 \overline{\boldsymbol{p}}_{ss} + 2b_1c_1\overline{\boldsymbol{p}}_{st} + c_1^2\overline{\boldsymbol{p}}_{tt}}_{\boldsymbol{k}_2} - d^2\underline{\boldsymbol{p}}^*_{vv}\big) = \boldsymbol{0}.$$

In a similar way, we can deal with the equations written for the homogeneous representation of rational patches, which were first published by Zheng, Wang and Liang [1995]. We assume that the homogeneous patches have polynomial parametrisations $\boldsymbol{P}$ and $\boldsymbol{P}^*$. The regularity of the rational patch $\boldsymbol{p}$ at the boundary to be shared with $\boldsymbol{p}^*$ implies the linear independence over $\mathbb{R}[\cdot]$ of three curves: the boundary curve $\overline{\boldsymbol{P}}$ and the partial derivatives $\overline{\boldsymbol{P}}_s$ and $\overline{\boldsymbol{P}}_t$ at the boundary. To simplify we assume that the junction function r_0 is a constant, 1; this assumption only restricts the set of pairs of homogeneous patches representing the smoothly joined rational patches to the pairs having a common boundary curve.

Theorem 3.6 *The existence of the junction functions $r_0 = 1, r_1, \ldots, r_n, s_1, \ldots, s_n$ and $t_1, \ldots, t_n$, satisfying Equations (3.7) for $j = 1, \ldots, n$, is equivalent to the existence of polynomials $a_1, \ldots, a_n$,*

[4] Equations for a junction of patches of class G^2 were first published by Kahmann [1983], who derived them from geometric considerations.

$b_1, \dots, b_n, c_1, \dots, c_n$ *and* d*, which satisfy the equations*

$$a_j\overline{\boldsymbol{P}} + b_j\overline{\boldsymbol{P}}_s + c_j\overline{\boldsymbol{P}}_t + d\big(\boldsymbol{K}_j - d^{2j-2}\underline{\boldsymbol{P}}^*_{vj}\big) = \boldsymbol{0}, \tag{3.12}$$

where

$$\boldsymbol{K}_j = \sum_{k=1}^{j}\sum_{h=0}^{k}\hat{A}_{jkh}\overline{\boldsymbol{P}}_{s^h t^{k-h}},$$

$$\hat{A}_{jkh} = \binom{k}{j}\sum_{\substack{i=k,\dots,j\\ k=1\Rightarrow i\neq j}}\binom{j}{i}\times$$

$$\times\sum_{\substack{m_1+\dots+m_k=i\\ m_1,\dots,m_k>0}}\frac{i!}{k!m_1!\dots m_k!}a_{j-i}b_{m_1}\dots b_{m_h}c_{m_{h+1}}\dots c_{m_k}d^{k-1}$$

and (by convention[5]*)* $a_0 = 1/d$.

For $j = 1$ and $j = 2$ Equation (3.12) has the form

$$a_1\overline{\boldsymbol{P}} + b_1\overline{\boldsymbol{P}}_s + c_1\overline{\boldsymbol{P}}_t - d\,\underline{\boldsymbol{P}}^*_v = \boldsymbol{0},$$

$$a_2\overline{\boldsymbol{P}} + b_2\overline{\boldsymbol{P}}_s + c_2\overline{\boldsymbol{P}}_t + d\big(\underbrace{2a_1b_1\overline{\boldsymbol{P}}_s + 2a_1c_1\overline{\boldsymbol{P}}_t + b_1^2\overline{\boldsymbol{P}}_{ss} + 2b_1c_1\overline{\boldsymbol{P}}_{st} + c_1^2\overline{\boldsymbol{P}}_{tt}}_{\boldsymbol{K}_2} - d^2\underline{\boldsymbol{P}}^*_{vv}\big) = \boldsymbol{0}.$$

If geometric continuity of order higher than 2 is not needed, then the latter equation may be replaced by

$$a_2\overline{\boldsymbol{P}} + \tilde{b}_2\overline{\boldsymbol{P}}_s + \tilde{c}_2\overline{\boldsymbol{P}}_t + d\big(b_1^2\overline{\boldsymbol{P}}_{ss} + 2b_1c_1\overline{\boldsymbol{P}}_{st} + c_1^2\overline{\boldsymbol{P}}_{tt} - d^2\underline{\boldsymbol{P}}^*_{vv}\big) = \boldsymbol{0},$$

with arbitrary polynomials $\tilde{b}_2$ and $\tilde{c}_2$ instead of b_2 and c_2. We do not need the polynomials b_2 and c_2, unless we are going to find the curve $\boldsymbol{K}_3$, which appears in the equation for the G^3 continuity of the junction of patches.

The transition from Equations (3.5) and (3.7) to (3.11) and (3.12) is the first step towards practical constructions of smooth junctions of patches. Suppose that having a patch $\boldsymbol{p}$ we need to construct a Bézier patch $\boldsymbol{p}^*$, that would make a junction of class G^n with $\boldsymbol{p}$. The problem reduces to finding the polynomial curves $\underline{\boldsymbol{p}}^*_v, \dots, \underline{\boldsymbol{p}}^*_{vn}$, i.e., the cross-boundary derivatives of the patch $\boldsymbol{p}^*$, which may then be obtained by solving an interpolation problem defined with these curves—which is particularly easy to do with the Bézier representation of the patch. But before turning to practical constructions, we shall take a closer look at the theoretical properties of the equations and their solutions.

By Theorem 3.1, taking the polynomial $d = \mathcal{F}(\overline{\boldsymbol{p}}_s, \overline{\boldsymbol{p}}_t)$ makes it possible to obtain *all* solutions (i.e., curves $\underline{\boldsymbol{p}}^*_{vj}$) of Equations (3.11). However, the polynomial d in practical constructions

[5]Each term of $\hat{A}_{jkh}$ with the factor a_0 has also the factor d, and the two factors cancel each other out.

is a constant, which may be 1 without any further loss of generality. Strictly speaking, though constructions found in literature use various notations and formulations of conditions of geometric continuity, their results are usually possible to obtain with the polynomial $d = 1$ in Equations (3.11). There are a number of reasons for taking this restriction.

The first reason is that finding a basis of the module $\mathcal{M}(\overline{\boldsymbol{p}}_s, \overline{\boldsymbol{p}}_t)$, which enables constructing *all* polynomial patches $\boldsymbol{p}^*$ joined smoothly with $\boldsymbol{p}$, is in general a complex and costly numerical procedure (see Kiciak [1996]). The second reason is that *almost always* this module has a basis whose elements are the curves $\overline{\boldsymbol{p}}_s$ and $\overline{\boldsymbol{p}}_t$, and using this procedure is unnecessary. This is the case when the highest factor of the two curves, $\mathcal{F}(\overline{\boldsymbol{p}}_s, \overline{\boldsymbol{p}}_t)$, is a constant, and here we discover the third reason: the polynomial $\mathcal{F}(\overline{\boldsymbol{p}}_s, \overline{\boldsymbol{p}}_t)$ is the highest common factor of three minors of the matrix $[\overline{\boldsymbol{p}}_s, \overline{\boldsymbol{p}}_t] \in \mathbb{R}[\cdot]^{3\times 2}$. Its set of (real and complex) zeros is the intersection of the sets of zeros of the minors. If $\mathcal{F}(\overline{\boldsymbol{p}}_s, \overline{\boldsymbol{p}}_t) \neq \text{const}$, then all minors have at least one common zero, but an arbitrarily small perturbation of the patch $\boldsymbol{p}$ may change the minors in such a way that their zeros become different. Then, solutions of Equations (3.11) possible to obtain only with d not being a constant disappear. To explain this phenomenon, consider the problem of computing a common point of two lines in the three-dimensional space. An arbitrarily small perturbation may turn intersecting lines into skew lines without a common point. Thus, constructions without the restriction $d = \text{const}$ are unstable.

The fourth reason of taking $d = \text{const}$ is relevant for junctions smoother than G^1. By Theorem 3.4, the set of curves $\boldsymbol{s}$ satisfying an equation of the form (3.9) may be empty. Let $n > 1$. The curve $\boldsymbol{k}_n$ is determined by cross-boundary derivatives of the patch $\boldsymbol{p}$ and by the junction functions taken to construct solutions of Equations (3.11) for $j < n$ (i.e., cross-boundary derivatives of order less than n of the patch $\boldsymbol{p}^*$). If the curve $\boldsymbol{k}_n$ has been fixed with a polynomial $d \neq \text{const}$, then any curve which might appear in the place of $d^{2n-2}\underline{\boldsymbol{p}}^*_{v^n}$ may not be divisible by the polynomial d^{2n-2}. The polynomial $d = 1$ relieves us of this trouble. And the fifth reason is that there is an easy way around the restriction $d = \text{const}$ using auxiliary patches (developing the analogy to the problem of finding the common point of lines in space, this approach corresponds to choosing a plane in the space and using a representation of the lines in this plane; if the lines intersect, small perturbations of such a representation cannot make their common point disappear). This technique is described in the next section.

Constructions of a rational patch $\boldsymbol{p}^*$ forming a smooth surface with a rational patch $\boldsymbol{p}$ are related to the module $\mathcal{M}(\overline{\boldsymbol{P}}, \overline{\boldsymbol{P}}_s, \overline{\boldsymbol{P}}_t) \subset \mathbb{R}[\cdot]^4$ determined by the boundary curve and the partial derivatives of the homogeneous patch $\boldsymbol{P}$ representing $\boldsymbol{p}$. The reasons for taking the polynomial $d = 1$ in practical constructions of homogeneous patches $\boldsymbol{P}^*$ are the same as in the non-rational case.

Let us try to gain some intuition about the modules discussed here. If the parametrisation $\boldsymbol{p}(s,t)$ is made of polynomials $x(s,t)$, $y(s,t)$ and $z(s,t)$, then the curves $\overline{\boldsymbol{p}}_s$, $\overline{\boldsymbol{p}}_t$ are described by restrictions (to $t = t_0$) of the partial derivatives, which we denote $\overline{x}_s, \overline{y}_s, \overline{z}_s, \overline{x}_t, \overline{y}_t, \overline{z}_t$. The curve $\overline{\boldsymbol{n}}$ whose points are normal vectors of the patch $\boldsymbol{p}$ at (the interesting part of) its boundary is

$$\overline{\boldsymbol{n}} = \begin{bmatrix} \overline{y}_s\overline{z}_t - \overline{y}_t\overline{z}_s \\ \overline{z}_s\overline{x}_t - \overline{z}_t\overline{x}_s \\ \overline{x}_s\overline{y}_t - \overline{x}_t\overline{y}_s \end{bmatrix}.$$

As we can see, the coordinates of the curve $\overline{\boldsymbol{n}}$ are the three minors of the matrix $[\overline{\boldsymbol{p}}_s, \overline{\boldsymbol{p}}_t]$, equipped with appropriate signs. Using the ordinary vector product in $\mathbb{R}^3$, we defined above a vector multiplication of curves in $\mathbb{R}[\cdot]^3$, which may be written briefly $\overline{\boldsymbol{n}} = \overline{\boldsymbol{p}}_s \wedge \overline{\boldsymbol{p}}_t$. The module $\mathcal{M}(\overline{\boldsymbol{p}}_s, \overline{\boldsymbol{p}}_t)$ is the set of all polynomial curves $\boldsymbol{p}_v^*$ such that for all $u \in \mathbb{R}$ the vectors $\overline{\boldsymbol{n}}(u)$ and $\boldsymbol{p}_v^*(u)$ are orthogonal (Fig. 3.4). Extending the notion of orthogonality, we can say that the module $\mathcal{M}(\overline{\boldsymbol{p}}_s, \overline{\boldsymbol{p}}_t)$ is the set of all polynomial curves orthogonal to the curve describing the normal vector at the boundary of the patch $\boldsymbol{p}$. Thus, this module is a representation of tangent planes at all points of the boundary of $\boldsymbol{p}$.

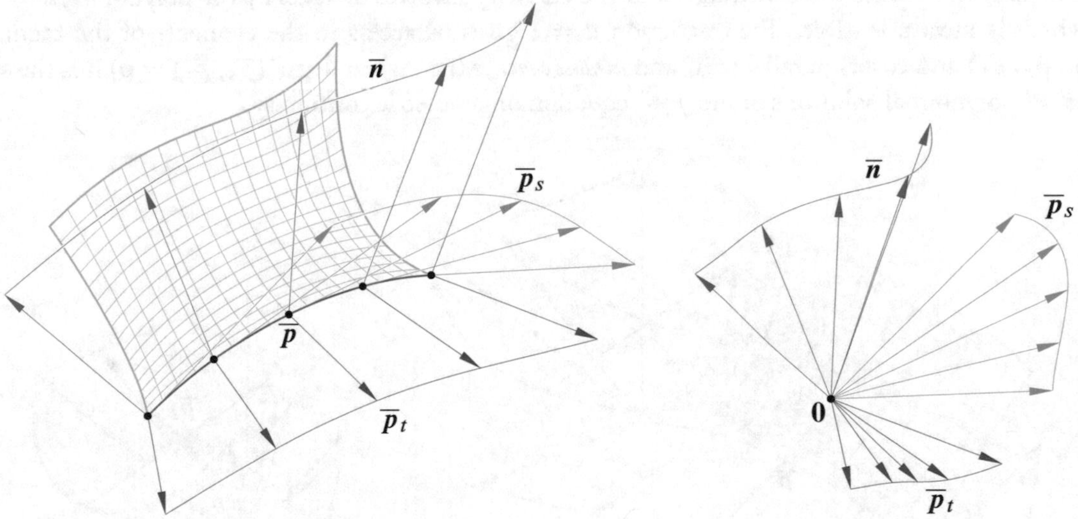

Figure 3.4: Curves $\overline{\boldsymbol{p}}_s$, $\overline{\boldsymbol{p}}_t$ and $\overline{\boldsymbol{n}}$

The degree of the curve $\overline{\boldsymbol{n}}$ is not higher than the sum of degrees of the curves $\overline{\boldsymbol{p}}_s$ and $\overline{\boldsymbol{p}}_t$, which in turn are determined by the degree of the patch $\boldsymbol{p}$. If $\deg \mathcal{F}(\overline{\boldsymbol{p}}_s, \overline{\boldsymbol{p}}_t) > 0$, then we can divide the curve $\overline{\boldsymbol{n}}$ by this polynomial and obtain a polynomial curve of a lower degree which also describes the normal vector at the boundary of the patch $\boldsymbol{p}$. Somewhat informally we can say that if the normal vector of the patch $\boldsymbol{p}$ may be described by a curve of lower degree than $\deg \overline{\boldsymbol{p}}_s + \deg \overline{\boldsymbol{p}}_t$, then the shape of the patch $\boldsymbol{p}$ is "simpler" than it might be within the limitations determined by the degree of $\boldsymbol{p}$. In the extreme case, if $\deg \mathcal{F}(\overline{\boldsymbol{p}}_s, \overline{\boldsymbol{p}}_t) = \deg \overline{\boldsymbol{p}}_s + \deg \overline{\boldsymbol{p}}_t$, then the normal vector at the boundary of $\boldsymbol{p}$ has a constant direction (as it may be described with a curve of degree 0), and then there exists a plane tangent to the patch $\boldsymbol{p}$ at *all* points of its boundary curve. The conclusion drawn from the considerations above is that by assuming $d = 1$, we give up only the patches $\boldsymbol{p}^*$ possible to construct when the patch $\boldsymbol{p}$ has a simpler shape than its degree allows.

At the end of Section 3.2 we stated that if the cross-boundary derivatives up to the order $n-1$ are fixed so as to achieve the G^{n-1} continuity of junction of the patch $\boldsymbol{p}^*$ with $\boldsymbol{p}$, then at each point of the common boundary the set of vectors $\underline{\boldsymbol{p}}^*_{v^n}(u)$ satisfying Equation (3.2) for $j = n$ is a coset of the space $\mathbb{R}^3$ parallel to the tangent plane at that point. The coset of the module $\mathbb{R}[\cdot]^3$ parallel to the submodule $\mathcal{M}(\overline{\boldsymbol{p}}_s, \overline{\boldsymbol{p}}_t)$ whose elements are polynomial curves satisfying (3.11) for $j = n$, if it exists (i.e., is non-empty), represents all these cosets of $\mathbb{R}^3$.

By substituting $d = 1$ in (3.11) we go back to Equations (3.5), but now all functions $s_1 = b_1, \ldots, s_n = b_n$ and $t_1 = c_1, \ldots, t_n = c_n$ are polynomials. By choosing b_1 and c_1 we fix the curve $\underline{\boldsymbol{p}}^*_v \in \mathcal{L}(\overline{\boldsymbol{p}}_s, \overline{\boldsymbol{p}}_t)$. The higher order cross-boundary derivatives of $\boldsymbol{p}^*$ are chosen from the cosets of the module $\mathbb{R}[\cdot]^3$ parallel to the submodule $\mathcal{L}(\overline{\boldsymbol{p}}_s, \overline{\boldsymbol{p}}_t)$ and containing $\boldsymbol{k}_j$. Each of the cosets is non-empty, because the construction of the curve $\boldsymbol{k}_j$ involves division by the polynomial $d = 1$, which is always feasible. The restriction $d = 1$ gives us access to the elements of the module $\mathcal{L}(\overline{\boldsymbol{p}}_s, \overline{\boldsymbol{p}}_t)$ and cosets parallel to it, and *almost always* (i.e., when $\deg \mathcal{F}(\overline{\boldsymbol{p}}_s, \overline{\boldsymbol{p}}_t) = 0$) it is the set of *all* polynomial solutions of the j-th equation of geometric continuity.

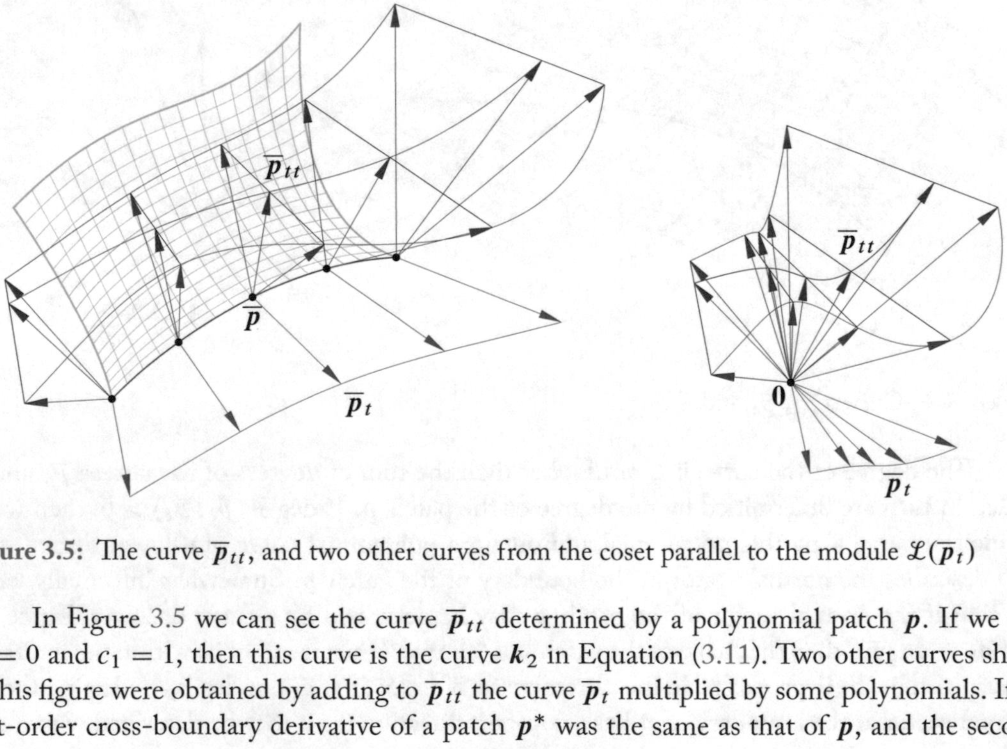

Figure 3.5: The curve $\overline{\boldsymbol{p}}_{tt}$, and two other curves from the coset parallel to the module $\mathcal{L}(\overline{\boldsymbol{p}}_t)$

In Figure 3.5 we can see the curve $\overline{\boldsymbol{p}}_{tt}$ determined by a polynomial patch $\boldsymbol{p}$. If we take $b_1 = 0$ and $c_1 = 1$, then this curve is the curve $\boldsymbol{k}_2$ in Equation (3.11). Two other curves shown in this figure were obtained by adding to $\overline{\boldsymbol{p}}_{tt}$ the curve $\overline{\boldsymbol{p}}_t$ multiplied by some polynomials. If the first-order cross-boundary derivative of a patch $\boldsymbol{p}^*$ was the same as that of $\boldsymbol{p}$, and the second-order cross-boundary derivative of $\boldsymbol{p}^*$ was described by any of those curves, then the junction of the two patches would be of class G^2.

Let us take a look at one more restriction in addition to $d = 1$. By fixing the parameters s and u (taking $u = s$), we obtain curves on the patches $\boldsymbol{p}$ and $\boldsymbol{p}^*$. If all polynomials $b_1, \dots, b_{n-1}$ are zero, then any choice of $c_1, \dots, c_{n-1}$ (where $c_1 \neq 0$) gives us the curves $\boldsymbol{k}_1, \dots, \boldsymbol{k}_n$ being linear combinations (over the ring $\mathbb{R}[\cdot]$) of the cross-boundary derivatives $\overline{\boldsymbol{p}}_t, \dots, \overline{\boldsymbol{p}}_{t^n}$. Equation (3.11) is, then, equivalent to (2.12); hence, the junction of the two constant parameter curves on the patches $\boldsymbol{p}$ and $\boldsymbol{p}^*$ is of class G^n—and this takes place for each pair of constant parameter curves meeting at the common boundary of the patches. By taking $b_1 = 0$ we restrict the possible choice of cross-boundary derivative $\boldsymbol{p}_v^*$ to the module $\mathcal{L}(\overline{\boldsymbol{p}}_t)$, whose elements are products of the vector function $\overline{\boldsymbol{p}}_t$ and (all) polynomials of the variable s. Taking all polynomials $b_1, \dots, b_n$ equal to 0 restricts the cross-boundary derivatives up to order n of the patch $\boldsymbol{p}^*$ to the cosets parallel to the module $\mathcal{L}(\overline{\boldsymbol{p}}_t)$.

Smooth junctions of rational patches in homogeneous representations are related in a similar way with the module $\mathcal{M}(\overline{\boldsymbol{P}}, \overline{\boldsymbol{P}}_s, \overline{\boldsymbol{P}}_t)$, which is a submodule of the module $\mathbb{R}[\cdot]^4$. The elements of this module are polynomial curves $\underline{\boldsymbol{P}}_v^*$ such that for any given $u = s$ the vector $\underline{\boldsymbol{P}}_v^*(u) \in \mathbb{R}^4$ is an element of the hyperplane $\Pi(s) \subset \mathbb{R}^4$, representing the plane $\pi(s)$ tangent to the patch $\boldsymbol{p}$ at the point $\overline{\boldsymbol{p}}(s)$. Any coset of the module $\mathbb{R}[\cdot]^4$ parallel to $\mathcal{M}(\overline{\boldsymbol{P}}, \overline{\boldsymbol{P}}_s, \overline{\boldsymbol{P}}_t)$ represents cosets[6] of the space $\mathbb{R}^4$ parallel to the hyperplanes $\Pi(s)$. By taking the polynomial $d = 1$, we restrict our choice of cross-boundary derivatives of the patch $\boldsymbol{P}^*$ to the submodule $\mathcal{L}(\overline{\boldsymbol{P}}, \overline{\boldsymbol{P}}_s, \overline{\boldsymbol{P}}_t)$ and to the cosets parallel to this submodule. This restriction gives us the stability of the construction and a guarantee that the parallel cosets (from which we may choose the cross-boundary derivatives of order higher than 1) are non-empty.

Exercise. Suppose that the degree of the polynomial $\mathcal{F}(\overline{\boldsymbol{p}}_s, \overline{\boldsymbol{p}}_t)$ is equal to $\deg \overline{\boldsymbol{p}}_s + \deg \overline{\boldsymbol{p}}_t$. Find a basis of the module $\mathcal{M}(\overline{\boldsymbol{p}}_s, \overline{\boldsymbol{p}}_t)$ (hint: there is no need to use any general algorithm in this and the following exercises; just read carefully the text above and use your imagination).

Exercise. Find a basis of the module $\mathcal{M}(\overline{\boldsymbol{p}}_s, \overline{\boldsymbol{p}}_t)$ if the curve $\overline{\boldsymbol{p}}_s$ is planar and the junction of regular patches $\boldsymbol{p}$ and $\boldsymbol{p}^*$, being images of each other in the symmetric reflection with respect to the plane containing $\overline{\boldsymbol{p}}_s$, is of class G^1.

Exercise. Prove that a symmetric pair of smooth patches whose junction is of class G^1 (like in the previous exercise) forms a surface of class G^2.

Exercise. Discuss the consequences of taking $d = 1$ and $b_1 = \cdots = b_n = 0$ in Equations (3.12) and the form of junctions of rational patches possible to obtain with such a restriction, in analogy to the discussion of junctions of polynomial patches given above.

---

[6]We reserve the term "hyperplanes" to the linear vector subspaces of dimension $d - 1$ of the space $\mathbb{R}^d$. All linear subspaces contain the zero vector, and other cosets are obtained by parallel translations of the linear subspaces.

3.5 CONSTRUCTING PAIRS OF PATCHES

Now we are getting closer to practical constructions of pairs of patches whose junctions are of class G^1 and G^2. A higher order continuity is rarely needed and, if necessary, the constructions described below may be extended in the obvious way. We use Bézier representation of the patches.

A key element of the construction is the **auxiliary patch**, denoted by $\boldsymbol{p}$. One of its boundary curves (corresponding to $t = 0$ in all examples in this section) becomes the common curve (corresponding to $v = 0$) of the two patches $\boldsymbol{p}^*$ to be constructed. The tangent plane of the auxiliary patch at each point of this curve will be *their* tangent plane. If the final patches are supposed to form a surface of class G^2, then the auxiliary patch determines also their normal curvature in all directions. The cross-boundary derivatives of each element of the pair will be constructed with its own set of polynomial junction functions. Their degrees are limited by the maximal degree of the final patch in the way explained with examples below.

The first example is a construction of polynomial patches shown in Figure 3.6, whose junction is of class G^1. The degree of the auxiliary patch is (3, 1), and the degree of the final patches with respect to the parameter u is 5. The Bézier representation of the common boundary curve of the final patches is obtained by degree elevation (from $n = 3$ to $n^* = 5$, see Section A.2.3) of the boundary curve of the auxiliary patch. The first-order cross-boundary derivatives of the final patches are obtained using the formula

$$\underline{\boldsymbol{p}}_v^* = b_1\overline{\boldsymbol{p}}_s + c_1\overline{\boldsymbol{p}}_t.$$

The degree of the product of polynomial functions is the sum of degrees of the factors. We do not assume any degree reduction (due to cancellation of the terms of the highest degree), nor the possibility of dividing the expression on the right-hand side by any polynomial d of degree greater than 0; the reasons were explained in the previous section. The degree of the function $\overline{\boldsymbol{p}}_s$, which is the derivative of the boundary curve of the auxiliary patch, is $n - 1 = 2$, and the cross-boundary derivative $\overline{\boldsymbol{p}}_t$ is of degree 3. Hence, in our example the degrees of the polynomials b_1 and c_1 must not exceed 3 and 2 respectively to guarantee that the cross-boundary derivative $\underline{\boldsymbol{p}}_v^*$ is at most quintic. Two polynomials c_1 used to obtain the cross-boundary derivatives of the two final patches must have constant and opposite signs in the interval [0, 1] in order to avoid getting a cusp-like junction; one of the final patches makes a cusp-like junction with the auxiliary patch. Apart from that, the junction polynomials may be arbitrary.

The multiplication of polynomials and polynomial vector functions given in Bézier form, i.e., in Bernstein polynomial bases, may be done as described in Section A.2.3. Having the Bézier representations of the boundary curve and the cross-boundary derivative, it is possible to find the control points of the final patches by solving linear equations obtained from Formulae (A.17) and (A.19). This gives us two rows of the control net of each final patch. The other rows do not affect the continuity of the tangent plane at the common boundary. To obtain further rows of the control net so as to satisfy the interpolation conditions for higher order cross-boundary derivatives, considered in the examples below, we can use Formula (A.20).

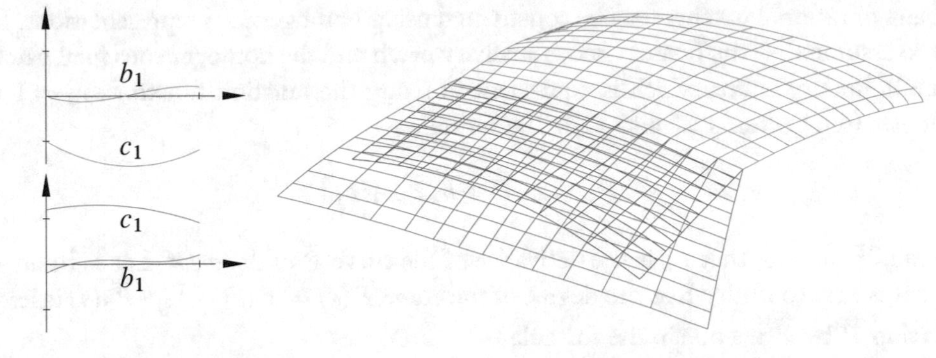

Figure 3.6: Junction of polynomial patches of class G^1 (the auxiliary patch is drawn in red)

The second example, a pair of patches whose junction is of class G^2, is shown in Figure 3.7. The auxiliary patch of degree (3, 2) is used to construct the boundary curve of the final patches and their first-order cross-boundary derivatives as in the previous example. The second-order cross-boundary derivatives are obtained using the formula

$$\boldsymbol{p}^*_{vv} = b_2\overline{\boldsymbol{p}}_s + c_2\overline{\boldsymbol{p}}_t + b_1^2\overline{\boldsymbol{p}}_{ss} + 2b_1c_1\overline{\boldsymbol{p}}_{st} + c_1^2\overline{\boldsymbol{p}}_{tt}.$$

It is assumed that the degree n^* of the final patches with respect to u is 7. The degree of $\overline{\boldsymbol{p}}_{ss}$ is 1, the degree of $\overline{\boldsymbol{p}}_s$ and $\overline{\boldsymbol{p}}_{st}$ is 2, and the degree of $\overline{\boldsymbol{p}}_t$, and $\overline{\boldsymbol{p}}_{tt}$ is 3. The rule that the degree of sum of polynomials is the highest degree of a term allows us to find the following limits for the degrees of junction polynomials: $\deg b_1 \leq 3$, $\deg c_1 \leq 2$, $\deg b_2 \leq 5$, $\deg c_2 \leq 4$. Note that to construct the first-order cross-boundary derivatives of degree 7, one might use polynomials b_1 and c_1 of degree 5 and 4, but the formula for the second-order derivative imposes tighter restrictions on their degrees. To obtain the final patches in Bézier form, one has to apply degree elevation to the boundary curve and to the first-order cross-boundary derivative, up to 7.

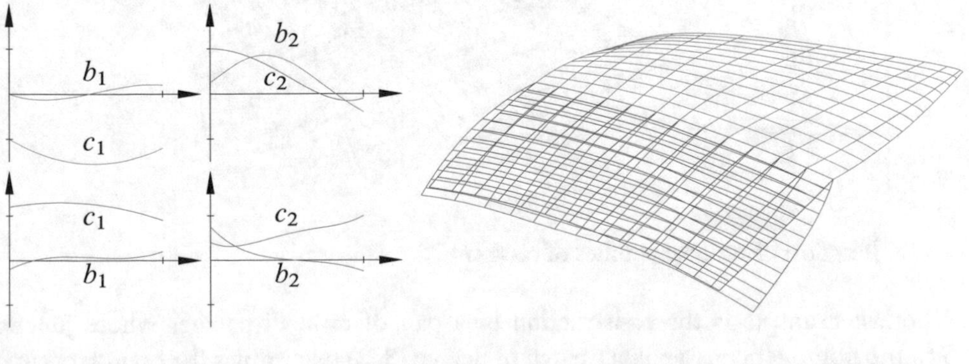

Figure 3.7: Junction of polynomial patches of class G^2

Pairs of rational patches may be constructed using homogeneous representations. It is convenient to assume that the homogeneous auxiliary patch and the homogeneous final patches have the same boundary curve, which is equivalent to fixing the junction functions $r_0 = 1$ for both final patches (see Sections 3.1 and 3.4). The formula

$$\underline{\boldsymbol{P}}_v^* = a_1\overline{\boldsymbol{P}} + b_1\overline{\boldsymbol{P}}_s + c_1\overline{\boldsymbol{P}}_t \tag{3.13}$$

may be modified; note that $\overline{\boldsymbol{P}}_s$ is the derivative of the curve $\overline{\boldsymbol{P}}$ of degree n. Let s_0 be an arbitrary number. It is easy to notice that the degree of the curve $\widehat{\boldsymbol{P}}(s) = \overline{\boldsymbol{P}}(s) - \frac{s-s_0}{n}\overline{\boldsymbol{P}}_s(s)$ is less than n. By replacing $\overline{\boldsymbol{P}}$ by $\widehat{\boldsymbol{P}}$ we obtain the formula

$$\underline{\boldsymbol{P}}_v^* = a_1\widehat{\boldsymbol{P}} + \hat{b}_1\overline{\boldsymbol{P}}_s + c_1\overline{\boldsymbol{P}}_t \tag{3.14}$$

with an arbitrary polynomial $\hat{b}_1$ instead of b_1. With this formula the limit for the degree of a_1 is increased by 1. It is best (because of symmetry of the construction) to take $s_0 = \frac{1}{2}$. Then, for a homogeneous patch $\boldsymbol{P}(s,t) = \sum_{i=0}^{n}\sum_{j=0}^{m} \boldsymbol{P}_{ij} B_i^n(s) B_j^m(t)$ one can find the following formulae: $\overline{\boldsymbol{P}}(s) = \sum_{i=0}^{n} \boldsymbol{P}_{i0} B_i^n(s)$ and $\widehat{\boldsymbol{P}}(s) = \sum_{i=0}^{n-1} \frac{1}{2}(\boldsymbol{P}_{i0} + \boldsymbol{P}_{i+1,0}) B_i^{n-1}(s)$.

Remark. Because the leading coefficients of the curves $\overline{\boldsymbol{P}}$ and $\overline{\boldsymbol{P}}_s$ in the power basis are vectors having the same direction, the triple $\{\overline{\boldsymbol{P}}, \overline{\boldsymbol{P}}_s, \overline{\boldsymbol{P}}_t\}$ cannot be a basis of minimal degree of the module $\mathcal{L}(\overline{\boldsymbol{P}}, \overline{\boldsymbol{P}}_s, \overline{\boldsymbol{P}}_t)$ (see Theorem 3.3). Replacing $\overline{\boldsymbol{P}}$ by $\widehat{\boldsymbol{P}}$ usually produces such a basis.

An example is shown in Figure 3.8. Here, the degree of the auxiliary patch is (3, 1) and the degree of both final patches with respect to the parameter u is 5. The degrees of the polynomials a_1, $\hat{b}_1$ and c_1 are 3, 3 and 2 respectively.

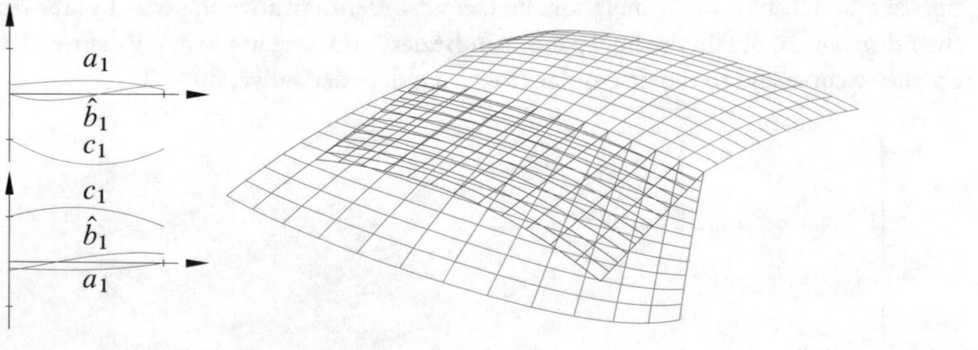

Figure 3.8: Junction of rational patches of class G^1

The last example is the construction of a pair of rational patches whose junction is of class G^2. The homogeneous auxiliary patch of degree (3, 2) determines the boundary curve of the final homogeneous patches, and their first-order cross-boundary derivatives may be constructed just as in the previous example. Again, we assume that the degree of the final patches with respect to u is 7. The second-order cross-boundary derivatives are obtained using the formula

$$\underline{\boldsymbol{P}}^*_{vv} = a_2\widehat{\boldsymbol{P}} + \hat{b}_2\overline{\boldsymbol{P}}_s + \tilde{c}_2\overline{\boldsymbol{P}}_t + b_1^2\overline{\boldsymbol{P}}_{ss} + 2b_1c_1\overline{\boldsymbol{P}}_{st} + c_1^2\overline{\boldsymbol{P}}_{tt}. \tag{3.15}$$

In the construction of the first-order cross-boundary derivative a polynomial $\hat{b}_1$ was used instead of b_1; we need the latter to construct the second-order cross-boundary derivative. There is $b_1(u) = \hat{b}_1(u) - \frac{u-1/2}{n}a_1(u)$. From the formula for the second-order cross-boundary derivative and the assumptions about the degrees of the auxiliary and final patches, we obtain the following restrictions for the degrees of junction polynomials: $\deg b_1 \le 3$, $\deg c_1 \le 2$, $\deg a_2 \le 5$, $\deg \hat{b}_2 \le 5$, $\deg \tilde{c}_2 \le 4$. To satisfy these restrictions we have to take $\deg a_1 \le 2$ and $\deg \hat{b}_1 \le 3$. However, using the original Formula (3.13) instead of (3.14) to obtain the first-order cross-boundary derivative allows us to take the polynomial a_1 of degree up to 4. An example of rational patches, represented by homogeneous patches whose cross-boundary derivatives were constructed using Formulae (3.13) and (3.15), is shown in Figure 3.9.

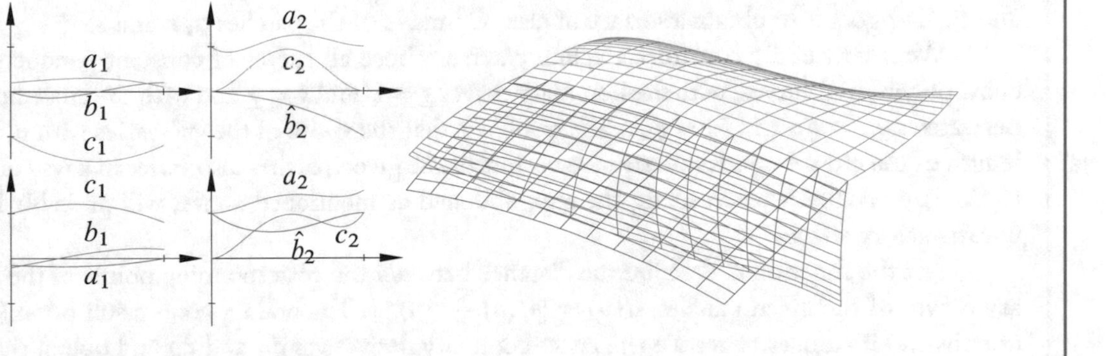

Figure 3.9: Junction of rational patches of class G^2

As promised, we got closer to practical constructions. However, the road from a smooth *pair* of patches to a smooth surface made of *many* patches is still long and bumpy. The main obstacle on this road is the necessity of satisfying the compatibility conditions, and the entire Chapter 4 is dedicated to this problem. In Chapter 5, a method of constructing pairs of patches using auxiliary patches, developed from the idea described here, allows us to obtain smooth surfaces made of many patches.

3.6 APPROXIMATING SMOOTH JUNCTIONS

As we have seen in the previous section, polynomial patches obtained with polynomial junction functions often have a high degree, which is undesirable in practical applications. In CAD applications it may be acceptable to give up the true geometric continuity in order to lower the degree of surfaces; it suffices that a non-smooth surface approximates a smooth surface within a given tolerance. After all, physical objects, even those obtained with the best manufacturing technologies, have rough surfaces, which one can see under a microscope.

Below, quasi-smooth junctions of spline patches are considered. Reducing the shape defects below a given tolerance threshold is possible due to the approximation properties of splines considered in Section A.8.4. We discuss two cases that are important in practice: joining a spline patch to given patches along their constant parameter (boundary) curves and joining a spline patch to a given trimmed patch.

3.6.1 JOINING PATCHES ALONG CONSTANT PARAMETER CURVES

We consider two bicubic B-spline patches denoted by $\boldsymbol{q}$ and $\boldsymbol{r}$. The domains of the parametrisations $\boldsymbol{q}(u, v)$ and $\boldsymbol{r}(u, v)$ are the rectangles $[c, d] \times [a, b]$ and $[e, f] \times [a, b]$ respectively; the range of the parameter v for both patches is the same interval $[a, b]$, though the sequences of "v" knots for the two patches may be different. Our purpose is to fill the gap between the boundary curves $\overline{\boldsymbol{q}}(v) = \boldsymbol{q}(d, v)$ and $\underline{\boldsymbol{r}}(v) = \boldsymbol{r}(e, v)$ with another spline patch, $\boldsymbol{s}$, whose domain is $[0, 1] \times [a, b]$ and whose two boundary curves, $\underline{\boldsymbol{s}}(v) = \boldsymbol{s}(0, v)$ and $\overline{\boldsymbol{s}}(v) = \boldsymbol{s}(1, v)$, coincide with the curves $\overline{\boldsymbol{q}}$ and $\underline{\boldsymbol{r}}$. Our goal is to obtain a surface of class G^1 made of the patches $\boldsymbol{q}$, $\boldsymbol{r}$ and $\boldsymbol{s}$.

We might easily construct a spline patch $\boldsymbol{s}$ whose all curves of constant parameter v are cubic polynomial curves, with the boundary curves $\underline{\boldsymbol{s}} = \overline{\boldsymbol{q}}$ and $\overline{\boldsymbol{s}} = \underline{\boldsymbol{r}}$ and with the cross-boundary derivatives $\underline{\boldsymbol{s}}_u = \overline{\boldsymbol{q}}_u$ and $\overline{\boldsymbol{s}}_u = \underline{\boldsymbol{r}}_u$.[7] The trouble is that the width of the gap varies with v, and the length of the cross-boundary derivative vectors of the given patches also varies in a way unrelated to the gap width. A patch filling the gap, obtained as mentioned above, will probably have an unsatisfactory shape.

Let the function d describe the distance between the corresponding points of the boundary curves of the given patches: $d(v) = \|\overline{\boldsymbol{q}}(v) - \underline{\boldsymbol{r}}(v)\|_2$. To obtain a good result often (though not always) it suffices to rescale the cross-boundary derivatives $\overline{\boldsymbol{q}}_u$ and $\underline{\boldsymbol{r}}_u$ and obtain the cross-boundary derivatives of the patch $\boldsymbol{s}$ such that $\|\underline{\boldsymbol{s}}_u(v)\|_2 = \|\overline{\boldsymbol{s}}_u(v)\|_2 = d(v)$ for all $v \in [a, b]$. The construction may be done by taking

$$\underline{\boldsymbol{s}}_u(v) = c_1(v)\overline{\boldsymbol{q}}_u(v) \qquad \text{and} \qquad \overline{\boldsymbol{s}}_u(v) = g_1(v)\underline{\boldsymbol{r}}_u(v), \tag{3.16}$$

where

$$c_1(v) = \frac{d(v)}{\|\overline{\boldsymbol{q}}_u(v)\|_2} \qquad \text{and} \qquad g_1(v) = \frac{d(v)}{\|\underline{\boldsymbol{q}}_u(v)\|_2}.$$

Formulae (3.16) are variants of (3.3), with the zero junction functions multiplying the derivative of the boundary curve. The scaling factors c_1 and g_1 are the junction functions multiplying the cross-boundary derivatives. They are transcendental because of the square root in the formula used to define the norm. An example is shown in Figure 3.10. To obtain spline vector functions

[7] The interpolation conditions described here define a unique patch—it is obtained just like the patch $\boldsymbol{p}_2$ in the construction of a bicubically blended Coons patch, see Section A.9.2.

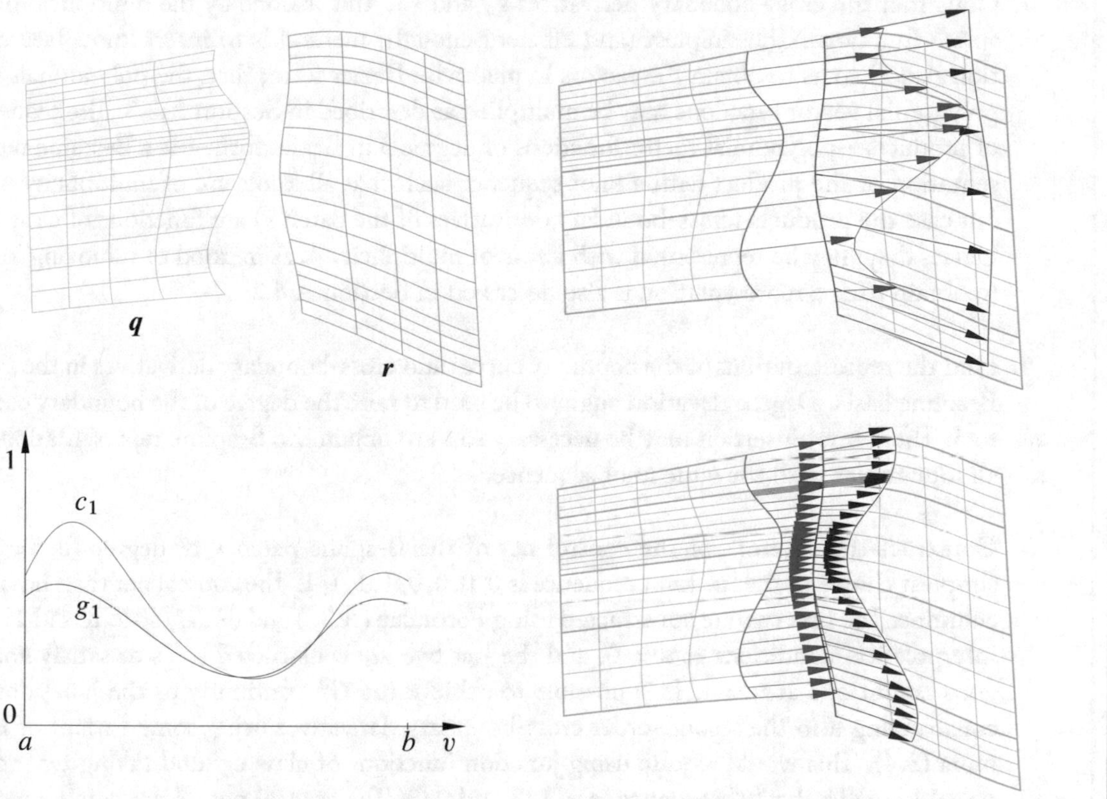

Figure 3.10: Rescaling cross-boundary derivatives of bicubic spline patches

which describe cross-boundary derivatives, and then a spline patch filling the gap, we have two ways.

The first way is to use spline functions c_1 and g_1 which may, but do not have to, approximate the transcendental functions defined above. As we are interested in G^1 continuity, we need the cross-boundary derivatives of class C^1; hence, the splines c_1 and g_1 must be at least quadratic. *If* they are quadratic, then the cross-boundary derivatives of the patch $\boldsymbol{s}$ are quintic and the degree of the patch $\boldsymbol{s}$ with respect to v will be 5. Though such a degree is not prohibitively high, the construction is rather complicated. It consists of the steps outlined below:

1. Extract (from the B-spline representations of the patches $\boldsymbol{q}$ and $\boldsymbol{r}$) the boundary curves $\overline{\boldsymbol{q}}$, $\underline{\boldsymbol{r}}$ and the cross-boundary derivatives $\overline{\boldsymbol{q}}_u$ and $\underline{\boldsymbol{r}}_u$, represented as cubic B-spline curves (see Section A.5).

2. Choose the junction functions (quadratic splines) c_1 and g_1 of class C^1.

3. Construct the cross-boundary derivatives $\underline{s}_u$ and $\overline{s}_u$. This is done by the multiplication of spline functions. The simplest (and efficient enough) method is to insert knots (see Section A.4.2) so as to obtain the factors in piecewise Bézier form; then the polynomials and polynomial vector functions may be multiplied as described in Section A.2.3. The sequence of products—polynomial vector functions of degree 5 in Bézier form—is a B-spline representation of the product with a knot sequence such that all knots are of multiplicity 6. In our case the products (cross-boundary derivatives of the patch $\boldsymbol{s}$) are functions of class C^1; hence, they may be represented with knots of multiplicity 4. A method of removing knots to obtain such a representation is also described in Section A.4.2.

4. Find the representations of the boundary curves and cross-boundary derivatives in the same B-spline basis. Degree elevation ought to be used to raise the degree of the boundary curves to 5. Then, knot insertion may be necessary so as to obtain the B-spline representations of all four curves with the same knot sequence.

5. Construct the columns of the control net of the B-spline patch $\boldsymbol{s}$ of degree $(3, 5)$. The simplest choice of the "u" knot sequence is $0, 0, 0, 0, 1, 1, 1, 1$. The control net then has four columns. The first two are constructed using Formulae (A.36) and (A.37) so as to satisfy the interpolation conditions at $u = 0$, and the last two are constructed so as to satisfy analogous conditions at $u = 1$. It is possible to achieve the G^2 continuity of the junctions by constructing also the second-order cross-boundary derivatives using some variant of Formula (3.4). This would require using junction functions of class C^2 and taking two additional knots in the "u" sequence, e.g. $1/3$ and $2/3$. The control net of the patch $\boldsymbol{s}$ would then have six columns. The third and fourth column should be obtained so as to interpolate the second-order cross-boundary derivatives at $u = 0$ and $u = 1$.

The second way is a much simpler construction whose result is a bicubic patch filling the gap. The price to pay is the discontinuity of the normal vector at the common boundaries of the patches. This discontinuity may be made arbitrarily small. To make the construction, we choose a sequence of interpolation knots in the interval $[a, b]$. This sequence should contain all knots of the "v" sequences of the patches $\boldsymbol{q}$ and $\boldsymbol{r}$. The presence of additional knots results in smaller discontinuities of the normal vector.

For each interpolation knot v_i we can compute the vectors $\overline{\boldsymbol{q}}_u(v_i)$ and $\underline{\boldsymbol{r}}_u(v_i)$, and the values of the junction functions. If a surface of class G^1 is the goal, then the junction functions only have to have continuous first-order derivatives but may otherwise be arbitrary (even transcendental). The cross-boundary derivative vectors at the knot v_i are obtained by a simple multiplication: $\underline{\boldsymbol{s}}_u(v_i) = c_1(v_i)\overline{\boldsymbol{q}}_u(v_i)$, $\overline{\boldsymbol{s}}_u(v_i) = g_1(v_i)\underline{\boldsymbol{r}}_u(v_i)$. The cross-boundary derivatives $\underline{\boldsymbol{s}}_u$ and $\overline{\boldsymbol{s}}_u$ are replaced by cubic spline curves of interpolation; we denote them $\underline{\boldsymbol{s}}_u^*$ and $\overline{\boldsymbol{s}}_u^*$. By knot insertion we can obtain the B-spline representations of all four cubic curves, $\underline{\boldsymbol{s}}$, $\overline{\boldsymbol{s}}$, $\underline{\boldsymbol{s}}_u^*$ and $\overline{\boldsymbol{s}}_u^*$, with the same knot sequence. It becomes the "v" knot sequence of the bicubic patch $\boldsymbol{s}^*$ whose junctions with

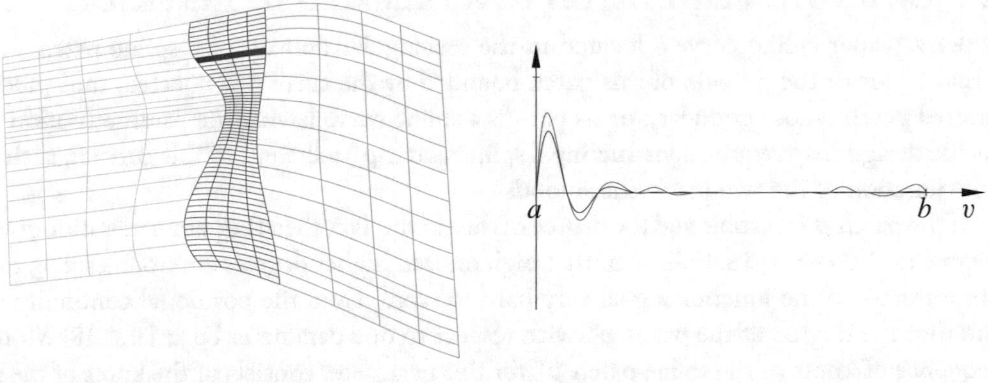

Figure 3.11: A bicubic patch filling the gap and the graph of discontinuities of the normal vector

the patches $\boldsymbol{q}$ and $\boldsymbol{r}$ are formally of class G^0, but the angle between the normal vectors of the patches at each common point is small.

The last statement needs an explanation. With the junction functions whose derivatives satisfy the Lipschitz condition, we obtain the products (vector functions) $\underline{\boldsymbol{s}}_u = c_1\overline{\boldsymbol{q}}_u$ and $\overline{\boldsymbol{s}}_u = g_1\underline{\boldsymbol{r}}_u$ whose derivatives also satisfy the Lipschitz condition; if the junction functions are of class C^2, so are these vector functions. Depending on the end conditions imposed on the curves of interpolation $\underline{\boldsymbol{s}}_u^*$ and $\overline{\boldsymbol{s}}_u^*$, we may apply Theorem A.19 (see the exercise on page 206) or Theorem A.20. By these theorems, increasing the density of interpolation knots, i.e., reducing the maximal distance h between consecutive knots, results in decreasing the approximation error, which has an estimate proportional to h^2 or even h^3. Hence, if the discontinuity of the normal vector (which, for any fixed $v \in [a, b]$, does not exceed the angle between the vectors $\underline{\boldsymbol{s}}_u(v)$ and $\underline{\boldsymbol{s}}_u^*(v)$ or $\overline{\boldsymbol{s}}_u(v)$ and $\overline{\boldsymbol{s}}_u^*(v)$) is too big, we can apply the simple remedy of choosing a denser sequence of interpolation knots.

An example of the result may be seen in Figure 3.11, where 17 interpolation knots in the interval $[a, b]$ were used. The graphs on the right side show the sine of the angle between the normal vectors of the patches $\boldsymbol{q}$ and $\boldsymbol{s}$, and $\boldsymbol{r}$ and $\boldsymbol{s}$ at their common points. This angle does not exceed $0°5'43''$ and $0°4'24''$ respectively. Doubling the density of knots reduced these discontinuities to $0°1'10''$ and $0°0'24''$ respectively.

Formulae (3.16) are special cases of (3.3) with the function $s_1 = 0$. It is possible to take non-zero junction functions which multiply the derivatives of the common curves in order to improve the results. It is also possible to obtain a junction of bicubic spline patches with small discontinuities of curvature; developing the details of such a construction is left as an exercise for the reader. In the next section, we tackle a more challenging problem, which is impossible to solve with spline patches of low degrees.

3.6.2 JOINING A SPLINE PATCH TO A TRIMMED SPLINE PATCH

Consider a planar spline curve $\boldsymbol{c}$ located in the rectangular domain of a spline patch $\boldsymbol{p}$. Suppose that a part of the domain of this patch bounded by the curve $\boldsymbol{c}$ is rejected, thus producing a **trimmed patch** whose boundary, or its part, is a spline curve having the parametrisation $\boldsymbol{p} \circ \boldsymbol{c}$. A specific design may require constructing a spline patch $\boldsymbol{p}^*$ adherent to this curve in such a way that the junction of the two patches is smooth.

If the patch $\boldsymbol{p}$ is bicubic and the degree of the curve $\boldsymbol{c}$ is 3, then the parametrisation $\boldsymbol{p} \circ \boldsymbol{c}$ has the degree $(3 + 3) \cdot 3 = 18$. Splines of that high or even higher degrees are troublesome, making the smoothness of the junction a goal very hard to score. Even the positional continuity would require that the degree of the patch $\boldsymbol{p}^*$ with respect to one parameter be at least 18. Moreover, the sequence of knots of the spline patch $\boldsymbol{p}^*$ for this parameter consists of the knots of the spline curve $\boldsymbol{c}$ and knots corresponding to intersections of this curve with the lines $u = u_i$ and $v = v_j$, where u_i and v_j are knots of the spline patch $\boldsymbol{p}$; the latter have to be found by solving nonlinear algebraic equations.

For the reasons given above, it makes sense to give up even the positional continuity of the junction. Instead of the patch $\boldsymbol{p}^*$ of an impractically high degree, whose junction with the trimmed patch $\boldsymbol{p}$ is of class G^1 or G^2, it is possible to construct a bicubic B-spline patch $\hat{\boldsymbol{p}}^*$ whose boundary approximates the boundary of the trimmed patch within a given tolerance. However, to develop such a construction we need to include in our theoretical considerations the patch $\boldsymbol{p}^*$ and to recognise the conditions which have to be satisfied by the patch $\boldsymbol{p}$ and by the trimming curve $\boldsymbol{c}$. If these conditions are satisfied, there exists a patch $\boldsymbol{p}^*$ whose junction with $\boldsymbol{p}$ at the boundary curve obtained by trimming is of class G^1 or G^2. The patch $\hat{\boldsymbol{p}}^*$ being the result of the construction approximates $\boldsymbol{p}^*$ and its junction with $\boldsymbol{p}$ may be said to be of class "quasi G^1" or "quasi G^2".

The idea of the construction is to obtain the boundary curve and one or two cross-boundary derivatives of the patch $\hat{\boldsymbol{p}}^*$, and then to construct this patch by solving an interpolation problem. The cross-boundary derivatives of the patch $\hat{\boldsymbol{p}}^*$ are constructed using the partial derivatives of $\boldsymbol{p}$ at the points of the curve $\boldsymbol{c}$ and junction functions.

Given a bicubic spline patch $\boldsymbol{p}$ trimmed by a cubic curve $\boldsymbol{c}$ it is possible to obtain a junction with the patch $\boldsymbol{p}^*$ of class G^1. To obtain the curvature continuity we need the parametrisation $\boldsymbol{p}$ of class C^4 and $\boldsymbol{c}$ of class C^3; the reasons are explained later. Therefore, the construction of the bicubic patch $\boldsymbol{p}^*$ making a junction of class G^2 with $\boldsymbol{p}$ assumes that the degree of the curve $\boldsymbol{c}$ is 4 and the patch $\boldsymbol{p}$ is biquintic. The curve $\boldsymbol{p} \circ \boldsymbol{c}$ being the boundary of the patch $\boldsymbol{p}^*$, has the degree $(5 + 5) \cdot 4 = 40$, which is prohibitive in practice.

The example in Figure 3.12 is a biquintic B-spline patch with a hole made by removing from the rectangular domain an area bounded by a closed quartic B-spline curve. The idea of the construction is explained using this example. The patch $\hat{\boldsymbol{p}}^*$, which we are going to construct, is a deformed cylinder whose boundary consists of two closed curves; one of them will be fitted to the boundary of the hole. After introducing junction functions we can write formulae to describe

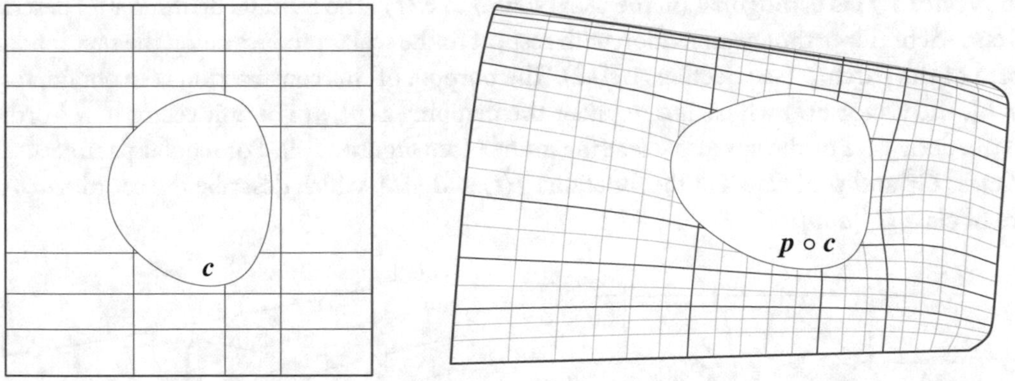

Figure 3.12: A trimmed biquintic B-spline patch

the boundary curve and two cross-boundary derivatives of the patch $\boldsymbol{p}^*$. The three curves will be evaluated at a number of points (interpolation knots) in the domain of $\boldsymbol{c}$ and then three cubic spline curves of interpolation will be constructed. The three curves are the boundary and cross-boundary derivatives of the bicubic patch $\hat{\boldsymbol{p}}^*$, which will be obtained by solving the interpolation problem for these curves. As in the previous section, Theorem A.20 will play an essential role in ensuring that by choosing a sufficiently dense sequence of knots, we can reduce the relevant discontinuities below the assumed tolerance threshold.

Let an interval $[a, b]$ be the domain of the parametrisation $\boldsymbol{c}$. This interval is also the domain of the composition $\boldsymbol{p} \circ \boldsymbol{c}$, which is a parametrisation of the boundary of the hole in the patch $\boldsymbol{p}$, of the junction functions s_1, t_1, s_2 and t_2, and of the cross-boundary derivatives of the patch $\boldsymbol{p}^*$. It will also be the interval for the parameter u of the B-spline patch $\hat{\boldsymbol{p}}^*$ being the result of the construction.

It is convenient to construct the junction functions in two steps. In the first step we use the differential $\mathrm{D}\boldsymbol{p}|_{\boldsymbol{c}(t)}$ of the parametrisation $\boldsymbol{p}$ at the point $\boldsymbol{c}(t)$ (see Section A.10.2). The curve $\boldsymbol{c}$ is regular, i.e., for any $t \in [a, b]$ the vector $\boldsymbol{c}'(t) = [x(t), y(t)]^T$ is non-zero. The image $\mathrm{D}\boldsymbol{p}|_{\boldsymbol{c}(t)}\big(\boldsymbol{c}'(t)\big)$ of the derivative of $\boldsymbol{c}$ at t is the derivative vector of the parametrisation $\boldsymbol{p} \circ \boldsymbol{c}$ of the boundary of the trimmed patch $\boldsymbol{p}$. Using the first fundamental form $\boldsymbol{I}_{\boldsymbol{c}(t)}$, we define for all $t \in [a, b]$ the vector

$$\boldsymbol{v}(t) = \frac{1}{\sqrt{\boldsymbol{I}_{\boldsymbol{c}(t)}\big(\boldsymbol{w}(t), \boldsymbol{w}(t)\big)}}\boldsymbol{w}(t) = \begin{bmatrix} g(t) \\ h(t) \end{bmatrix},$$

where

$$\boldsymbol{w}(t) = \boldsymbol{z}(t) - \frac{\boldsymbol{I}_{\boldsymbol{c}(t)}\big(\boldsymbol{c}'(t), \boldsymbol{z}(t)\big)}{\boldsymbol{I}_{\boldsymbol{c}(t)}\big(\boldsymbol{c}'(t), \boldsymbol{c}'(t)\big)}\boldsymbol{c}'(t), \qquad \boldsymbol{z}(t) = \begin{bmatrix} -y(t) \\ x(t) \end{bmatrix}. \tag{3.17}$$

The vector $\boldsymbol{z}(t)$ is orthogonal (in the usual sense) to $\boldsymbol{c}'(t)$. The formula defining $\boldsymbol{v}(t)$ describes the Gram–Schmidt orthonormalisation with respect to the scalar product being the first fundamental form of the patch $\boldsymbol{p}$ (see Section A.10.2). The purpose of this construction is to obtain, for all $t \in [a, b]$, the vector $\boldsymbol{v}(t)$ whose image under the mapping $\mathrm{D}\boldsymbol{p}|_{\boldsymbol{c}(t)}$ is a unit vector in $\mathbb{R}^3$ orthogonal to the derivative of the curve $\boldsymbol{p} \circ \boldsymbol{c}$ at the point t (see Figure 3.13). For regular parametrisations $\boldsymbol{c}$ of class C^3 and $\boldsymbol{p}$ of class C^4 the functions $g(t)$ and $h(t)$, which describe the coordinates of $\boldsymbol{v}(t)$, are of class $C^2[a, b]$.

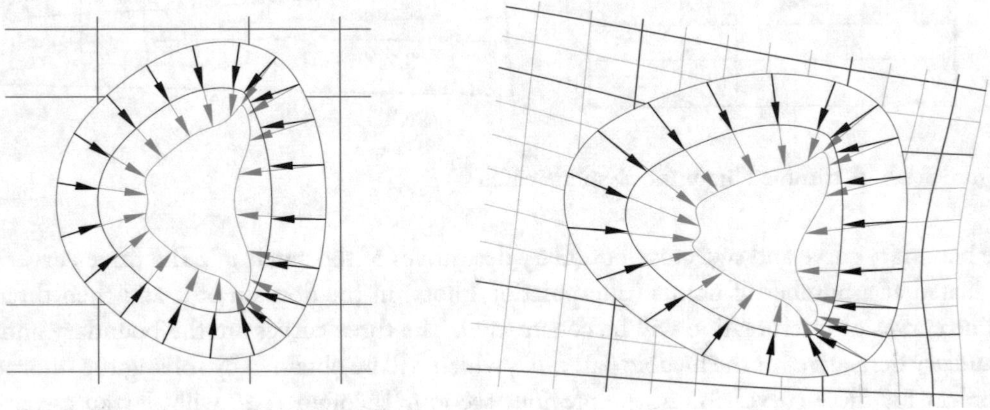

Figure 3.13: Vectors $\boldsymbol{z}(t)$ (green) and $\boldsymbol{v}(t)$ (black) at points of the curve $\boldsymbol{c}$, and their images under the mappings $\mathrm{D}\boldsymbol{p}|_{\boldsymbol{c}(t)}$

The first-order cross-boundary derivative of the patch $\boldsymbol{p}^*$ is obtained using the formula

$$\underline{\boldsymbol{p}}_v^* = s_1 \overline{\boldsymbol{p}}_u + t_1 \overline{\boldsymbol{p}}_v,$$

where the vector functions $\overline{\boldsymbol{p}}_u$ and $\overline{\boldsymbol{p}}_v$ describe the partial derivatives of the patch $\boldsymbol{p}$ at the points of the curve $\boldsymbol{c}$. If the junction functions s_1 and t_1 are taken as

$$s_1(t) = l_1(t)g(t), \quad t_1(t) = l_1(t)h(t),$$

then the length of the cross-boundary derivative vector $\underline{\boldsymbol{p}}_v^*(t)$ is equal to $|l_1(t)|$; the function l_1 must be of class $C^2[a, b]$, it must not have zeros and its sign must be chosen so as to avoid producing a cusp-like junction (the correct sign of l_1 is determined by the orientation of the curve $\boldsymbol{c}$). Modifying the function l_1 is a convenient method of controlling the length of the cross-boundary derivative vectors of the patch $\boldsymbol{p}^*$. The vector $\underline{\boldsymbol{p}}_v^*(t)$ chosen in this way remains perpendicular to the boundary curve of the trimmed patch, which is often desirable. However, sometimes it may be necessary to take

$$s_1(t) = l_1(t)\big(g(t) + m_1(t)x(t)\big), \quad t_1(t) = l_1(t)\big(h(t) + m_1(t)y(t)\big),$$

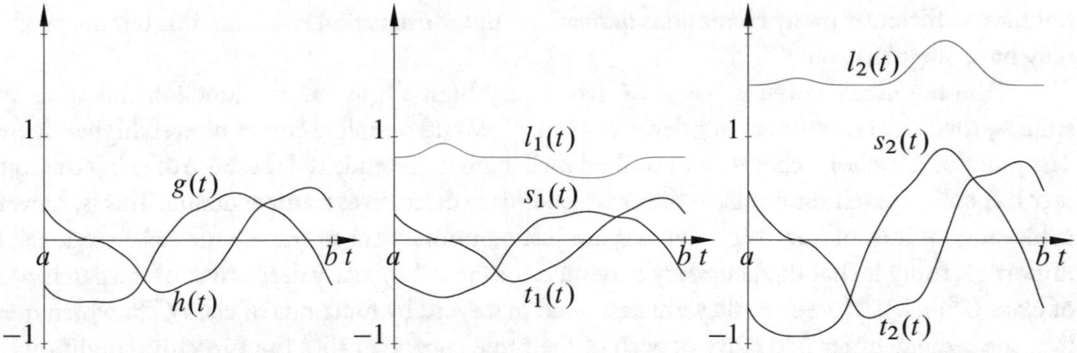

Figure 3.14: Coordinates of the vector $\boldsymbol{v}(t)$ and junction functions s_1, t_1, s_2 and t_2 obtained with the scaling function l_1 and l_2 shown on the graph, and $m_1 = m_2 = 0$

where the functions x and y describe the coordinates of $\boldsymbol{c}'(t)$; in this way the pair of junction functions (s_1, t_1) is replaced by the pair of functions (l_1, m_1), which is more convenient to use; the function l_1 controls the length of the component of $\boldsymbol{p}_v^*$ in the direction perpendicular to the boundary of $\boldsymbol{p}$, while m_1 influences the "slant".[8]

The second-order cross-boundary derivative of the patch $\boldsymbol{p}^*$ is obtained from the formula

$$\underline{\boldsymbol{p}_{vv}^*} = s_2\overline{\boldsymbol{p}}_u + t_2\overline{\boldsymbol{p}}_v + s_1^2\overline{\boldsymbol{p}}_{uu} + 2s_1t_1\overline{\boldsymbol{p}}_{uv} + t_1^2\overline{\boldsymbol{p}}_{vv}.$$

It is convenient to introduce auxiliary functions l_2 and m_2 and take

$$s_2(t) = l_2(t)\big(g(t) + m_2(t)x(t)\big), \quad t_2(t) = l_2(t)\big(h(t) + m_2(t)y(t)\big).$$

The functions l_1, m_1, l_2 and m_2 must have the second-order derivative continuous, and if the curve $\boldsymbol{c}$ is closed (as in our example), they must have a periodic extension of class C^2 on $\mathbb{R}$—cubic splines are appropriate in this application; see Section A.8.

Looking at the formula for the second-order cross-boundary derivative, we can see the reason for choosing the degree 5 for the patch $\boldsymbol{p}$ and 4 for the curve $\boldsymbol{c}$: the patch $\boldsymbol{p}^*$ is (by assumption) a tensor product patch. The first- and second-order cross-boundary derivatives of a tensor product patch of class C^2 are of class C^2. To ensure the continuity of the cross-boundary derivatives of $\boldsymbol{p}^*$, we need the continuity of the second-order derivatives of the second-order partial derivatives of the patch $\boldsymbol{p}$, which must, therefore, be of class C^4. Such a spline patch, consisting of more than one polynomial piece, must be of degree at least $(5, 5)$. Also, because of taking the derivative in Formula (3.17), we need the curve $\boldsymbol{c}$ of class C^3 to obtain the junction functions of class C^2; such a spline curve must be at least quartic. If the patch $\boldsymbol{p}$ or the curve $\boldsymbol{c}$ do

[8]Actually, if the functions s_1 and t_1 are constructed as described here, $\|\boldsymbol{c}'(t)\|_2 m_1(t)$ is the tangent of the angle between the normal plane of the curve $\boldsymbol{p} \circ \boldsymbol{c}$ and the cross-boundary derivative vector of the patch $\boldsymbol{p}^*$.

not have sufficiently many continuous derivatives, the construction is still feasible but the patch $\boldsymbol{p}^*$ may have singularities.

The boundary curve $\underline{\boldsymbol{p}}^* = \boldsymbol{p} \circ \boldsymbol{c}$ has a very high degree. If the junction functions were splines, then the cross-boundary derivatives of $\boldsymbol{p}^*$ would be spline curves of even higher degrees. The junction functions chosen as described earlier are transcendental (because of using the square root). A spline patch cannot have the cross-boundary derivatives transcendental. This is, however, irrelevant; splines of very high degrees are just as impractical as transcendental functions. The important thing is that the boundary curve and the cross-boundary derivatives of the patch $\boldsymbol{p}^*$ are of class $C^2[a,b]$. Moreover, they are described piecewise by functions of class C^∞, which means that the second-order derivative of each of the three curves satisfies the Lipschitz condition (i.e., for each of them there exists a finite constant L_2, considered in Theorem A.20). Therefore, it makes sense to choose a number of interpolation knots in the interval $[a,b]$. For each knot u_i, we can compute the point $\underline{\boldsymbol{p}}^*(u_i) = \boldsymbol{p}(\boldsymbol{c}(u_i))$ and two cross-boundary derivative vectors, $\underline{\boldsymbol{p}}^*_v(u_i)$ and $\underline{\boldsymbol{p}}^*_{vv}(u_i)$. This data may be used to construct three cubic B-spline curves of interpolation which will become the boundary curve of $\hat{\boldsymbol{p}}^*$ and its first and second cross-boundary derivatives. Due to Theorem A.20, by choosing a sufficiently dense set of knots in the interval $[a,b]$, we can decrease the approximation errors—not just for the curves, but also for their derivatives up to the second order. This is exactly what we need to obtain a patch $\hat{\boldsymbol{p}}^*$ whose junction with the trimmed patch $\boldsymbol{p}$ has arbitrarily small discontinuities of curvature.

The cubic spline curve of interpolation to the points $\boldsymbol{p}(\boldsymbol{c}(u_i))$ will be denoted by $\underline{\hat{\boldsymbol{p}}}^*$; the curves of interpolation to the cross-boundary derivatives of the first and second order at the knots u_i will be denoted by $\underline{\hat{\boldsymbol{p}}}^*_v$ and $\underline{\hat{\boldsymbol{p}}}^*_{vv}$ respectively. The way of finding their B-spline representations is described in Section A.8.3. We have

$$\underline{\hat{\boldsymbol{p}}}^*(t) = \sum_{i=0}^{N+2} \boldsymbol{k}_{i0} N_i^3(t), \quad \underline{\hat{\boldsymbol{p}}}^*_v(t) = \sum_{i=0}^{N+2} \boldsymbol{k}_{i1} N_i^3(t), \quad \underline{\hat{\boldsymbol{p}}}^*_{vv}(t) = \sum_{i=0}^{N+2} \boldsymbol{k}_{i2} N_i^3(t),$$

where $N+1$ is the number of interpolation knots in the interval $[a,b]$; in our example they are equidistant. After finding these curves, it is time to construct the patch $\hat{\boldsymbol{p}}^*$. The parameter t of the curve $\boldsymbol{c}$ is now identified with the parameter u of the patch $\hat{\boldsymbol{p}}^*$.

The knot sequence for the curves $\underline{\hat{\boldsymbol{p}}}^*$, $\underline{\hat{\boldsymbol{p}}}^*_v$ and $\underline{\hat{\boldsymbol{p}}}^*_{vv}$ becomes the "u" knot sequence for the bicubic patch $\hat{\boldsymbol{p}}^*$. It is convenient to choose the "v" sequence, $v_0, \ldots, v_M$, with $v_1 = v_2 = v_3$, because the control points in the first three *rows* of control net of the B-spline patch

$$\hat{\boldsymbol{p}}^*(u,v) = \sum_{i=0}^{N+2} \sum_{j=0}^{M-4} \boldsymbol{d}_{ij} N_i^3(u) N_j^3(v)$$

may be computed as described in Section A.5. The other control points of $\hat{\boldsymbol{p}}^*$ may be chosen arbitrarily.

Results of the construction may be seen in Figure 3.15; the first patch, shown on the left side, was obtained with curves having $N+1 = 17$ interpolation knots, while the second patch (on

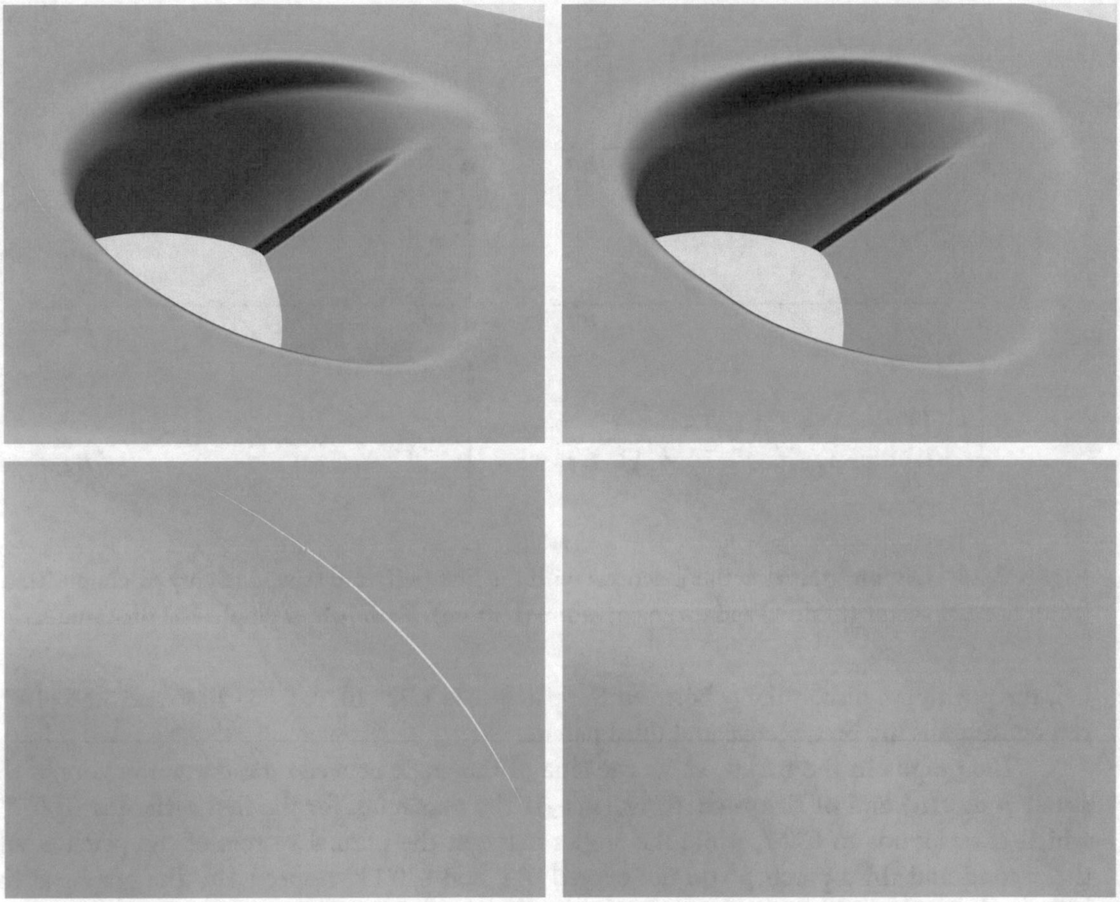

Figure 3.15: Junctions of bicubic B-spline patches with a trimmed patch

the right side) used $N+1=33$ interpolation knots. The magnified picture on the left side below reveals the positional discontinuity in the former case, and the same magnification is insufficient to expose the discontinuity in the latter case. Pictures of the patch obtained with $N+1=65$ knots are indistinguishable from pictures of the second patch. The texture on the surface is used to show the distribution of the mean curvature, which is mapped on the hue of the surface points. Due to the Linkage Curve Theorem (see Section 6.6), the mean curvature image is a visualisation tool powerful enough to examine geometric continuity of the second order.

Figure 3.16 shows graphs of approximation errors for the first and second patch. The topmost graphs show the distance between the points of our theoretical goal, the curve $\boldsymbol{p}^*(u, v_3) = \boldsymbol{p}(\boldsymbol{c}(u))$, and the corresponding points of the boundary curve of the bicubic patch, $\hat{\boldsymbol{p}}^*(u, v_3)$. The diameter of the patch $\boldsymbol{p}$ is about 3; the maximal distance between the corresponding points,

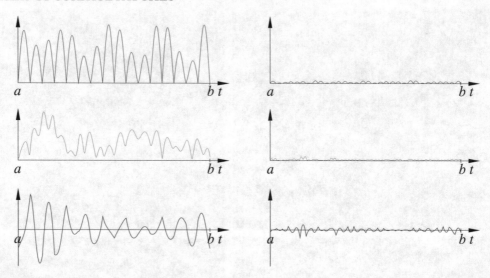

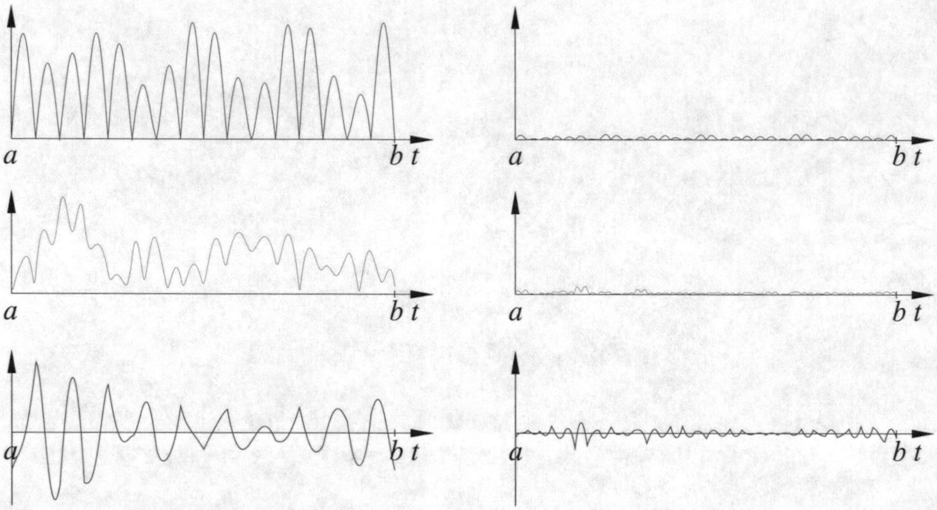

Figure 3.16: Discontinuities at the junctions with the first (left) and second (right) patch: positional (top), normal vector (middle) and mean curvature (bottom). Each pair of graphs has the same scale

i.e., the positional discontinuity between the patches, is $1.22 \cdot 10^{-3}$, $5.85 \cdot 10^{-5}$ and $3.66 \cdot 10^{-6}$, respectively for the first, second and third patch.

The graphs in the middle show the sine of the angle between the normal vectors of the patch $\boldsymbol{p}$ at $\boldsymbol{c}(u)$ and of the patch $\hat{\boldsymbol{p}}^*$ at (u, v_3). The maximum for the first surface is 0.00749, which corresponds to $0°26'$, while the angles between the normal vectors of the patch $\boldsymbol{p}$ and the second and third patch $\hat{\boldsymbol{p}}^*$ do not exceed $0°2'$ and $0°0'13''$ respectively. The graphs at the bottom show the difference between the mean curvatures of the patch $\boldsymbol{p}$ and the first and second patch $\hat{\boldsymbol{p}}^*$ at the corresponding points of their boundary curves. The mean curvature of the patch $\boldsymbol{p}$ at the trimming boundary ranges from -0.361 to $+0.6365$, and the maximal difference between the mean curvature of this patch and the mean curvature of the first, second and third patch $\hat{\boldsymbol{p}}^*$ is 0.145, 0.0313 and 0.0085 respectively.

As we can see, increasing the density of the interpolation knots results in a significant reduction of all three discontinuities. If the surface made of the patches $\boldsymbol{p}$ and $\hat{\boldsymbol{p}}^*$ were a part of the design of a physical object, then the accuracy obtained with $N = 32$ (recall: the maximal positional discontinuity less than 59μm for an object whose diameter is about 3m) should be sufficient for most manufacturing technologies.

CHAPTER 4

Compatibility conditions

A smooth surface usually consists of a number of parametric patches, and it has points at which several patches meet. Such a point is also a common point of the common boundary curves of the patches. There are conditions which must be satisfied by derivatives of the patches and their common boundary curves meeting at the common point. For a surface of class G^n these **compatibility conditions** involve derivatives up to the order $2n$, and they cause trouble in constructions of smooth surfaces. In this chapter we study the compatibility conditions in detail, in order to understand them and to learn how to satisfy them in practical constructions.

4.1 HAHN'S SCHEME OF FILLING POLYGONAL HOLES

In [1988] Hahn outlined a general method of filling polygonal holes in surfaces made of tensor product patches. This method, with various modifications, was implemented in numerous constructions developed later. An outline of this outline given below will serve us as a reference point in the analysis of compatibility conditions made in this chapter and as a framework for constructions described in the next chapter.

A given surface with a hole is made of smooth regular tensor product patches having common boundary curves. Each pair of patches having a common curve may be reparametrised so that their rectangular domains have a common edge and the parametrisation of the surface made of the two patches over the union of the two rectangles is of class C^n. The boundary of the hole consists of k smooth curves (made of boundary curves of the patches making the surface). The goal is to construct k tensor product patches which would fill the hole; the junctions between the new patches and the given ones and between any two new patches having a common boundary are supposed to be of class G^n.

The first step of the construction is to find the cross-boundary derivatives of the given patches surrounding the hole, up to the order n (Fig. 4.1a). The second step is to choose the common corner of the patches to be constructed, i.e., the "central point" of the filling surface, and vectors which will be the first-order derivatives of the boundary curves of the final patches at this point (Fig. 4.1b). These vectors, which must be coplanar (they determine the tangent plane of the surface at the central point), are related to what we call a partition of the full angle, which later in this chapter will be the subject of extensive study. Then in the third step (Fig. 4.1c) derivatives up to the order $2n$ of *one* of the patches filling the hole at the central point are fixed. By reparametrisation of this patch (and using the generalised Fàa di Bruno's formula, see Section A.11), it is possible to obtain the partial derivatives up to the order $2n$ of all the other patches at the central

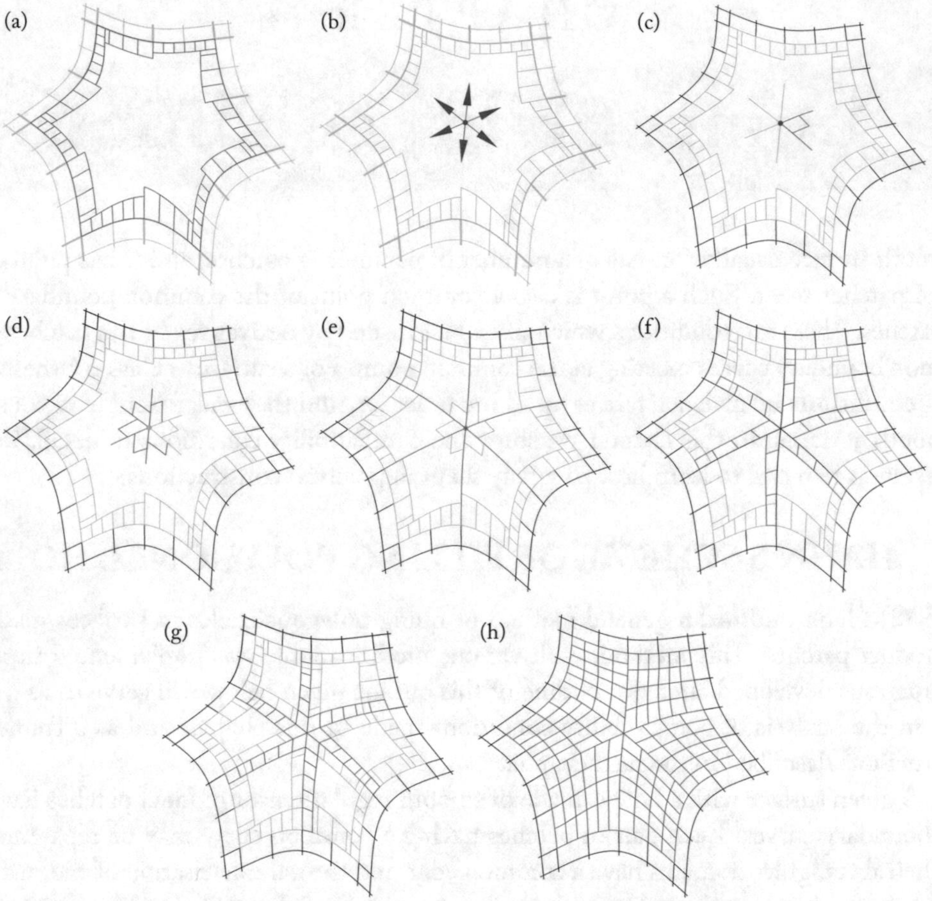

Figure 4.1: Hahn's scheme of filling a polygonal hole, $k = 6$

point (Fig. 4.1d). Then, by solving Hermite interpolation problems, we can construct the curves between the central point and the points in the middle of the edges of the hole (Fig. 4.1e); these curves will be the common curves of the patches. The next step is to construct auxiliary patches along these curves; the auxiliary patches determine planes tangent to the final patches along the curves (Fig. 4.1f) and, if a higher order geometric continuity is the goal, also the normal curvatures and attributes of the surface determined by higher order cross-boundary derivatives.

Hahn suggested constructing directly cross-boundary derivatives of one of the patches adjacent to each of the common curves and using them to construct the cross-boundary derivatives of the other patch using junction functions. In this way the final patches are constructed in a non-symmetrical way (Fig. 4.1g). Having the cross-boundary derivatives along all four boundary curves, we can obtain the final patches (Fig. 4.1h) as Coons patches (Section A.9)—bicubically blended, if the surface is of class G^1, biquintically if G^2, etc.

Some assumptions made in Hahn's scheme seem too restrictive; in particular, the idea of fixing the partial derivatives up to the order $2n$ at the central point of one final patch, and then computing the partial derivatives of the other final patches. Later, we shall weaken this restriction as much as possible. We shall also take a look at the possibility of fixing the common curves *before* proceeding to the construction of mixed partial derivatives at the common point, as it seems to be a tempting alternative.

4.2 COMPATIBILITY CONDITIONS AT A COMMON CORNER

Consider two regular auxiliary patches, $\boldsymbol{q}(s,u)$ and $\boldsymbol{r}(v,t)$, to which we fit a patch $\boldsymbol{p}(u,v)$ being a part of the surface to be constructed. If smooth junctions of the patch $\boldsymbol{p}$ with $\boldsymbol{q}$ and $\boldsymbol{r}$ are to be obtained, then the auxiliary patches must comply to some restrictions. The most obvious one is that if the patch $\boldsymbol{p}$ has a common corner with $\boldsymbol{q}$ and $\boldsymbol{r}$, then the auxiliary patches have the common corner. This is not enough if smooth junctions of the patches are to be constructed; if a surface of class G^n, where $n \geq 1$, is to be obtained, then the three patches must have a common tangent plane at their common corner; also higher order derivatives are involved. In this section, we study in detail the compatibility conditions for one pair of auxiliary patches, $\boldsymbol{q}$ and $\boldsymbol{r}$. In the next section, we deal with the bigger problem, showing how to construct a set of compatible auxiliary patches in order to obtain final patches surrounding a common corner.

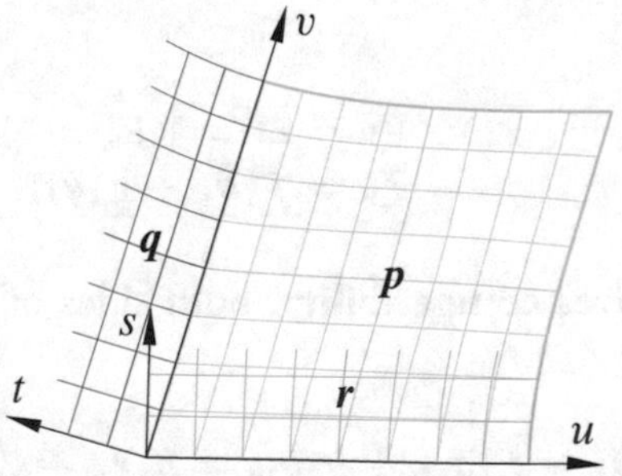

Figure 4.2: A patch and two auxiliary patches with a common corner

Two boundary curves of the patch $\boldsymbol{p}$ will be denoted by $\overline{\boldsymbol{p}}$ and $\underline{\boldsymbol{p}}$. The former is the curve of constant parameter[1] $u = u_b$ and it coincides with the boundary curve of the patch $\boldsymbol{q}$, denoted by $\underline{\boldsymbol{q}}$ (there is $\underline{\boldsymbol{q}}(v) = \boldsymbol{q}(v,0)$); the latter is the curve of constant parameter $v = v_b$ and it coincides with the curve $\overline{\boldsymbol{r}}(u) = \boldsymbol{r}(u,0)$ (see Figure 4.2).

[1]We assume that the patch $\boldsymbol{p}$ has a rectangular domain, $[u_0,u_1] \times [v_0,v_1]$ and u_b is either u_0 or u_1; similarly, the symbol v_b will denote v_0 or v_1. In practice we often take the default domain of a Bézier patch, with $u_0 = v_0 = 0$ and $u_1 = v_1 = 1$. The boundary curves of the patches $\boldsymbol{q}(v,t)$ and $\boldsymbol{r}(u,s)$ correspond to $t = 0$ and $s = 0$.

As we shall see, apart from the partial derivatives of the patches the compatibility conditions also involve junction functions and their derivatives up to the order n. We rewrite Equations (3.1), (3.3) and (3.4) for the junctions of the patch $\boldsymbol{p}$ with $\boldsymbol{q}$ and $\boldsymbol{r}$; both junctions have their own junction functions, which we denote f_1, g_1, f_2, g_2 (whose argument is the parameter v), and b_1, c_1, b_2, c_2 (with the argument u) respectively:

$$\underline{\boldsymbol{p}} = \overline{\boldsymbol{r}}, \tag{4.1}$$
$$\overline{\boldsymbol{p}} = \underline{\boldsymbol{q}}, \tag{4.2}$$
$$\underline{\boldsymbol{p}}_v = b_1\overline{\boldsymbol{r}}_u + c_1\overline{\boldsymbol{r}}_s, \tag{4.3}$$
$$\overline{\boldsymbol{p}}_u = f_1\underline{\boldsymbol{q}}_v + g_1\underline{\boldsymbol{q}}_t, \tag{4.4}$$
$$\underline{\boldsymbol{p}}_{vv} = b_2\overline{\boldsymbol{r}}_u + c_2\overline{\boldsymbol{r}}_s + b_1^2\overline{\boldsymbol{r}}_{uu} + 2b_1c_2\overline{\boldsymbol{r}}_{us} + c_1^2\overline{\boldsymbol{r}}_{ss}, \tag{4.5}$$
$$\overline{\boldsymbol{p}}_{uu} = f_2\underline{\boldsymbol{q}}_v + g_2\underline{\boldsymbol{q}}_t + f_1^2\underline{\boldsymbol{q}}_{vv} + 2f_1g_2\underline{\boldsymbol{q}}_{vt} + f_1^2\underline{\boldsymbol{q}}_{tt}. \tag{4.6}$$

If the junctions are supposed to be of class G^1, then only the first four equations are relevant; more equations are needed for junctions of class G^3 or higher. Now we calculate derivatives (with respect to u) of the curves $\underline{\boldsymbol{p}}$, $\underline{\boldsymbol{p}}_v$ and $\underline{\boldsymbol{p}}_{vv}$ at $u = u_b$, and, similarly, derivatives (with respect to v) of $\overline{\boldsymbol{p}}$, $\overline{\boldsymbol{p}}_u$ and $\overline{\boldsymbol{p}}_{uu}$ at $v = v_b$. The double restriction (to $u = u_b$, $v = v_b$) gives us equations with the vectors in $\mathbb{R}^3$, denoted by under- *and* overlining. Underlined symbols of the junction functions and their derivatives denote their values at u_b or v_b. The equations express

positional continuity: the patches have a common corner, $\underline{\overline{\boldsymbol{p}}} = \underline{\overline{\boldsymbol{q}}} = \underline{\overline{\boldsymbol{r}}}$;

first-order derivatives compatibility: the sides of the equations below are equal to $\underline{\overline{\boldsymbol{p}}}_v$ and $\underline{\overline{\boldsymbol{p}}}_u$, respectively:

$$\underline{\overline{\boldsymbol{q}}}_v = \underline{b}_1\underline{\overline{\boldsymbol{r}}}_u + \underline{c}_1\underline{\overline{\boldsymbol{r}}}_s, \tag{4.7}$$
$$\underline{\overline{\boldsymbol{r}}}_u = \underline{f}_1\underline{\overline{\boldsymbol{q}}}_v + \underline{g}_1\underline{\overline{\boldsymbol{q}}}_t; \tag{4.8}$$

mixed second-order derivatives compatibility: both sides of the equation below are equal to $\underline{\overline{\boldsymbol{p}}}_{uv}$:

$$\underline{f}_1'\underline{\overline{\boldsymbol{q}}}_v + \underline{g}_1'\underline{\overline{\boldsymbol{q}}}_t + \underline{f}_1\underline{\overline{\boldsymbol{q}}}_{vv} + \underline{g}_1\underline{\overline{\boldsymbol{q}}}_{vt} = \underline{b}_1'\underline{\overline{\boldsymbol{r}}}_u + \underline{c}_1'\underline{\overline{\boldsymbol{r}}}_s + \underline{b}_1\underline{\overline{\boldsymbol{r}}}_{uu} + \underline{c}_1\underline{\overline{\boldsymbol{r}}}_{us}. \tag{4.9}$$

The equations above suffice if the junctions of patches are of class G^1. With junctions of class G^2 we have further equations:

second-order derivatives compatibility: the sides of the equations

$$\underline{\overline{\boldsymbol{q}}}_{vv} = \underline{b}_2\underline{\overline{\boldsymbol{r}}}_u + \underline{c}_2\underline{\overline{\boldsymbol{r}}}_s + \underline{b}_1^2\underline{\overline{\boldsymbol{r}}}_{uu} + 2\underline{b}_1\underline{c}_1\underline{\overline{\boldsymbol{r}}}_{us} + \underline{c}_1^2\underline{\overline{\boldsymbol{r}}}_{ss}, \tag{4.10}$$
$$\underline{\overline{\boldsymbol{r}}}_{uu} = \underline{f}_2\underline{\overline{\boldsymbol{q}}}_v + \underline{g}_2\underline{\overline{\boldsymbol{q}}}_t + \underline{f}_1^2\underline{\overline{\boldsymbol{q}}}_{vv} + 2\underline{f}_1\underline{g}_1\underline{\overline{\boldsymbol{q}}}_{vt} + \underline{g}_1^2\underline{\overline{\boldsymbol{q}}}_{tt} \tag{4.11}$$

are equal to $\underline{\overline{\boldsymbol{p}}}_{vv}$ and $\underline{\overline{\boldsymbol{p}}}_{uu}$;

mixed third-order partial derivatives compatibility: the sides of the equations below are equal to $\underline{\bar{\boldsymbol{p}}}_{uvv}$ and $\underline{\bar{\boldsymbol{p}}}_{uuv}$:

$$\begin{aligned}
&\underline{f}_1''\underline{\bar{\boldsymbol{q}}}_v + \underline{g}_1''\underline{\bar{\boldsymbol{q}}}_t + 2\underline{f}_1'\underline{\bar{\boldsymbol{q}}}_{vv} + 2\underline{g}_1'\underline{\bar{\boldsymbol{q}}}_{vt} + \underline{f}_1\underline{\bar{\boldsymbol{q}}}_{vvv} + \underline{g}_1\underline{\bar{\boldsymbol{q}}}_{vvt} = \\
&\qquad \underline{b}_2'\underline{\bar{\boldsymbol{r}}}_u + \underline{c}_2'\underline{\bar{\boldsymbol{r}}}_s + (2\underline{b}_1\underline{b}_1' + \underline{b}_2)\underline{\bar{\boldsymbol{r}}}_{uu} + (2\underline{b}_1'\underline{c}_1 + 2\underline{b}_1\underline{c}_1' + \underline{c}_2)\underline{\bar{\boldsymbol{r}}}_{us} + \\
&\qquad 2\underline{c}_1\underline{c}_1'\underline{\bar{\boldsymbol{r}}}_{ss} + \underline{b}_1^2\underline{\bar{\boldsymbol{r}}}_{uuu} + 2\underline{b}_1\underline{c}_1\underline{\bar{\boldsymbol{r}}}_{uus} + \underline{c}_1^2\underline{\bar{\boldsymbol{r}}}_{uss},
\end{aligned} \tag{4.12}$$

$$\begin{aligned}
&\underline{b}_1''\underline{\bar{\boldsymbol{r}}}_u + \underline{c}_1''\underline{\bar{\boldsymbol{r}}}_s + 2\underline{b}_1'\underline{\bar{\boldsymbol{r}}}_{uu} + 2\underline{c}_1'\underline{\bar{\boldsymbol{r}}}_{us} + \underline{b}_1\underline{\bar{\boldsymbol{r}}}_{uuu} + \underline{c}_1\underline{\bar{\boldsymbol{r}}}_{uus} = \\
&\qquad \underline{f}_2'\underline{\bar{\boldsymbol{q}}}_v + \underline{g}_2'\underline{\bar{\boldsymbol{q}}}_t + (2\underline{f}_1\underline{f}_1' + \underline{f}_2)\underline{\bar{\boldsymbol{q}}}_{vv} + (2\underline{f}_1'\underline{g}_1 + 2\underline{f}_1\underline{g}_1' + \underline{g}_2)\underline{\bar{\boldsymbol{q}}}_{vt} + \\
&\qquad 2\underline{g}_1\underline{g}_1'\underline{\bar{\boldsymbol{q}}}_{tt} + \underline{f}_1^2\underline{\bar{\boldsymbol{q}}}_{vvv} + 2\underline{f}_1\underline{g}_1\underline{\bar{\boldsymbol{q}}}_{vvt} + \underline{g}_1^2\underline{\bar{\boldsymbol{q}}}_{vtt};
\end{aligned} \tag{4.13}$$

mixed fourth-order partial derivatives compatibility: both sides of the equation below are equal to $\underline{\bar{\boldsymbol{p}}}_{uuvv}$:

$$\begin{aligned}
&\underline{f}_2''\underline{\bar{\boldsymbol{q}}}_v + \underline{g}_2''\underline{\bar{\boldsymbol{q}}}_t + 2(\underline{f}_1\underline{f}_1'' + \underline{f}_1'^2 + \underline{f}_2')\underline{\bar{\boldsymbol{q}}}_{vv} + \\
&2(\underline{f}_1''\underline{g}_1 + 2\underline{f}_1'\underline{g}_1' + \underline{f}_1\underline{g}_1'' + \underline{g}_2')\underline{\bar{\boldsymbol{q}}}_{vt} + 2(\underline{g}_1\underline{g}_1'' + \underline{g}_1'^2)\underline{\bar{\boldsymbol{q}}}_{tt} + \\
&(4\underline{f}_1\underline{f}_1' + \underline{f}_2)\underline{\bar{\boldsymbol{q}}}_{vvv} + (4\underline{f}_1'\underline{g}_1 + 4\underline{f}_1\underline{g}_1' + \underline{g}_2)\underline{\bar{\boldsymbol{q}}}_{vvt} + 4\underline{g}_1\underline{g}_1'\underline{\bar{\boldsymbol{q}}}_{vtt} + \\
&\underline{f}_1^2\underline{\bar{\boldsymbol{q}}}_{vvvv} + 2\underline{f}_1\underline{g}_1\underline{\bar{\boldsymbol{q}}}_{vvvt} + \underline{g}_1^2\underline{\bar{\boldsymbol{q}}}_{vvtt} = \\
&\qquad \underline{b}_2''\underline{\bar{\boldsymbol{r}}}_u + \underline{c}_2''\underline{\bar{\boldsymbol{r}}}_s + 2(\underline{b}_1\underline{b}_1'' + \underline{b}_1'^2 + \underline{b}_2')\underline{\bar{\boldsymbol{r}}}_{uu} + \\
&\qquad 2(\underline{b}_1''\underline{c}_1 + 2\underline{b}_1'\underline{c}_1' + \underline{b}_1\underline{c}_1'' + \underline{c}_2')\underline{\bar{\boldsymbol{r}}}_{us} + 2(\underline{c}_1\underline{c}_1'' + \underline{c}_1'^2)\underline{\bar{\boldsymbol{r}}}_{ss} + \\
&\qquad (4\underline{b}_1\underline{b}_1' + \underline{b}_2)\underline{\bar{\boldsymbol{r}}}_{uuu} + (4\underline{b}_1'\underline{c}_1 + 4\underline{b}_1\underline{c}_1' + \underline{c}_2)\underline{\bar{\boldsymbol{r}}}_{uus} + 4\underline{c}_1\underline{c}_1'\underline{\bar{\boldsymbol{r}}}_{uss} + \\
&\qquad \underline{b}_1^2\underline{\bar{\boldsymbol{r}}}_{uuuu} + 2\underline{b}_1\underline{c}_1\underline{\bar{\boldsymbol{r}}}_{uuus} + \underline{c}_1^2\underline{\bar{\boldsymbol{r}}}_{uuss}.
\end{aligned} \tag{4.14}$$

These equations may draw some people into hopelessness. Ugly they look, but the situation is not yet desperate. Each equation above is a system of three scalar equations. With the auxiliary patches $\boldsymbol{q}$ and $\boldsymbol{r}$ fixed, we have their derivative vectors fixed and we can solve these equations with respect to the values of the junction functions and their derivatives.

In Equation (4.7) we have two unknown variables, $\underline{b}_1$ and $\underline{c}_1$. Note that the vector $\underline{\bar{\boldsymbol{q}}}_v$ must be a linear combination of $\underline{\bar{\boldsymbol{r}}}_u$ and $\underline{\bar{\boldsymbol{r}}}_s$, otherwise the system is inconsistent. Similarly, the vector $\underline{\bar{\boldsymbol{r}}}_u$ must be a linear combination of $\underline{\bar{\boldsymbol{q}}}_v$ and $\underline{\bar{\boldsymbol{q}}}_t$ to obtain a consistent system of equations (4.8) with unknown $\underline{f}_1$ and $\underline{g}_1$. It turns out that all four vectors, $\underline{\bar{\boldsymbol{q}}}_v$, $\underline{\bar{\boldsymbol{q}}}_t$, $\underline{\bar{\boldsymbol{r}}}_u$ and $\underline{\bar{\boldsymbol{r}}}_s$, must span a two-dimensional subspace of $\mathbb{R}^3$ which determines the plane tangent to all three patches at their common corner.

After solving Equations (4.7) and (4.8) we may substitute the solutions, $\underline{b}_1$, $\underline{c}_1$, $\underline{f}_1$ and $\underline{g}_1$, in the equations following these two. In Equation (4.9) we have four unknown variables, $\underline{b}_1'$, $\underline{c}_1'$, $\underline{f}_1'$ and $\underline{g}_1'$, and, thus, this system of equations, if consistent, is indefinite. After rewriting this equation in the form

$$\underline{f}_1'\underline{\bar{\boldsymbol{q}}}_v + \underline{g}_1'\underline{\bar{\boldsymbol{q}}}_t - \underline{b}_1'\underline{\bar{\boldsymbol{r}}}_u - \underline{c}_1'\underline{\bar{\boldsymbol{r}}}_s = \underline{b}_1\underline{\bar{\boldsymbol{r}}}_{uu} + \underline{c}_1\underline{\bar{\boldsymbol{r}}}_{us} - \underline{f}_1\underline{\bar{\boldsymbol{q}}}_{vv} - \underline{g}_1\underline{\bar{\boldsymbol{q}}}_{vt}$$

we notice that this system is consistent if and only if the right-hand side vector is an element of the two-dimensional subspace spanned by $\underline{\bar{\boldsymbol{q}}}_v$, $\underline{\bar{\boldsymbol{q}}}_t$, $\underline{\bar{\boldsymbol{r}}}_u$ and $\underline{\bar{\boldsymbol{r}}}_s$.

In constructions of surfaces of class G^1 the above is enough; we obtain the values and first-order derivatives of the junction functions at the point corresponding to the common corner of the patches. Similar compatibility conditions for the other corners of the patch $\boldsymbol{p}$ give us the values and derivatives of the junction functions at points corresponding to those corners. It is, then, natural to obtain the junction functions as polynomials satisfying the Hermite interpolation conditions. Having the junction functions and the cross-boundary derivatives of the auxiliary patches surrounding the (yet unknown) patch $\boldsymbol{p}$, we can construct the cross-boundary derivatives of this patch using (4.3) and (4.4), and then construct the patch $\boldsymbol{p}$ itself; in the simplest case, it may be a bicubically blended Coons patch determined by the boundary curves (taken from the auxiliary patches) and the cross-boundary derivatives.

If a surface of class G^2 is to be constructed, we need the values of four more junction functions, b_2, c_2, f_2 and g_2, and the derivatives of all eight junction functions up to the order two. By solving Equations (4.10) and (4.11) we obtain $\underline{b}_2, \underline{c}_2, \underline{f}_2$ and $\underline{g}_2$. Then, we can solve (4.12) to obtain $\underline{b}'_2, \underline{c}'_2, \underline{f}''_1$ and $\underline{g}''_1$, and (4.13) to obtain $\underline{f}'_2, \underline{g}'_2, \underline{b}''_1$ and $\underline{c}''_1$. Finally, from (4.14) we obtain $\underline{b}''_2$, $\underline{c}''_2$, $\underline{f}''_2$ and $\underline{g}''_2$. The last three equations are systems of three scalar equations with four unknown variables, and they are linear with respect to these variables. Each of them, if consistent, has infinitely many solutions. Having the values and derivatives of the junction functions satisfying the compatibility conditions for each corner of the patch $\boldsymbol{p}$, we can solve Hermite interpolation problems to obtain polynomials which we take for the junction functions, then we can construct the cross-boundary derivatives of the patch $\boldsymbol{p}$ and then construct this patch as a biquintically blended Coons patch.

It is worth taking a look at the compatibility conditions in an important particular case, when we need a solution such that $\underline{b}_1 = \underline{f}_1 = \underline{b}_2 = \underline{c}_2 = \underline{f}_2 = \underline{g}_2 = 0$. Then,

$$\underline{\bar{\boldsymbol{q}}}_v = \underline{c}_1\underline{\bar{\boldsymbol{r}}}_s, \tag{4.7$'$}$$

$$\underline{\bar{\boldsymbol{r}}}_u = \underline{g}_1\underline{\bar{\boldsymbol{q}}}_t, \tag{4.8$'$}$$

$$\underline{f}'_1\underline{\bar{\boldsymbol{q}}}_v + \underline{g}'_1\underline{\bar{\boldsymbol{q}}}_t + \underline{g}_1\underline{\bar{\boldsymbol{q}}}_{vt} = \underline{b}'_1\underline{\bar{\boldsymbol{r}}}_u + \underline{c}'_1\underline{\bar{\boldsymbol{r}}}_s + \underline{c}_1\underline{\bar{\boldsymbol{r}}}_{us}, \tag{4.9$'$}$$

$$\underline{\bar{\boldsymbol{q}}}_{vv} = \underline{c}_1^2\underline{\bar{\boldsymbol{r}}}_{ss}, \tag{4.10$'$}$$

$$\underline{\bar{\boldsymbol{r}}}_{uu} = \underline{g}_1^2\underline{\bar{\boldsymbol{q}}}_{tt}, \tag{4.11$'$}$$

$$\begin{aligned}&\underline{f}''_1\underline{\bar{\boldsymbol{q}}}_v + \underline{g}''_1\underline{\bar{\boldsymbol{q}}}_t + 2\underline{f}'_1\underline{\bar{\boldsymbol{q}}}_{vv} + 2\underline{g}'_1\underline{\bar{\boldsymbol{q}}}_{vt} + \underline{g}_1\underline{\bar{\boldsymbol{q}}}_{vvt} = \\ &\qquad \underline{b}'_2\underline{\bar{\boldsymbol{r}}}_u + \underline{c}'_2\underline{\bar{\boldsymbol{r}}}_s + 2\underline{b}'_1\underline{c}_1\underline{\bar{\boldsymbol{r}}}_{us} + 2\underline{c}_1\underline{c}'_1\underline{\bar{\boldsymbol{r}}}_{ss} + \underline{c}_1^2\underline{\bar{\boldsymbol{r}}}_{uss},\end{aligned} \tag{4.12$'$}$$

$$\begin{aligned}&\underline{b}''_1\underline{\bar{\boldsymbol{r}}}_u + \underline{c}''_1\underline{\bar{\boldsymbol{r}}}_s + 2\underline{b}'_1\underline{\bar{\boldsymbol{r}}}_{uu} + 2\underline{c}'_1\underline{\bar{\boldsymbol{r}}}_{us} + \underline{c}_1\underline{\bar{\boldsymbol{r}}}_{uus} = \\ &\qquad \underline{f}'_2\underline{\bar{\boldsymbol{q}}}_v + \underline{g}'_2\underline{\bar{\boldsymbol{q}}}_t + 2\underline{f}'_1\underline{g}_1\underline{\bar{\boldsymbol{q}}}_{vt} + 2\underline{g}_1\underline{g}'_1\underline{\bar{\boldsymbol{q}}}_{tt} + \underline{g}_1^2\underline{\bar{\boldsymbol{q}}}_{vtt},\end{aligned} \tag{4.13$'$}$$

$$\begin{aligned}&\underline{f}''_2\underline{\bar{\boldsymbol{q}}}_v + \underline{g}''_2\underline{\bar{\boldsymbol{q}}}_t + 2(\underline{f}'^2_1 + \underline{f}'_2)\underline{\bar{\boldsymbol{q}}}_{vv} + 2(\underline{f}''_1\underline{g}_1 + 2\underline{f}'_1\underline{g}'_1 + \underline{g}'_2)\underline{\bar{\boldsymbol{q}}}_{vt} + \\ &2(\underline{g}_1\underline{g}''_1 + \underline{g}'^2_1)\underline{\bar{\boldsymbol{q}}}_{tt} + 4\underline{f}'_1\underline{g}_1\underline{\bar{\boldsymbol{q}}}_{vvt} + 4\underline{g}_1\underline{g}'_1\underline{\bar{\boldsymbol{q}}}_{vtt} + \underline{g}_1^2\underline{\bar{\boldsymbol{q}}}_{vvtt} = \\ &\qquad \underline{b}''_2\underline{\bar{\boldsymbol{r}}}_u + \underline{c}''_2\underline{\bar{\boldsymbol{r}}}_s + 2(\underline{b}'^2_1 + \underline{b}'_2)\underline{\bar{\boldsymbol{r}}}_{uu} + 2(\underline{b}''_1\underline{c}_1 + 2\underline{b}'_1\underline{c}'_1 + \underline{c}'_2)\underline{\bar{\boldsymbol{r}}}_{us} + \\ &\qquad 2(\underline{c}_1\underline{c}''_1 + \underline{c}'^2_1)\underline{\bar{\boldsymbol{r}}}_{ss} + 4\underline{b}'_1\underline{c}_1\underline{\bar{\boldsymbol{r}}}_{uus} + 4\underline{c}_1\underline{c}'_1\underline{\bar{\boldsymbol{r}}}_{uss} + \underline{c}_1^2\underline{\bar{\boldsymbol{r}}}_{uuss}.\end{aligned} \tag{4.14$'$}$$

As we can see, in this case the derivatives of the boundary curves $\overline{r}$ and $\underline{q}$ of order higher than the desired order of geometric continuity are absent, and they have no influence on the consistency of these equations.

4.3 COMPATIBILITY CONDITIONS AROUND A POINT

Now we consider several patches surrounding their common corner. At first we use a straightforward method of the analysis, which suffices for the G^1 compatibility conditions. Then, we discuss a more powerful approach that works for compatibility conditions of arbitrary order. We illustrate this approach for the case of G^2, which is sufficient in the majority of practical applications.

4.3.1 G^1 COMPATIBILITY CONDITIONS

To obtain k patches which surround their common corner and form a smooth surface (Fig. 4.3b), we may first construct k auxiliary patches meeting at the common corner (Fig. 4.3a). Boundary curves of the auxiliary patches will be common boundary curves for the final patches; moreover, each auxiliary patch determines the tangent plane of two final patches at each point of their common boundary. Our goal is to find conditions necessary for the feasibility of the construction of regular final patches. We number the patches from 0 to $k-1$, and in the calculation below we always take the indices of the patches modulo k, though we drop the symbol "mod k" to shorten the formulae.

The common curve of the patches $\boldsymbol{p}_{i-1}$ and $\boldsymbol{p}_i$ will be denoted by $\boldsymbol{c}_i$, and its parameter is the variable u_i. The parameters of the patch $\boldsymbol{p}_i$ are the variables u_i and u_{i+1}. The auxiliary patches are denoted by $\boldsymbol{q}_0, \ldots, \boldsymbol{q}_{k-1}$ and the parameters of the patch $\boldsymbol{q}_i$ are the variables u_i and s_i. We assume that the common corner of all the patches is obtained by taking all variables $u_0, \ldots, u_{k-1}$ and $s_0, \ldots, s_{k-1}$ equal to zero. The under- and overlining is used in the same way as in Section 4.2. Thus, $\boldsymbol{c}_i$ is the curve of constant parameter $u_{i+1} = 0$ of the patch $\boldsymbol{p}_i$, the curve of constant parameter $u_{i-1} = 0$ of the patch $\boldsymbol{p}_{i-1}$, and the curve of constant parameter $s_i = 0$ of the patch $\boldsymbol{q}_i$.

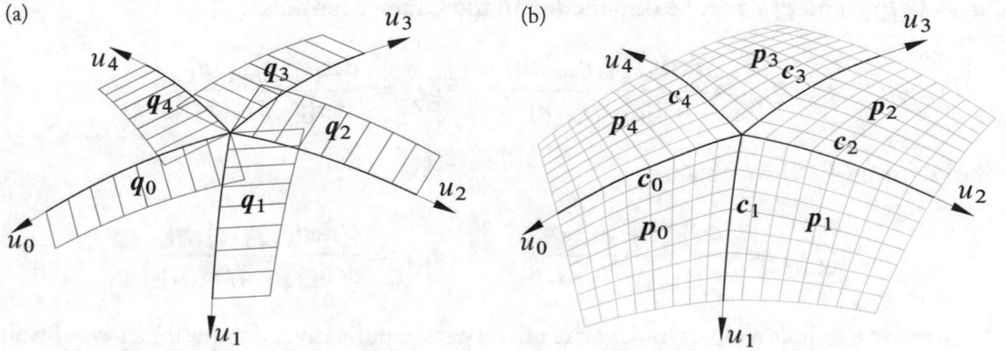

Figure 4.3: Symbols to write G^1 compatibility conditions around a common corner

In our notation

$$\boldsymbol{c}_i = \underline{\boldsymbol{p}}_i = \overline{\boldsymbol{p}}_{i-1} = \underline{\boldsymbol{q}}_i$$

and the common corner is the point

$$\overline{\boldsymbol{p}}_0 = \cdots = \overline{\boldsymbol{p}}_{k-1} = \overline{\boldsymbol{q}}_0 = \cdots = \overline{\boldsymbol{q}}_{k-1} = \underline{\boldsymbol{c}}_0 = \cdots = \underline{\boldsymbol{c}}_{k-1}.$$

As the situation gets more complicated, we need new symbols for the junction functions; we introduce them in the rewritten Formulae (4.3) and (4.4):

$$\underline{\boldsymbol{p}}_{i,u_{i+1}} = b_{i,1}\overline{\boldsymbol{q}}_{i,u_i} + c_{i,1}\overline{\boldsymbol{q}}_{i,s_i}, \tag{4.15}$$

$$\overline{\boldsymbol{p}}_{i,u_i} = f_{i,1}\underline{\boldsymbol{q}}_{i+1,u_{i+1}} + g_{i,1}\underline{\boldsymbol{q}}_{i+1,s_{i+1}}. \tag{4.16}$$

To simplify the notation, instead of $\overline{\boldsymbol{q}}_{i,s_i}$ and $\overline{\boldsymbol{q}}_{i,u_is_i}$ we shall write $\overline{\boldsymbol{q}}_{i,s}$ and $\overline{\boldsymbol{q}}_{i,us}$; the first subscript already indicates with respect to which variables the derivatives are taken. Consequently, the parameter of the curve $\boldsymbol{c}_i(=\underline{\boldsymbol{q}}_i)$ is u_i, and we shall write $\boldsymbol{c}_i'(=\underline{\boldsymbol{q}}_{i,u})$ and $\underline{\boldsymbol{c}}_i''(=\overline{\boldsymbol{q}}_{i,uu})$. With these symbols we may rewrite Equations (4.7) and (4.8) like this:

$$\underline{\boldsymbol{c}}_{i+1}' = \underline{b}_{i,1}\underline{\boldsymbol{c}}_i' + \underline{c}_{i,1}\overline{\boldsymbol{q}}_{i,s}, \tag{4.17}$$

$$\underline{\boldsymbol{c}}_i' = \underline{f}_{i,1}\underline{\boldsymbol{c}}_{i+1}' + \underline{g}_{i,1}\overline{\boldsymbol{q}}_{i+1,s}. \tag{4.18}$$

Let us recall that these are systems of three scalar equations with two unknown variables. These systems are consistent for *all* auxiliary patches if there exists a unit vector $\boldsymbol{n}$ orthogonal to the first-order partial derivatives of all auxiliary patches. The vector $\boldsymbol{n}$ is the normal vector of the patches at their common corner. To find $\underline{b}_{i,1}$ and $\underline{c}_{i,1}$, we replace (4.17) with

$$\underline{\boldsymbol{c}}_i'\underline{b}_{i,1} + \overline{\boldsymbol{q}}_{i,s}\underline{c}_{i,1} + \boldsymbol{n}d = \underline{\boldsymbol{c}}_{i+1}'.$$

The 3×3 matrix $[\underline{\boldsymbol{c}}_i', \overline{\boldsymbol{q}}_{i,s}, \boldsymbol{n}]$ is non-singular, so this system also has a unique solution such that $d = 0$; $\underline{b}_{i,1}$ and $\underline{c}_{i,1}$ may be obtained with the Cramer formulae:

$$\underline{b}_{i,1} = \frac{\det[\underline{\boldsymbol{c}}_{i+1}', \overline{\boldsymbol{q}}_{i,s}, \boldsymbol{n}]}{\det[\underline{\boldsymbol{c}}_i', \overline{\boldsymbol{q}}_{i,s}, \boldsymbol{n}]}, \qquad \underline{c}_{i,1} = \frac{\det[\underline{\boldsymbol{c}}_i', \underline{\boldsymbol{c}}_{i+1}', \boldsymbol{n}]}{\det[\underline{\boldsymbol{c}}_i', \overline{\boldsymbol{q}}_{i,s}, \boldsymbol{n}]}.$$

Similarly, from (4.18) we obtain

$$\underline{f}_{i,1} = \frac{\det[\underline{\boldsymbol{c}}_i', \overline{\boldsymbol{q}}_{i+1,s}, \boldsymbol{n}]}{\det[\underline{\boldsymbol{c}}_{i+1}', \overline{\boldsymbol{q}}_{i+1,s}, \boldsymbol{n}]}, \qquad \underline{g}_{i,1} = \frac{\det[\underline{\boldsymbol{c}}_{i+1}', \underline{\boldsymbol{c}}_i', \boldsymbol{n}]}{\det[\underline{\boldsymbol{c}}_{i+1}', \overline{\boldsymbol{q}}_{i+1,s}, \boldsymbol{n}]}.$$

Now we can take a closer look at the mixed partial derivatives compatibility condition. If we intend to fix the curves $\boldsymbol{c}_0, \ldots, \boldsymbol{c}_{k-1}$ before constructing the patches whose boundaries are made of these curves, then we need to recognise any condition which must be satisfied by these curves to ensure the consistency of the equations with unknown derivatives of the junction functions. Equation (4.9) for the patch $\boldsymbol{p}_i$ in our present notation is

$$\begin{aligned}\underline{f}'_{i,1}\overline{\underline{\boldsymbol{c}}}'_{i+1} + \underline{g}'_{i,1}\overline{\underline{\boldsymbol{q}}}_{i+1,s} + \underline{f}_{i,1}\overline{\underline{\boldsymbol{c}}}''_{i+1} + \underline{g}_{i,1}\overline{\underline{\boldsymbol{q}}}_{i+1,us} = \\ \underline{b}'_{i,1}\overline{\underline{\boldsymbol{c}}}'_i + \underline{c}'_{i,1}\overline{\underline{\boldsymbol{q}}}_{i,s} + \underline{b}_{i,1}\overline{\underline{\boldsymbol{c}}}''_i + \underline{c}_{i,1}\overline{\underline{\boldsymbol{q}}}_{i,us}.\end{aligned}$$

This is a system of three scalar equations with four unknown variables, i.e., derivatives of the junction functions at 0. It is consistent if the vector

$$\underline{f}_{i,1}\overline{\underline{\boldsymbol{c}}}''_{i+1} + \underline{g}_{i,1}\overline{\underline{\boldsymbol{q}}}_{i+1,us} - \underline{b}_{i,1}\overline{\underline{\boldsymbol{c}}}''_i - \underline{c}_{i,1}\overline{\underline{\boldsymbol{q}}}_{i,us}$$

is a linear combination of $\overline{\underline{\boldsymbol{c}}}'_i$, $\overline{\underline{\boldsymbol{q}}}_{i,s}$, $\overline{\underline{\boldsymbol{c}}}'_{i+1}$ and $\overline{\underline{\boldsymbol{q}}}_{i+1,s}$, i.e., if it is orthogonal to the vector $\boldsymbol{n}$. The scalar product of this vector and $\boldsymbol{n}$ must be 0; hence,

$$\underline{c}_{i,1}\langle\overline{\underline{\boldsymbol{q}}}_{i,us},\boldsymbol{n}\rangle - \underline{g}_{i,1}\langle\overline{\underline{\boldsymbol{q}}}_{i+1,us},\boldsymbol{n}\rangle = -\underline{b}_{i,1}\langle\overline{\underline{\boldsymbol{c}}}''_i,\boldsymbol{n}\rangle + \underline{f}_{i,1}\langle\overline{\underline{\boldsymbol{c}}}''_{i+1},\boldsymbol{n}\rangle. \tag{4.19}$$

We can gather Equations (4.19) for all patches surrounding the common corner, thus obtaining the system

$$\begin{bmatrix} \underline{c}_{0,1} & -\underline{g}_{0,1} & & \\ & \underline{c}_{1,1} & \ddots & \\ & & \ddots & -\underline{g}_{k-2,1} \\ -\underline{g}_{k-1,1} & & & \underline{c}_{k-1,1}\end{bmatrix}\begin{bmatrix}\langle\overline{\underline{\boldsymbol{q}}}_{0,us},\boldsymbol{n}\rangle \\ \vdots \\ \vdots \\ \langle\overline{\underline{\boldsymbol{q}}}_{k-1,us},\boldsymbol{n}\rangle\end{bmatrix} = \begin{bmatrix} -\underline{b}_{0,1} & \underline{f}_{0,1} & & \\ & -\underline{b}_{1,1} & \ddots & \\ & & \ddots & \underline{f}_{k-2,1} \\ \underline{f}_{k-1,1} & & & -\underline{b}_{k-1,1}\end{bmatrix}\begin{bmatrix}\langle\underline{\boldsymbol{c}}''_0,\boldsymbol{n}\rangle \\ \vdots \\ \vdots \\ \langle\underline{\boldsymbol{c}}''_{k-1},\boldsymbol{n}\rangle\end{bmatrix}. \tag{4.20}$$

Let the $k \times k$ matrices on the left- and right-hand side of (4.20) be denoted respectively by A and B. Matrices with non-zero coefficients distributed like this are called cyclic bidiagonal; it is easy to find their determinants:

$$\det A = \prod_{i=0}^{k-1}\underline{c}_{i,1} - \prod_{i=0}^{k-1}\underline{g}_{i,1}, \qquad \det B = (-1)^k\left(\prod_{i=0}^{k-1}\underline{b}_{i,1} - \prod_{i=0}^{k-1}\underline{f}_{i,1}\right).$$

Using the formulae for the coefficients of the matrix A, we find

$$\det A = \frac{\prod_{i=0}^{k-1}\det[\underline{\boldsymbol{c}}'_i,\underline{\boldsymbol{c}}'_{i+1},\boldsymbol{n}]}{\prod_{i=0}^{k-1}\det[\underline{\boldsymbol{c}}'_i,\overline{\underline{\boldsymbol{q}}}_{i,s},\boldsymbol{n}]} - \frac{\prod_{i=0}^{k-1}\det[\underline{\boldsymbol{c}}'_{i+1},\underline{\boldsymbol{c}}'_i,\boldsymbol{n}]}{\prod_{i=0}^{k-1}\det[\underline{\boldsymbol{c}}'_{i+1},\overline{\underline{\boldsymbol{q}}}_{i+1,s},\boldsymbol{n}]}.$$

The indices are taken modulo k; hence, the two fractions above have the same denominator. Also, we see that the numerators are products of the same factors with opposite signs. Thus, both fractions subtracted in our formula have the same absolute value; if k is odd, $\det A$ is twice greater than the first fraction (which is non-zero if $\underline{c}'_i$ and $\underline{c}'_{i+1}$ have different directions for all i), while for k even the two fractions cancel each other out, giving $\det A = 0$.

To complete the analysis of G^1 compatibility conditions, we consider the two cases separately, i.e., when k is odd and when k is even. Fixing the curves $\boldsymbol{c}_0, \ldots, \boldsymbol{c}_{k-1}$ and the first-order cross-boundary derivatives of the auxiliary patches at 0, i.e., the vectors $\overline{\boldsymbol{q}}_{0,s}, \ldots \overline{\boldsymbol{q}}_{k-1,s}$, determines the matrices A and B. The curves determine also the vectors $\overline{\underline{c}}''_0, \ldots, \overline{\underline{c}}''_{k-1}$.

If k is odd, then the matrix A is non-singular and System (4.20), with unknown quantities $\langle \underline{\boldsymbol{q}}_{0,us}, \boldsymbol{n}\rangle, \ldots, \langle \underline{\boldsymbol{q}}_{k-1,us}, \boldsymbol{n}\rangle$, has a unique solution for any vectors $\overline{\underline{c}}''_0, \ldots, \overline{\underline{c}}''_{k-1}$. If k is even, the matrix A is singular; arbitrary vectors $\overline{\underline{c}}''_0, \ldots, \overline{\underline{c}}''_{k-1}$ will usually give us an inconsistent system (4.20). Therefore, we need a consistency condition.

Let k be even. Any restriction on the vectors $\underline{\boldsymbol{c}}''_0, \ldots, \underline{\boldsymbol{c}}''_{k-1}$ does not depend on the choice of the vectors $\overline{\boldsymbol{q}}_{0,s}, \ldots, \overline{\boldsymbol{q}}_{k-1,s}$, which only have to be orthogonal to the vector $\boldsymbol{n}$ and such that each pair $\{\underline{\boldsymbol{c}}'_i, \overline{\boldsymbol{q}}_{i,s}\}$ is linearly independent. To simplify the calculation, we choose $\underline{\boldsymbol{q}}_{i,s} = \underline{\boldsymbol{c}}'_{i+1}$ for all i. Then, we have

$$\underline{b}_{i,1} = \frac{\det[\underline{\boldsymbol{c}}'_{i+1}, \underline{\boldsymbol{c}}'_{i+1}, \boldsymbol{n}]}{\det[\underline{\boldsymbol{c}}'_i, \underline{\boldsymbol{c}}'_{i+1}, \boldsymbol{n}]} = 0, \qquad \underline{c}_{i,1} = \frac{\det[\underline{\boldsymbol{c}}'_i, \underline{\boldsymbol{c}}'_{i+1}, \boldsymbol{n}]}{\det[\underline{\boldsymbol{c}}'_i, \underline{\boldsymbol{c}}'_{i+1}, \boldsymbol{n}]} = 1.$$

The system of equations (4.20) has the form $AX = BY$, and the matrices A and B with the coefficients as above are respectively

$$\begin{bmatrix} 1 & -\underline{g}_{0,1} & & \\ & 1 & \ddots & \\ & & \ddots & -\underline{g}_{k-2,1} \\ -\underline{g}_{k-1,1} & & & 1 \end{bmatrix}, \quad \begin{bmatrix} 0 & \underline{f}_{0,1} & & \\ & 0 & \ddots & \\ & & \ddots & \underline{f}_{k-2,1} \\ \underline{f}_{k-1,1} & & & 0 \end{bmatrix}.$$

For any non-singular $k \times k$ matrix C, the systems $AX = BY$ and $CAX = CBY$ are equivalent. Let C be the matrix, whose first $k-1$ rows are these of the identity matrix, and the last row consists of the numbers[2]

$$\underline{g}_{k-1,1}, \quad \underline{g}_{k-1,1}\underline{g}_{0,1}, \quad \underline{g}_{k-1,1}\underline{g}_{0,1}\underline{g}_{1,1}, \quad \ldots, \quad \underline{g}_{k-1,1}\underline{g}_{0,1}\underline{g}_{1,1}\cdots\underline{g}_{k-2,1}, \quad 1.$$

Then the first $k-1$ rows of the matrices A and CA are the same (these rows are linearly independent) and the last row of CA consists of zeros. The last row of the matrix CB consists of the numbers

$$\begin{aligned} &\underline{f}_{k-1,1}, \quad \underline{g}_{k-1,1}\underline{f}_{0,1}, \quad \underline{g}_{k-1,1}\underline{g}_{0,1}\underline{f}_{1,1}, \quad \ldots \\ &\quad \ldots, \quad \underline{g}_{k-1,1}\underline{g}_{0,1}\cdots\underline{g}_{k-4,1}\underline{f}_{k-3,1}, \quad \underline{g}_{k-1,1}\underline{g}_{0,1}\cdots\underline{g}_{k-3,1}\underline{f}_{k-2,1}. \end{aligned}$$

[2]The expressions, whose values these numbers are, were obtained as a result of symbolic Gaussian elimination of the matrix A.

System (4.20) is consistent if and only if the last coordinate of the vector CBY is equal to zero. After substituting the expressions

$$f_{i,1} = \frac{\det[\underline{c}_i', \underline{c}_{i+2}', \boldsymbol{n}]}{\det[\underline{c}_{i+1}', \underline{c}_{i+2}', \boldsymbol{n}]}, \qquad g_{i,1} = \frac{\det[\underline{c}_{i+1}', \underline{c}_i', \boldsymbol{n}]}{\det[\underline{c}_{i+1}', \underline{c}_{i+2}', \boldsymbol{n}]},$$

dividing the coefficients of the last row of CB by $\det[\underline{c}_{k-1}', \underline{c}_0', \boldsymbol{n}]$ and reordering, we obtain the following equation:

$$\sum_{i=0}^{k-1} (-1)^i \frac{\det[\underline{c}_{i-1}', \underline{c}_{i+1}', \boldsymbol{n}]}{\det[\underline{c}_{i-1}', \underline{c}_i', \boldsymbol{n}]\det[\underline{c}_i', \underline{c}_{i+1}', \boldsymbol{n}]} \langle \underline{c}_i'', \boldsymbol{n}\rangle = 0. \tag{4.21}$$

The number $\langle \underline{c}_i'', \boldsymbol{n}\rangle / \|\underline{c}_i'\|_2^2$ is the normal curvature of the patches $\boldsymbol{p}_{i-1}$ and $\boldsymbol{p}_i$ in the direction of the vector $\underline{c}_i'$ (see Formula (A.51)). Let the numbers $\langle \underline{\overline{q}}_{i,us}, \boldsymbol{n}\rangle$ be (somewhat imprecisely) called the twists of the auxiliary patches. We have proved that if k is odd, then there is no restriction for the second-order derivatives of the curves $\boldsymbol{c}_i$; System (4.20) is always consistent and it has a unique solution which describes the twists of the auxiliary patches. If k is even, then the normal curvatures determined by the curves $\boldsymbol{c}_i$ cannot be arbitrary; Equation (4.21) must be satisfied. If it is satisfied, then System (4.20) is consistent, but it has infinitely many solutions. One can arbitrarily fix the twist of one auxiliary patch, and then the twists of the other patches are uniquely determined. This degree of freedom makes it possible to construct surfaces of class G^1 known as monkey saddles (Fig. 4.4).

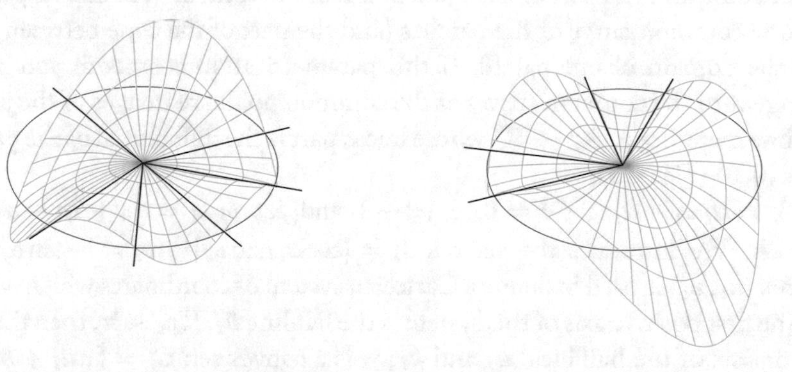

Figure 4.4: Piecewise quadratic monkey saddles for $k = 6$ and $k = 4$

It is worth noting that if the vectors $\underline{c}_{i-1}'$ and $\underline{c}_{i+1}'$ have the same direction, then the number $\langle \underline{c}_i'', \boldsymbol{n}\rangle$ in Equation (4.21) is multiplied by 0, and the vector $\underline{c}_i''$ may be arbitrary. Moreover, if $k = 4$ and the vectors $\underline{c}_0'$ and $\underline{c}_2'$ have the same direction (and opposite orientations), and a similar condition is satisfied by the vectors $\underline{c}_1'$ and $\underline{c}_3'$, then Equation (4.21) is $0 = 0$. System (4.20) is then consistent and it has infinitely many solutions; fixing arbitrarily the twist of one auxiliary patch determines the twists of the other patches.

4.3.2 PIECEWISE POLYNOMIAL FUNCTIONS IN $\mathbb{R}^2$

The G^1 compatibility conditions turned out to be tractable in an elementary way; we have dealt with them in the previous section. A direct go at the compatibility conditions of order 2 or higher, to my knowledge, has not yet led to a complete success in the form of a list of necessary and sufficient conditions for consistency of Equations (4.12)–(4.14) (see e.g. Hermann, Peters and Strotman [2012]). The approach used in this and following sections (Kiciak [1999]) focuses on the possibilities of constructing scalar functions of class C^n of two variables. A surface of class G^n may be obtained by choosing three such functions, assembling them into a vector function and applying a (piecewise) reparametrisation.

A smooth surface made of k regular patches surrounding a common point is always associated with k halflines tangent to the common boundary curves at that point. These halflines remain in the plane tangent to the surface, dividing it to k cones. The scalar functions which we are going to study are defined in the plane $\mathbb{R}^2$, similarly divided into k cones by halflines $\hbar_0, \dots, \hbar_{k-1}$ whose origin is the point $(0,0)$. These halflines will be called the **junction halflines.** A set of numbers

$$\Delta = \{\alpha_0, \dots, \alpha_{k-1}\colon \alpha_0 < \alpha_1 < \dots < \alpha_{k-1} < \alpha_0 + 2\pi\},$$

will be called a **partition of the full angle.** We assume that the number α_i is the angle of inclination of the halfline $\hbar_i$. We can choose a neighbourhood of the point $(0,0)$ (e.g. a sufficiently small disc) and consider such a parametrisation $\boldsymbol{p}$ of a part of the surface made of k patches around the common point that the part of each junction halfline contained in this neighbourhood is the domain of one common curve of the patches (and the part of the cone between two consecutive halflines is the domain of one patch). If this parametrisation is smooth and regular, then the halflines tangent to the common curves at the common point are images of the junction halflines under an affine mapping $\boldsymbol{g}\colon \mathbb{R}^2 \to \mathbb{R}^3$ whose linear part is the differential of the parametrisation $\boldsymbol{p}$ at the point $(0,0)$ (Fig. 4.5).

Let $\delta_i = \alpha_{i+1} - \alpha_i$ for $i = 0, \dots, k-2$ and let $\delta_{k-1} = \alpha_0 + 2\pi - \alpha_{k-1}$. Obviously, $\sum_{i=0}^{k-1} \delta_i = 2\pi$. We introduce the vectors $\boldsymbol{u}_i = [\cos\alpha_i, \sin\alpha_i]^T$, $\boldsymbol{v}_i = [-\sin\alpha_i, \cos\alpha_i]^T$. Each pair of vectors $(\boldsymbol{u}_i, \boldsymbol{v}_i)$ is used to define a Cartesian system of coordinates which will be denoted by x_i and y_i. The first positive axis of this system is the halfline $\hbar_i$. If $\delta_i < \pi$, then the cone C_i, whose boundary consists of the halflines[3] $\hbar_i$ and $\hbar_{i+1}$, is a convex set: $C_i = \{a\boldsymbol{u}_i + b\boldsymbol{u}_{i+1}\colon a, b \geq 0\}$. If $\delta_i = \pi$, then the cone $C_i = \{a\boldsymbol{u}_i + b\boldsymbol{v}_i\colon a \in \mathbb{R}, b \geq 0\}$ is a halfplane, and if $\delta_i > \pi$, then the cone is concave: $C_i = \{a\boldsymbol{u}_i + b\boldsymbol{u}_{i+1}\colon a \leq 0 \text{ or } b \leq 0\}$.

Suppose that in each cone C_i a function f_i of class C^{2n} is defined. If the function f whose restriction to each cone C_i is f_i is of class $C^n(\mathbb{R}^2)$, then the terms of Taylor expansions of the functions f_i satisfy some dependencies—the compatibility conditions. From now on, we restrict our attention to the polynomials obtained by cutting the Taylor expansions of the functions f_i

[3]The indices of junction halflines, cones etc. are taken modulo k, in the same way as in the previous section. In particular, $\alpha_k = \alpha_0 + 2\pi$, $\alpha_{-1} = \alpha_{k-1} - 2\pi$, $\delta_0 = \delta_k$ and $\delta_{-1} = \delta_{k-1}$.

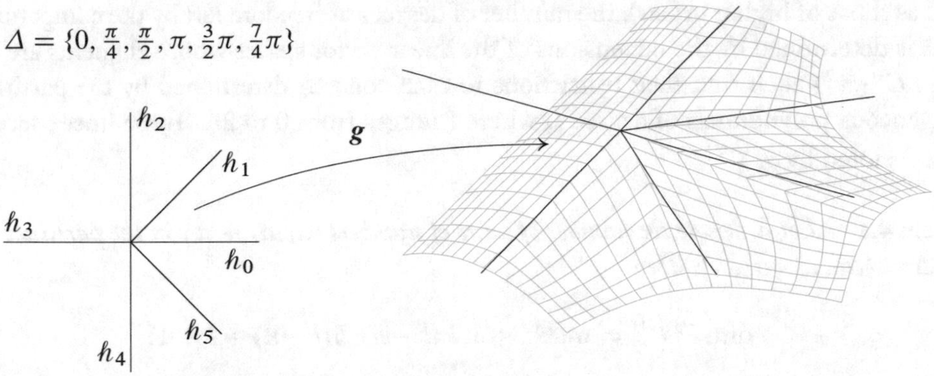

Figure 4.5: A partition of the full angle and halflines tangent to common curves

after the terms with derivatives of order $2n$. Thus, our subject of interest are piecewise polynomial functions of two variables, of degree up to $2n$, of class $C^n(\mathbb{R}^2)$. Such functions will be called **bivariate splines of degree** $2n$ **over the partition** Δ. The *knots* of these splines, i.e., places where different polynomials meet, are the junction halflines.

An algebraic polynomial of degree $2n$ is the sum of $2n+1$ homogeneous terms:

$$p(x,y) = a_{00} + (a_{10}x + a_{11}y) + (a_{20}x^2 + a_{21}xy + a_{22}y^2) + \cdots + \\ (a_{2n,0}x^{2n} + a_{2n,1}x^{2n-1}y + \cdots + a_{2n,2n}y^{2n}). \tag{4.22}$$

Let $\mathcal{H}^l$ denote the linear vector space whose elements are homogeneous polynomials of degree l of two variables. Any such a polynomial, expressed as a function of the Cartesian coordinates introduced above, is also a homogeneous polynomial of the variables x_i and y_i.

Now we consider functions $g_l(x,y)$ whose restrictions to the cones $C_0, \ldots, C_{k-1}$ are homogeneous polynomials of degree l of two variables. We can express the function g_l in terms of the polar coordinates (r, φ):

$$g_l(r\cos\varphi, r\sin\varphi) = r^l s_l(\varphi). \tag{4.23}$$

We shall take a closer look at the admissible functions s_l in the following section. At this point we notice that the restriction of the function g_l to any halfline in $\mathbb{R}^2$ with the origin at $(0,0)$ is either zero or a homogeneous polynomial of degree l.

The analysis of G^1 compatibility conditions in the previous section revealed that if the number $k \geq 3$ of the common curves meeting at the common point of patches is odd, then there are k degrees of freedom in a construction. For example, one can arbitrarily choose the normal curvatures of the patches in the directions tangent to the common curves (which fixes the numbers $\langle \underline{\boldsymbol{c}}_j'', \boldsymbol{n}\rangle$), and then the twists $\langle \underline{\overline{\boldsymbol{q}}}_{j,us}, \boldsymbol{n}\rangle$ of the auxiliary patches satisfying the system of equations (4.20) are unique. In general (which includes the case of G^1 compatibility conditions as well as those of higher orders), the number of degrees of freedom left by the compatibility conditions is determined by the dimensions of the linear vector spaces whose elements are functions of class $C^n(\mathbb{R}^2)$ such that their restrictions to each cone C_i determined by the partition Δ are homogeneous polynomials of degree l, where l ranges from 0 to $2n$. These linear vector spaces will be denoted by $\mathcal{H}_\Delta^{(l,n)}$.

Theorem 4.1 *Let h denote the number of pairs of numbers $\{\alpha_i, \alpha_i + \pi\}$ in the partition of the full angle $\Delta = \{\alpha_0, \dots, \alpha_{k-1}\}$. Then*

$$\dim \mathcal{H}_\Delta^{(l,n)} = \max\{l + 1, k(l - n), h(l - n) + l + 1\}.$$

Proof. The dimension of a linear vector space may be found by finding its basis and counting its elements. We shall construct a basis.

If a function $f \in \mathcal{H}_\Delta^{(l,n)}$ of class C^n in the cones C_{i-1} and C_i, neighbouring across the halfline h_i (whose points in the system of coordinates (x_i, y_i) have the y_i coordinate equal to 0) is described by polynomials p_{i-1} and p_i, then the difference of these polynomials must be divisible by the polynomial y_i^{n+1}. Hence, if their degree l is less than or equal to n, then there must be $p_{i-1} = p_i$. Therefore, if $l \le n$, then the space $\mathcal{H}_\Delta^{(l,n)}$ is the space $\mathcal{H}^l$ of homogeneous bivariate polynomials of degree l and its dimension is $l + 1$. If $l > n$, then $\mathcal{H}^l$ is a subspace of $\mathcal{H}_\Delta^{(l,n)}$. We can take the polynomials $x^l, x^{l-1}y, \dots, xy^{l-1}, y^l$ as the first $l + 1$ elements of a basis of this space.

The functions $x_i^m y_i^{l-m}$ are homogeneous polynomials of degree l of the coordinates x, y in $\mathbb{R}^2$. Let $y_{i,+} = \max\{0, y_i\}$. Consider a pair of numbers $\{\alpha_i, \alpha_j\} \subset \Delta$ such that $\alpha_j = \alpha_i + \pi$. The union of corresponding haflines h_i and h_j is a line. One can easily check that the functions $x_i^m y_{i,+}^{l-m}$ (see Figure 4.6), where $m \in \{0, \dots, l - n - 1\}$, are described by homogeneous polynomials of degree l in each cone and they are of class $C^n(\mathbb{R}^2)$. In general, with h lines made of the halflines determined by the partition Δ, we have $h(l - n)$ functions of class $C^n(\mathbb{R}^2)$, equal to zero

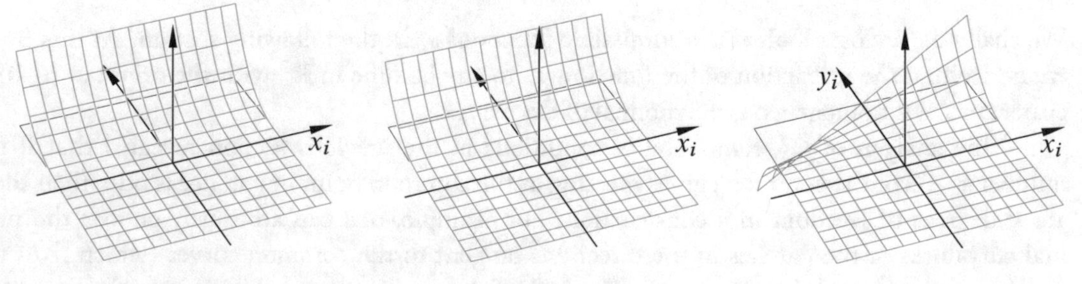

Figure 4.6: Truncated power functions of two variables: $y_{i,+}^3$, $y_{i,+}^4$ and $x_i y_{i,+}^3$

in halfplanes bounded by the junction halflines. The proof that these functions, together with the polynomials found earlier, form a linearly independent set is left as an exercise.

Now we are looking for more functions in the space $\mathcal{H}_{\Delta}^{(l,n)}$ linearly independent from the ones found above. We construct an auxiliary partition $\tilde{\Delta}$. To obtain it, for each pair $\{\alpha_i, \alpha_j\} \subset \Delta$ such that $\alpha_j = \alpha_i + \pi$, we reject α_j. In what is left, we replace each number $\alpha_j > \alpha_0 + \pi$ by $\alpha_j - \pi$. The new partition consists of $k - h$ numbers, which we renumber to obtain an increasing sequence: $\tilde{\alpha}_0 < \cdots < \tilde{\alpha}_{k-h-1}$. The halfline with the origin at $(0,0)$ whose inclination angle is the number $\tilde{\alpha}_j$ will be denoted by $\tilde{h}_j$; these halflines divide the plane $\mathbb{R}^2$ into cones $\tilde{C}_j$, all but one of which are convex. The concave cone $\tilde{C}_{k-h-1}$ is bounded by the halflines $\tilde{h}_0$ and $\tilde{h}_{k-h-1}$.

Due to the inequality $\tilde{\alpha}_{k-h-1} < \tilde{\alpha}_0 + \pi$, there exists a line ℓ in $\mathbb{R}^2$ that does not pass through the point $(0,0)$ and that intersects all halflines $\tilde{h}_j$. In the space $\mathcal{H}_{\tilde{\Delta}}^{(l,n)}$ we are looking for functions equal to 0 in the concave cone $\tilde{C}_{k-h-1}$. The restrictions of such functions to the line ℓ are univariate splines of class C^n, of degree at most l, and equal to zero in the intersection of the line ℓ and the cone $\tilde{C}_{k-h-1}$. By fixing an affine mapping $\ell \to \mathbb{R}$ we obtain a coordinate system on ℓ; we take the numbers corresponding to the intersections of the halflines h_i with ℓ for *knots*; to obtain functions of class C^n we assume that these knots have multiplicities $l - n$. Thus, we have a non-decreasing sequence of $(l-n)(k-h)$ knots, and if their number is greater than $l+1$, then we can define $(l-n)(k-h) - l - 1$ linearly independent B-spline functions of degree l on the line ℓ (Fig. 4.7).

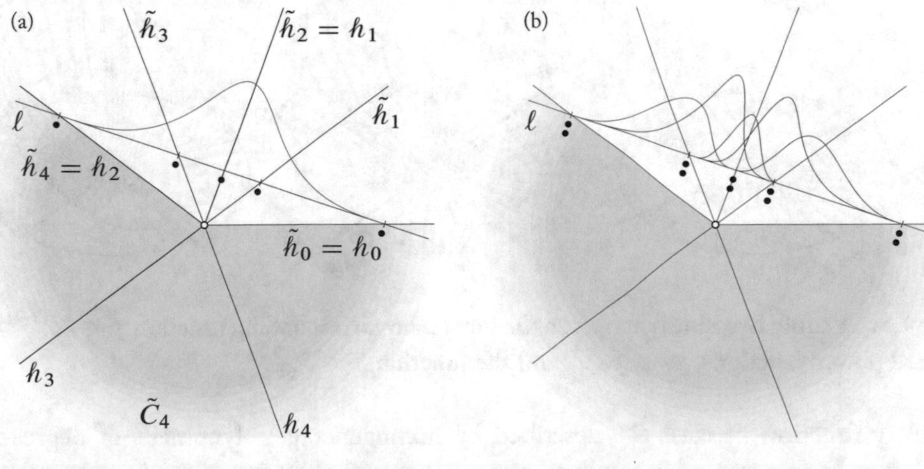

Figure 4.7: B-spline functions of class C^2 on the line ℓ: (a) cubic and (b) quartic

Let $\tilde{f}\colon \mathbb{R}^2 \to \mathbb{R}$ be a function equal to 0 in the concave cone $\tilde{C}_{k-h-1}$, whose restrictions to *all* halflines in $\mathbb{R}^2$ having the origin at $(0,0)$ are either zero, or homogeneous polynomials of degree l, and whose restriction to the line ℓ is one of the B-spline functions described above. The function $\tilde{f}$ is an element of the space $\mathcal{H}_{\tilde{\Delta}}^{(l,n)}$. We can use it to construct a function $f \in \mathcal{H}_{\Delta}^{(l,n)}$

as follows. If $\tilde{\alpha}_i \notin \Delta$, then the halfline $\tilde{h}_i$ is located inside some cone C_m determined by the partition Δ. The number $\alpha_j = \tilde{\alpha}_i + \pi$ is an element of Δ. In both cones, $\tilde{C}_{i-1}$ and $\tilde{C}_i$ (adjacent to the halfline $\tilde{h}_i$), the function f is described by the same homogeneous polynomial, while $\tilde{f}$ is not. The difference of the two polynomials, which describe $\tilde{f}$ on the opposite sides of the halfline $\tilde{h}_i$, is a homogeneous polynomial p_j of degree l divisible by y_j^{n+1}. So, we replace this factor of p_j by the truncated power function $y_{j,+}^{n+1}$ and we add the function $p_{j,+}$ obtained in this way to the function $\tilde{f}$, thus getting a function of class $C^n(\mathbb{R}^2)$ which has a seam along the halfline h_j but not along $\tilde{h}_i$. After doing this for *all* numbers $\tilde{\alpha}_i$ absent in Δ, we obtain a function f, which is a new element of our basis of the space $\mathcal{H}_{\Delta}^{(l,n)}$. The example in Figure 4.8 is a continuation of the one in Figure 4.7a; to better visualise the functions, lines of constant polar coordinates φ, r instead of lines of constant variables x, y are drawn on the graphs.

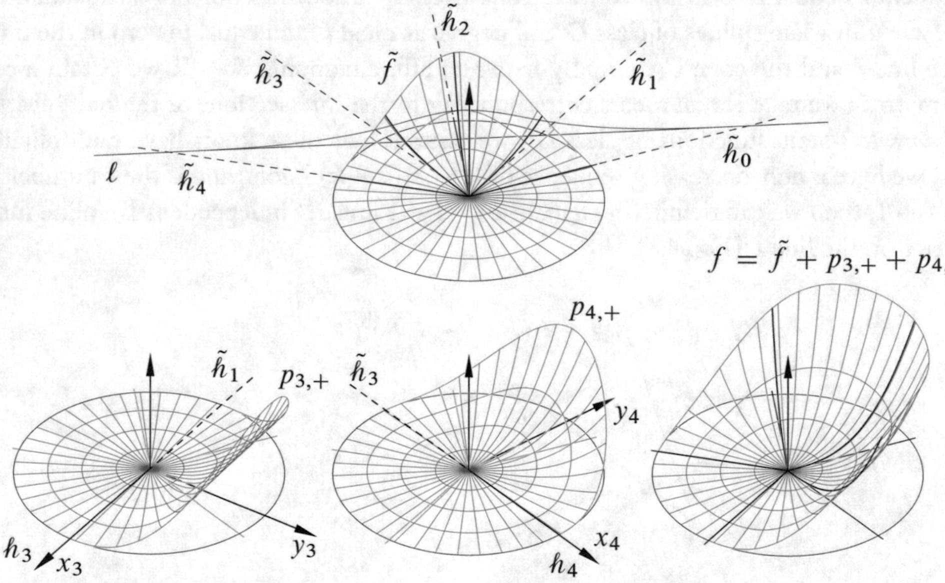

Figure 4.8: A cubic B-spline function on the line ℓ, the corresponding function $\tilde{f} \in \mathcal{H}_{\tilde{\Delta}}^{(3,2)}$, two cubic truncated power functions, $p_{3,+}$, $p_{4,+}$, and the function $f \in \mathcal{H}_{\Delta}^{(3,2)}$

Any function of class C^n described by homogeneous polynomials of degree l in each cone C_i may be represented in a unique way as a linear combination of the functions found above, which, therefore, form a basis. To complete the proof, it suffices to reorder the expression $d_1 + d_2 + d_3$, where $d_1 = l + 1$ is the number of polynomials in our basis, $d_2 = \max\{0, h(l-n)\}$ is the number of basis functions whose sets of zeros are the halfplanes bounded by lines made of halflines h_i, h_j, and $d_3 = \max\{0, (l-n)(k-h) - l - 1\}$ is the number of functions constructed with the use of B-spline functions on the line ℓ. □

The theorem just proved reveals *all* degrees of freedom left by the G^n compatibility conditions around a common corner of k patches. The direct algebraic sum[4]

$$\mathcal{P}_{\Delta}^{(n)} = \mathcal{H}_{\Delta}^{(0,n)} \oplus \cdots \oplus \mathcal{H}_{\Delta}^{(2n,n)}$$

is a linear vector space whose elements are functions of class $C^n(\mathbb{R}^2)$: bivariate splines of degree $2n$ over the partition Δ. In a construction of a surface we can pick a triple of such splines and form a vector function $\boldsymbol{f}\colon \mathbb{R}^2 \to \mathbb{R}^3$. Then, for each cone C_i we can take the function $\boldsymbol{f}_i$ made of the three polynomials which describe the function $\boldsymbol{f}$ in C_i, and change the variables (x, y) to (x_i, y_i). In this way we obtain the auxiliary patches (or rather their partial derivatives up to the order $2n$ at the common point), which determine derivatives of the common boundary curves and which may be used to construct the cross-boundary derivatives of patches forming a surface of class G^n.

Now we take a closer look at the functions obtained by extending the B-spline functions defined on the line ℓ to the entire plane. In the proof of Theorem 4.1 such functions, denoted by $\tilde{f}$, were used to construct $(k-h)(l-n)-l-1$ elements of a basis of the space $\mathcal{H}_{\Delta}^{(l,n)}$. We can assume that $0 < \tilde{\alpha}_0 < \cdots < \tilde{\alpha}_{k-h-1} < \pi$ (we can satisfy this condition by rotating the junction halflines) and then all junction halflines intersect the linc $y = 1$. The definition below is a modification of Definition A.5.

Definition 4.2 *Let $\ell = \{(x, y)\colon x \in \mathbb{R},\ y = 1\}$ and let $u_i \ldots, u_{i+n+1}$ be a non-decreasing sequence of numbers (knots). An* ***extended normalised B-spline function*** *of degree n with these knots is given by the formula*

$$E_i^n(x, y) = \begin{cases} (-1)^{n+1}(u_{i+n+1} - u_i) f_{x,y}^n[u_i, \ldots, u_{i+n+1}], & \textit{if } y > 0, \\ 0 & \textit{else,} \end{cases} \tag{4.24}$$

where $f_{x,y}^n[u_i, \ldots, u_{i+n+1}]$ is the divided difference of order $n+1$ of the function

$$f_{x,y}^n(u) \stackrel{\text{def}}{=} (x - uy)_+^n = \begin{cases} (x - uy)^n & \textit{if } x \ge uy, \\ 0 & \textit{else.} \end{cases}$$

It is obvious that $E_i^n(x, 1) = N_i^n(x)$ for all $x \in \mathbb{R}$, i.e., the restriction of the function E_i^n to the line ℓ is a normalised B-spline function of degree n. Moreover, the restriction of E_i^n to any

[4]The **algebraic sum** of linear subspaces $X_1, \ldots, X_m$ of a real linear vector space V is the subspace $X = \{\sum_{i=1}^m a_i \boldsymbol{x}_i : a_i \in \mathbb{R}, \boldsymbol{x}_i \in X_i\} \subset V$. The algebraic sum is **direct** if the union of any bases of the subspaces $X_1, \ldots, X_m$ is a linearly independent set of vectors in the space V.

halfline in $\mathbb{R}^2$ with the origin at $(0,0)$, i.e., the function $E_i^n(tx, ty)$ of the variable $t \geq 0$, is either zero, or a homogeneous polynomial of degree n.

Property 4.3 *The functions E_i^n satisfy the following (extended Mansfield–de Boor–Cox) recursive formula:*

$$E_i^0(x,y) = \begin{cases} 1 & \text{for } y > 0 \text{ and } x \in [u_i y, u_{i+1} y), \\ 0 & \text{else,} \end{cases} \tag{4.25}$$

$$E_i^n(x,y) = \frac{x - u_i y}{u_{i+n} - u_i} E_i^{n-1}(x,y) + \frac{u_{i+n+1} y - x}{u_{i+n+1} - u_{i+1}} E_{i+1}^{n-1}(x,y) \quad \text{for } n > 0. \tag{4.26}$$

Property 4.4 *Let $n > 0$. If $y > 0$ and $x \neq u_j y$, where u_j is a knot of multiplicity n, or $y \leq 0$, then the partial derivatives of the function E_i^n are described by the formulae*

$$\frac{\partial}{\partial x} E_i^n(x,y) = \frac{n}{u_{i+n} - u_i} E_i^{n-1}(x,y) - \frac{n}{u_{i+n+1} - u_{i+1}} E_{i+1}^{n-1}(x,y), \tag{4.27}$$

$$\frac{\partial}{\partial y} E_i^n(x,y) = \frac{n u_{i+n+1}}{u_{i+n+1} - u_{i+1}} E_{i+1}^{n-1}(x,y) - \frac{n u_i}{u_{i+n} - u_i} E_i^{n-1}(x,y). \tag{4.28}$$

Exercise. Prove Properties 4.3 and 4.4. Hint: it may be done with almost the same calculations as in the proofs of Properties A.7 and A.12.

Exercise. Show how to modify Algorithm A.7 (p. 180) to evaluate the extended B-spline functions and their derivatives.

Elements of a basis of the space $\mathcal{H}_\Delta^{(l,n)}$ are obtained from the functions E_i^l of class C^n. With the partition $\tilde{\Delta} = \{\tilde{\alpha}_0, \ldots, \tilde{\alpha}_{k-h-1}\}$ such that $0 < \tilde{\alpha}_0 < \cdots < \tilde{\alpha}_{k-h-1} < \pi$ and the line ℓ chosen as in Definition 4.2, the knots are

$$u_{(l-n)i+j} = \cot \tilde{\alpha}_{k-h-1-i} \quad \text{for } i = 0, \ldots, k-h-1,\ j = 0, \ldots, l-n-1.$$

Their numbering is established so as to obtain a non-decreasing sequence. Each knot in it appears $l - n$ times (see Figure 4.7). Such a sequence guarantees that the B-spline functions N_i^l on the line ℓ and the functions E_i^l in $\mathbb{R}^2$ are of class C^n.

The examples below show how elements of the spaces $\mathcal{H}_\Delta^{(l,n)}$ may be obtained from the functions E_i^l, for (l,n) equal to $(2,1)$, $(3,2)$ and $(4,2)$, i.e., in all cases relevant for constructions of surfaces of class G^1 and G^2. We divide the set of indices of the knots, $S = \{0, \ldots, (l-n)(k-h)-1\}$ to two subsets; the first, denoted by S^+, consists of the indices j such that the number $\tilde{\alpha}_m$, which determines the knot $u_j = \cot \tilde{\alpha}_m$, is an element of the original partition Δ, and the other indices are elements of the second subset, S^-.

If $l = n + 1$, then all knots are of multiplicity 1 and the divided difference in Formula (4.24) is a linear combination of the values of $f^l_{x,y}(u)$ at the knots; hence,

$$E^l_i(x, y) = (-1)^{l+1}(u_{i+l+1} - u_i)\sum_{j=i}^{i+l+1} c_{ij} f^l_{x,y}(u_j),$$

for some numbers c_{ij}. The element of the space $\mathcal{H}^{(l,l-1)}_\Delta$ corresponding to this function is

$$f(x, y) = \begin{cases} (-1)^{l+1}(u_{i+l+1} - u_i) \displaystyle\sum_{j\in\{i,\dots,i+l+1\}\cap S^+} c_{ij} f^l_{x,y}(u_i) & \text{if } y > 0, \\ (-1)^l(u_{i+l+1} - u_i) \displaystyle\sum_{j\in\{i,\dots,i+l+1\}\cap S^-} c_{ij} f^l_{x,y}(u_i) & \text{else.} \end{cases}$$

If $l = n + 2$, then the knots which determine the function E^l_i are of multiplicity 1 or 2; for $l = 4$, $n = 2$ there is either $u_i = u_{i+1} < u_{i+2} = u_{i+3} < u_{i+4} = u_{i+5}$, or $u_i < u_{i+1} = u_{i+2} < u_{i+3} = u_{i+4} < u_{i+5}$. The divided difference in the definition of the function E^4_i involves the value of the function $f^4_{x,y}(u)$ at each knot, and if the knot appears twice, then it also involves the first-order derivative, equal to $-4yf^3_{x,y}(u)$. We can write[5]

$$E^4_i(x, y) = -(u_{i+5} - u_i)\sum_{j=0}^{3}\bigl(c_{i,i+2j} f^4_{x,y}(u_{i+2j}) - 4yd_{i,i+2j} f^3_{x,y}(u_{i+2j})\bigr),$$

and the function f, which is the corresponding element of the space $\mathcal{H}^{(4,2)}_\Delta$, is given by

$$f(x, y) = \begin{cases} -(u_{i+5} - u_i) \displaystyle\sum_{\substack{j=0,\dots,3,\\ i+2j\in\{i,\dots,i+5\}\cap S^+}} \bigl(c_{i,i+2j} f^4_{x,y}(u_{i+2j}) - 4yd_{i,i+2j} f^3_{x,y}(u_{i+2j})\bigr) & \text{if } y > 0, \\ (u_{i+5} - u_i) \displaystyle\sum_{\substack{j=0,\dots,3,\\ i+2j\in\{i,\dots,i+5\}\cap S^-}} \bigl(c_{i,i+2j} f^4_{x,y}(u_{i+2j}) - 4yd_{i,i+2j} f^3_{x,y}(u_{i+2j})\bigr) & \text{else.} \end{cases}$$

Explicit formulae for the coefficients c_{ij} and d_{ij} may be found using the definition of divided differences; this is left as an exercise for readers determined and patient enough.[6] These formulae, however, are not very attractive or convenient. Instead, it is better to do the following exercise:

Exercise. Describe how Algorithms A.1 and A.2 (see Section A.1.1) may be used to compute numerically the coefficients c_{ij} and d_{ij}.

[5] Note that if $u_i = u_{i+1}$, then $c_{i,i+6} = d_{i,i+6} = 0$ and the upper limit for j in this and the following formula should be 2; on the other hand, if $u_i < u_{i+1}$, then $d_{i,i} = d_{i,i+6} = 0$.

[6] I once was.

4.3.3 CONNECTION WITH TRIGONOMETRIC SPLINES

Let $p(x, y)$ be a polynomial of two variables described by Formula (4.22). The mapping $\boldsymbol{c}\colon (r, \varphi) \to (x, y)$, given by the formula

$$\boldsymbol{c}(r, \varphi) \stackrel{\text{def}}{=} (r \cos\varphi, r \sin\varphi),$$

describes a transition from polar coordinates (r, φ) to the Cartesian coordinates (x, y) in $\mathbb{R}^2$. The composition of the function p with the mapping $\boldsymbol{c}$ may be described by the formula

$$p(x, y) = p\big(\boldsymbol{c}(r, \varphi)\big) = r^0 T_0(\varphi) + r^1 T_1(\varphi) + r^2 T_2(\varphi) + \cdots + r^{2n} T_{2n}(\varphi). \tag{4.29}$$

The functions $T_0, \ldots, T_{2n}$ are trigonometric polynomials of degrees $0, \ldots, 2n$ of the variable φ. Consider a term $r^l T_l(\varphi) = p_l(x, y)$ of the sum above. It is a homogeneous polynomial of degree l of the variables x, y. If l is even, then $p_l(-x, -y) = p_l(x, y)$ for all $x, y \in \mathbb{R}$. If l is odd, then $p_l(-x, -y) = -p_l(x, y)$. These properties are equivalent to $T_l(\varphi + \pi) = T_l(\varphi)$ in the former case and $T_l(\varphi + \pi) = -T_l(\varphi)$ in the latter case, for all φ. Hence, the formula for the trigonometric polynomials T_l in (4.29) depends on the parity of l:

$$T_l(\varphi) = \begin{cases} a_{l,0} + \displaystyle\sum_{j=1}^{l/2} (a_{l,2j} \cos 2j\varphi + b_{l,2j} \sin 2j\varphi) & \text{if } l \text{ is even,} \\ \displaystyle\sum_{j=0}^{(l-1)/2} \big(a_{l,2j+1} \cos(2j+1)\varphi + b_{l,2j+1} \sin(2j+1)\varphi\big) & \text{if } l \text{ is odd.} \end{cases}$$

Exercise. Find homogeneous algebraic polynomials p_{lj} and q_{lj} of two variables, such that $p_{lj}(x, y) = r^l \cos j\varphi$ and $q_{lj}(x, y) = r^l \sin j\varphi$, for $l = 0, \ldots, 4$ and $j = 0, \ldots, l$.

Now we consider bivariate splines of degree $2n$ over a fixed partition Δ. By composing such a function g of class $C^n(\mathbb{R}^2)$ with the mapping $\boldsymbol{c}$, we obtain

$$g(x, y) = r^0 s_0(\varphi) + r^1 s_1(\varphi) + r^2 s_2(\varphi) + \cdots + r^{2n} s_{2n}(\varphi). \tag{4.30}$$

The function $s_l(\varphi)$ is a **trigonometric spline** of degree l. It is a periodic function, with the period 2π; it is of class $C^n(\mathbb{R})$ and its restriction to each interval (α_i, α_{i+1}) is an even or odd (depending on the parity of l) trigonometric polynomial. Such trigonometric splines form a linear vector space, which we denote by $\mathcal{T}_\Delta^{(l,n)}$. There is a one-to-one correspondence between these trigonometric splines and bivariate splines in $\mathbb{R}^2$ over the partition Δ, described in each cone C_i by homogeneous polynomials of degree l; the mapping obtained as before by transition to the polar coordinates and taking $r = 1$ is an isomorphism of the spaces $\mathcal{H}_\Delta^{(l,n)}$ and $\mathcal{T}_\Delta^{(l,n)}$ (see Figure 4.9). An obvious consequence of Theorem 4.1 is

Corollary 4.5 *The dimension of the space $\mathcal{T}_\Delta^{(l,n)}$ is the same as the dimension of the space $\mathcal{H}_\Delta^{(l,n)}$, i.e.,*

$$\dim \mathcal{T}_\Delta^{(l,n)} = \max\{l + 1, k(l - n), h(l - n) + l + 1\},$$

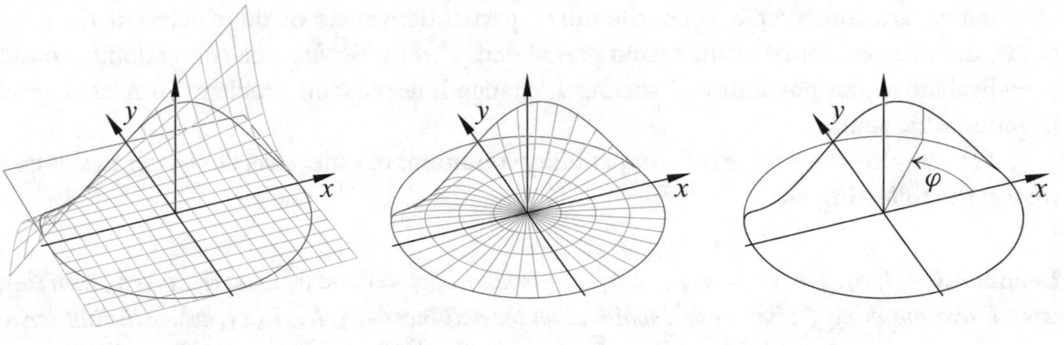

Figure 4.9: The correspondence between a bivariate spline and a trigonometric spline

where k is the number of elements of the partition of the full angle Δ and h is the number of pairs $\{\alpha_i, \alpha_i + \pi\}$ *in it.*

If $l \leq n$, then all elements of the space $\mathcal{H}_\Delta^{(l,n)}$ are homogeneous polynomials of degree l and, accordingly, all elements of the space $\mathcal{T}_\Delta^{(l,n)}$ are even or odd trigonometric polynomials. By differentiating the composition of the function g with the mapping $\boldsymbol{c}$, we obtain a connection between the partial derivatives at the point $(0,0)$ of polynomials which describe g in the cones C_i and the values and derivatives of the trigonometric splines in Formula (4.30). In this way, we obtain the derivatives with respect to the variables x_i, y_i, i.e., the Cartesian coordinates in $\mathbb{R}^2$ defined with the frames $(\boldsymbol{u}_i, \boldsymbol{v}_i)$ in Section 4.3.2. In particular, the polynomial q_i, which describes the function g in the cone C_i, has the following value and derivatives[7] involved in the G^2 compatibility conditions:

$$\begin{aligned}
\overline{\underline{q}}_i &= s_0(\alpha_i), & \overline{\underline{q}}_{i,x_i} &= s_1(\alpha_i), & \overline{\underline{q}}_{i,y_i} &= s_1'(\alpha_i),\\
\overline{\underline{q}}_{i,x_ix_i} &= 2s_2(\alpha_i), & \overline{\underline{q}}_{i,x_iy_i} &= s_2'(\alpha_i), & \overline{\underline{q}}_{i,y_iy_i} &= 2s_2(\alpha_i) + s_2''(\alpha_i),\\
\overline{\underline{q}}_{i,x_ix_ix_i} &= 6s_3(\alpha_i), & \overline{\underline{q}}_{i,x_ix_iy_i} &= 2s_3'(\alpha_i), & \overline{\underline{q}}_{i,x_iy_iy_i} &= 3s_3(\alpha_i) + s_3''(\alpha_i),\\
\overline{\underline{q}}_{i,x_ix_ix_ix_i} &= 24s_4(\alpha_i), & \overline{\underline{q}}_{i,x_ix_ix_iy_i} &= 6s_4'(\alpha_i), & \overline{\underline{q}}_{i,x_ix_iy_iy_i} &= 8s_4(\alpha_i) + 2s_4''(\alpha_i).
\end{aligned}$$

Similarly, we can find derivatives appearing in higher order compatibility conditions.

If a smooth surface made of patches $\boldsymbol{p}_0, \ldots, \boldsymbol{p}_{k-1}$ surrounding a common corner is parametrised in such a way that the intersection of the cone C_i with a neighbourhood of the point $(0,0)$ is the domain of the parametrisation of the patch $\boldsymbol{p}_i$ then (parts of) the junction halflines are the domains of the common boundary curves of those patches. The partial derivatives of the parametrisation in the directions of the junction halflines (i.e., with respect to x_i) are derivatives of the common curves. As we can see from the formulae above, fixing the derivatives of the curves fixes also the values of the trigonometric splines at the points α_i.[8] If the com-

[7] As before, the under- and overlining denotes the value or derivative at the common point of domains of auxiliary and final patches, whose construction we have in mind all the time.

[8] This applies also to higher order derivatives; in general, there is $\overline{\underline{q}}_{i,x_i^m} = m!s_m(\alpha_i)$.

mon curves are constructed *before* the mixed partial derivatives of the patches at the common point, then the existence of the mixed partial derivatives satisfying the compatibility conditions is equivalent to the possibility of solving Lagrange interpolation problems in related spaces of trigonometric splines.

To simplify the analysis of compatibility conditions of order two, done later in this chapter, we use the following facts:

Lemma 4.6 *If* $\alpha_{i-1} < \alpha_i < \alpha_{i+1} < \alpha_{i-1} + \pi$ *and* $\tilde{\alpha}_{i-1} < \tilde{\alpha}_i < \tilde{\alpha}_{i+1} < \tilde{\alpha}_{i-1} + \pi$*, then there exists a linear mapping* $\boldsymbol{f}\colon \mathbb{R}^2 \to \mathbb{R}^2$*, which maps the halflines* h_{i-1}, h_i, h_{i+1}*, whose inclination angles are* $\alpha_{i-1}, \alpha_i, \alpha_{i+1}$*, to the halflines* $\tilde{h}_{i-1}, \tilde{h}_i, \tilde{h}_{i+1}$*, whose inclination angles are* $\tilde{\alpha}_{i-1}, \tilde{\alpha}_i, \tilde{\alpha}_{i+1}$ *respectively.*

Proof. The halflines h_{i-1} and h_{i+1} bound a convex cone in $\mathbb{R}^2$, and so do the halflines $\tilde{h}_{i-1}$ and $\tilde{h}_{i+1}$. Due to the inequalities $\alpha_{i-1} < \alpha_i < \alpha_{i+1}$ and $\tilde{\alpha}_{i-1} < \tilde{\alpha}_i < \tilde{\alpha}_{i+1}$, these cones contain the halflines h_i and $\tilde{h}_i$.

The vector $\boldsymbol{u}_i = [\cos\alpha_i, \sin\alpha_i]^T$ is an element of the halfline h_i and, as the halflines h_{i-1} and h_{i+1} are not collinear, it is the sum of two vectors, say, $\boldsymbol{a} \in h_{i-1}$ and $\boldsymbol{b} \in h_{i+1}$ (Fig. 4.10). Similarly, there exist vectors $\tilde{\boldsymbol{a}} \in \tilde{h}_{i-1}$ and $\tilde{\boldsymbol{b}} \in \tilde{h}_{i+1}$ whose sum is the vector $\tilde{\boldsymbol{u}}_i = [\cos\tilde{\alpha}_i, \sin\tilde{\alpha}_i]^T \in \tilde{h}_i$. The mapping $\boldsymbol{f}$ is represented by the matrix $[\tilde{\boldsymbol{a}}, \tilde{\boldsymbol{b}}][\boldsymbol{a}, \boldsymbol{b}]^{-1}$. □

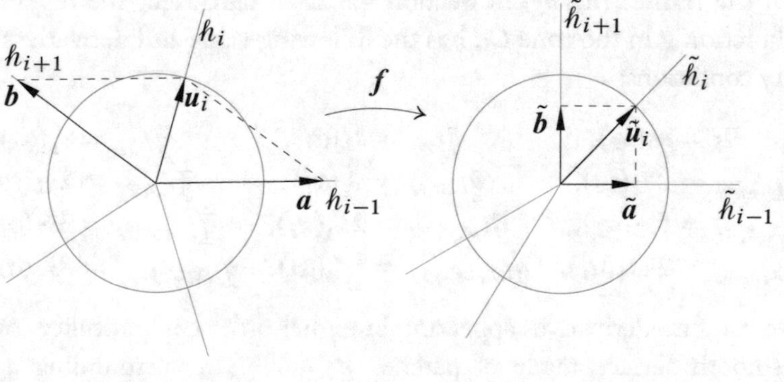

Figure 4.10: Transforming a partition of the full angle

Lemma 4.7 *Let* $\Delta = \{\alpha_0, \ldots, \alpha_{k-1}\}$ *and* $\tilde{\Delta} = \{\tilde{\alpha}_0, \ldots, \tilde{\alpha}_{k-1}\}$ *be partitions of the full angle such that for* $i \in \{0, \ldots, k-1\}$ *there is*

$$\cos\tilde{\alpha}_i = \frac{p_i}{\sqrt{p_i^2 + q_i^2}}, \quad \sin\tilde{\alpha}_i = \frac{q_i}{\sqrt{p_i^2 + q_i^2}},$$

where $p_i = a\cos\alpha_i + b\sin\alpha_i$, $q_i = c\cos\alpha_i + d\sin\alpha_i$, *and* a, b, c, d *are numbers such that* $ad - bc \neq 0$. *The number* h *of pairs* $\{\alpha_i, \alpha_i + \pi\} \in \Delta$ *is equal to the number of pairs* $\{\tilde{\alpha}_i, \tilde{\alpha}_i + \pi\} \in \tilde{\Delta}$. *The spaces* $\mathcal{T}_{\Delta}^{(l,n)}$ *and* $\mathcal{T}_{\tilde{\Delta}}^{(l,n)}$ *are isomorphic.*

Proof. The matrix $\left[\begin{smallmatrix} a & b \\ c & d \end{smallmatrix}\right]$ describes a one-to-one linear transformation $\boldsymbol{f}\colon \mathbb{R}^2 \to \mathbb{R}^2$. Let $\tilde{\hbar}_0, \ldots, \tilde{\hbar}_{k-1}$ be the images of the halflines $\hbar_0, \ldots, \hbar_{k-1}$, whose inclination angles are $\alpha_0, \ldots, \alpha_{k-1}$. Then, the numbers $\tilde{\alpha}_i$ are the inclination angles of the halflines $\tilde{\hbar}_i$. A one-to-one correspondence between the elements of the spaces $\mathcal{T}_{\Delta}^{(l,n)}$ and $\mathcal{T}_{\tilde{\Delta}}^{(l,n)}$ may be obtained as follows: given a trigonometric spline $s \in \mathcal{T}_{\Delta}^{(l,n)}$, we convert the function $r^l s(\varphi)$ of the polar coordinates (r, φ) in $\mathbb{R}^2$ to the function of Cartesian coordinates (x, y). Thus, we obtain a bivariate spline $p \in \mathcal{H}_{\Delta}^{(l,n)}$. The composition $\tilde{p} = p \circ \boldsymbol{f}^{-1}$ is a bivariate spline of degree l over the partition $\tilde{\Delta}$, which is an element of the space $\mathcal{H}_{\tilde{\Delta}}^{(l,n)}$. As $\boldsymbol{f}$ is a bijection, this mapping of $\mathcal{H}_{\Delta}^{(l,n)}$ to $\mathcal{H}_{\tilde{\Delta}}^{(l,n)}$ is an isomorphism of linear vector spaces. By returning from Cartesian to polar coordinates and fixing $r = 1$, we obtain a trigonometric spline $\tilde{s} \in \mathcal{T}_{\tilde{\Delta}}^{(l,n)}$. □

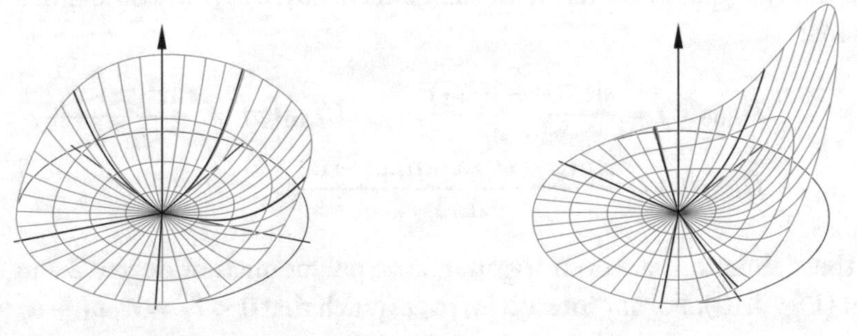

Figure 4.11: Corresponding bivariate splines over two partitions of the full angle; the partitions shown here are those from Figure 4.10

4.3.4 TRIGONOMETRIC SPLINES AND G^1 COMPATIBILITY CONDITIONS

Although the direct analysis of the G^1 compatibility conditions in Section 4.3.1 gave us a complete solution of the problem, in particular revealing *the weakest* restrictions for the derivatives of the common curves at the common point, below we see how these things look from the trigonometric spline perspective.

A quadratic bivariate spline of class C^1 over a partition $\Delta = \{\alpha_0, \ldots, \alpha_{k-1}\}$ in trigonometric form has only three terms:

$$g(x, y) = s_0 + r s_1(\varphi) + r^2 s_2(\varphi).$$

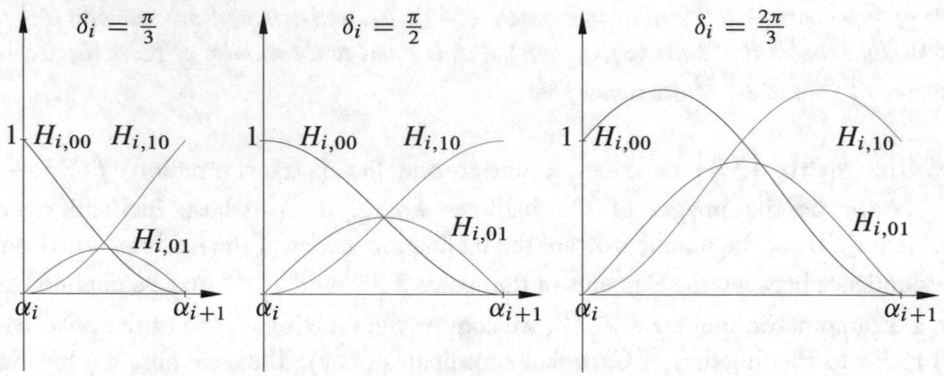

Figure 4.12: "Local" bases of even quadratic trigonometric polynomials

The function s_0 is a constant, equal to $g(0, 0)$. The function s_1 is an odd trigonometric polynomial of degree 1 and it determines the plane tangent to the graph of the function g at the central point. To deal with the last (quadratic) term, for each interval $[\alpha_i, \alpha_{i+1}]$ whose length is δ_i, we introduce three functions:

$$H_{i,00}(\varphi) = \frac{\sin^2(\varphi - \alpha_{i+1})}{\sin^2 \delta_i}, \qquad H_{i,10}(\varphi) = \frac{\sin^2(\varphi - \alpha_i)}{\sin^2 \delta_i},$$
$$H_{i,01}(\varphi) = \frac{\sin(\varphi - \alpha_i)\sin(\alpha_{i+1} - \varphi)}{\sin \delta_i}.$$

Obviously, these functions are even trigonometric polynomials of degree 2 and they are linearly independent (Fig. 4.12). For any interval $[\alpha_i, \alpha_{i+1}]$ such that $0 < \delta_i = \alpha_{i+1} - \alpha_i < \pi$, these functions exist.

Any trigonometric spline function considered below in each interval $[\alpha_i, \alpha_{i+1}]$ is a trigonometric polynomial being a linear combination of the elements of the "local" basis shown above. There is

$$\begin{aligned} H_{i,00}(\alpha_i) &= 1, & H_{i,10}(\alpha_i) &= 0, & H_{i,01}(\alpha_i) &= 0, \\ H_{i,00}(\alpha_{i+1}) &= 0, & H_{i,10}(\alpha_{i+1}) &= 1, & H_{i,01}(\alpha_{i+1}) &= 0. \end{aligned}$$

Due to the above, the trigonometric polynomial

$$p_i(\varphi) = a_i H_{i,00}(\varphi) + a_{i+1} H_{i,10}(\varphi) + b_i H_{i,01}(\varphi),$$

regardless of the coefficient b_i, satisfies the interpolation conditions $p_i(\alpha_i) = a_i$ and $p_i(\alpha_{i+1}) = a_{i+1}$. The value of the trigonometric polynomial

$$p_{i-1}(\varphi) = a_{i-1} H_{i-1,00}(\varphi) + a_i H_{i-1,10}(\varphi) + b_{i-1} H_{i-1,01}(\varphi)$$

at α_i is also a_i. Therefore, for arbitrary numbers $a_0, \ldots, a_{k-1}$ and $b_0, \ldots, b_{k-1}$ we can define a spline function s_2 whose restriction to each interval $[\alpha_i, \alpha_{i+1}]$ is the trigonometric polynomial p_i; this function is continuous and there is $s_2(\alpha_i) = a_i$ for all i.

We need a function s_2 of class C^1. A simple calculation gives us

$$H'_{i,00}(\alpha_i) = \frac{\sin 2(\varphi - \alpha_{i+1})}{\sin^2 \delta_i}, \quad H'_{i,10}(\alpha_i) = 0, \quad H'_{i,01}(\alpha_i) = 1,$$
$$H'_{i,00}(\alpha_{i+1}) = 0, \quad H'_{i,10}(\alpha_{i+1}) = \frac{\sin 2(\varphi - \alpha_i)}{\sin^2 \delta_i}, \quad H'_{i,01}(\alpha_{i+1}) = -1.$$

From the above we obtain the derivatives of the functions p_{i-1} and p_i at α_i:

$$p'_{i-1}(\alpha_i) = \frac{\sin 2\delta_{i-1}}{\sin^2 \delta_{i-1}} a_i - b_{i-1}, \quad p'_i(\alpha_i) = -\frac{\sin 2\delta_i}{\sin^2 \delta_i} a_i + b_i.$$

By rearranging the equation $p'_{i-1}(\alpha_i) = p'_i(\alpha_i)$, we obtain the following:

$$b_{i-1} + b_i = \left(\frac{\sin 2\delta_{i-1}}{\sin^2 \delta_{i-1}} + \frac{\sin 2\delta_i}{\sin^2 \delta_i} \right) a_i.$$

By standard trigonometric identities, we have

$$\frac{\sin 2\delta_{i-1}}{\sin^2 \delta_{i-1}} + \frac{\sin 2\delta_i}{\sin^2 \delta_i} = 2(\cot \delta_{i-1} + \cot \delta_i) = \frac{2\sin(\delta_{i-1} + \delta_i)}{\sin \delta_{i-1} \sin \delta_i}.$$

Hence, the system of equations expressing the continuity of derivative of the function s_2 at all points $\alpha_0, \ldots, \alpha_{k-1}$ has the form

$$\begin{bmatrix} 1 & 1 & & \\ & 1 & \ddots & \\ & & \ddots & 1 \\ 1 & & & 1 \end{bmatrix} \begin{bmatrix} b_0 \\ \vdots \\ \vdots \\ b_{k-1} \end{bmatrix} = 2 \begin{bmatrix} (\cot \delta_0 + \cot \delta_1) a_1 \\ \vdots \\ \vdots \\ (\cot \delta_{k-1} + \cot \delta_0) a_0 \end{bmatrix}. \tag{4.31}$$

The determinant of the matrix on the left-hand side is 2 if k is odd and 0 if k is even. Hence, if k is odd, then for any given values of the function s_2 at $\alpha_0, \ldots, \alpha_{k-1}$, the interpolation problem in the space $\mathcal{T}_\Delta^{(2,1)}$ has a unique solution. If k is even, then the rank of this matrix is $k-1$ and the consistency condition for this system may be expressed by one linear equation being the linear combination of Equations (4.31) with the coefficients $-\frac{1}{2}, \frac{1}{2}, \ldots, -\frac{1}{2}, \frac{1}{2}$:

$$0 = \sum_{i=0}^{k-1} (-1)^i (\cot \delta_{i-1} + \cot \delta_i) a_i.$$

We can rewrite it in the form

$$\sum_{i=0}^{k-1}(-1)^i \frac{\sin(\delta_{i-1}+\delta_i)}{\sin\delta_{i-1}\sin\delta_i} s_2(\alpha_i) = 0. \tag{4.32}$$

Exercise. Compare Equation (4.32) with (4.21) to verify that both equations describe the same consistency condition for any even number of common curves meeting at a point. Note that $\det[\boldsymbol{a}, \boldsymbol{b}, \boldsymbol{n}]$ is the scalar product of the vectors $\boldsymbol{a} \wedge \boldsymbol{b}$ and $\boldsymbol{n}$; if the vectors $\boldsymbol{a}$ and $\boldsymbol{b}$ are orthogonal to a unit vector $\boldsymbol{n}$, then this determinant is equal to the product of their lengths and the sine of the (signed) angle between $\boldsymbol{a}$ and $\boldsymbol{b}$.

We may see the contents of this section as a test for the trigonometric spline approach—the analysis gave us results equivalent to those of the direct method. Though the direct method seems more natural, to my knowledge, in the case of compatibility conditions of order 2 or higher, it leads only to partial results; in particular, it does not reveal *all* degrees of freedom in constructing the common curves of the patches. We tackle this problem in the next section.

4.3.5 TRIGONOMETRIC SPLINES AND G^2 COMPATIBILITY CONDITIONS

Just as in the previous section, we consider bivariate splines over a fixed partition Δ which determine derivatives of k patches surrounding a common point. For a surface of class G^2 we need quartic splines of class C^2. In trigonometric form, such a spline is

$$g(x, y) = s_0 + r s_1(\varphi) + r^2 s_2(\varphi) + r^3 s_3(\varphi) + r^4 s_4(\varphi).$$

The function $s_0 = g(0, 0)$ is a constant; the function s_1, which is an odd trigonometric polynomial of degree 1, determines the tangent plane of the graph of the function g, while the even trigonometric polynomial s_2 determines the second-order derivatives of g at $(0, 0)$. A vector function $\boldsymbol{p}$ made of three bivariate splines is a parametrisation of a surface in $\mathbb{R}^3$; if it is regular, then the terms of degree 0, 1 and 2 determine the common point of the patches, their tangent plane at that point and the osculating paraboloid of the surface.

The functions s_3 and s_4, which appear in the last two terms, are trigonometric splines of class $C^2(\mathbb{R})$; s_3 is an odd cubic trigonometric spline, while s_4 is an even quartic trigonometric spline. The dimensions of the linear vector spaces $\mathcal{T}_\Delta^{(3,2)}$ and $\mathcal{T}_\Delta^{(4,2)}$, isomorphic to the spaces $\mathcal{H}_\Delta^{(3,2)}$ and $\mathcal{H}_\Delta^{(4,2)}$, depend on the partition Δ in the way described by Theorem 4.1 (see Section 4.3.2) and Corollary 4.5 (see Section 4.3.3).

G^2 compatibility conditions and odd cubic trigonometric splines

To study the degrees of freedom and restrictions on the third- and fourth-order derivatives of the common curves of the patches at the common point, we shall make an analysis of interpolation

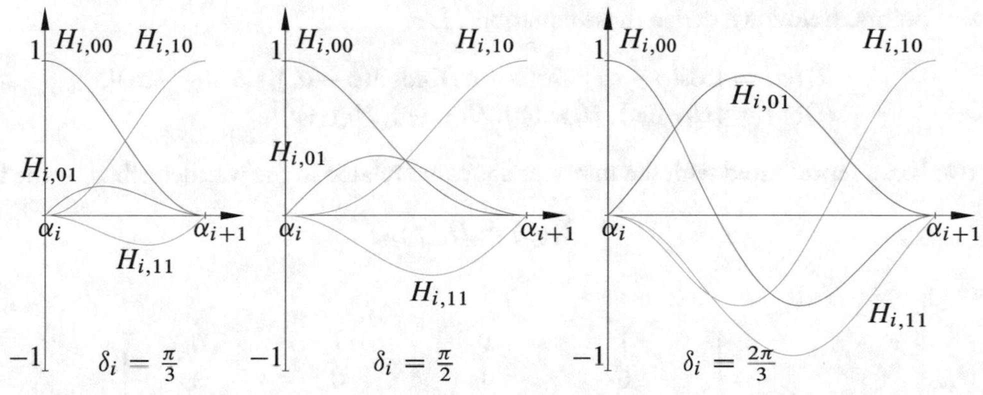

Figure 4.13: "Local" bases of the space of odd cubic trigonometric polynomials

problems in the spaces $\mathcal{T}_\Delta^{(3,2)}$ and $\mathcal{T}_\Delta^{(4,2)}$. We begin with the space $\mathcal{T}_\Delta^{(3,2)}$, whose dimension, by Corollary 4.5, is $\max\{k, 4+h\}$, where h is the number of pairs $\{\alpha_i, \alpha_i + \pi\}$ in Δ. Note that always $h \le \lfloor k/2 \rfloor$; hence, if $k \ge 7$, then $\dim \mathcal{T}_\Delta^{(3,2)} = k$.

To analyse interpolation problems in $\mathcal{T}_\Delta^{(3,2)}$, we construct special bases of the space of odd cubic trigonometric polynomials; for an interval $[\alpha_i, \alpha_{i+1}]$, we need four functions, $H_{i,00}, H_{i,01}, H_{i,10}, H_{i,11}$, such that

$$\begin{bmatrix} H_{i,00}(\alpha_i) & H'_{i,00}(\alpha_i) & H_{i,00}(\alpha_{i+1}) & H'_{i,00}(\alpha_{i+1}) \\ H_{i,01}(\alpha_i) & H'_{i,01}(\alpha_i) & H_{i,01}(\alpha_{i+1}) & H'_{i,01}(\alpha_{i+1}) \\ H_{i,10}(\alpha_i) & H'_{i,10}(\alpha_i) & H_{i,10}(\alpha_{i+1}) & H'_{i,10}(\alpha_{i+1}) \\ H_{i,11}(\alpha_i) & H'_{i,11}(\alpha_i) & H_{i,11}(\alpha_{i+1}) & H'_{i,11}(\alpha_{i+1}) \end{bmatrix}$$

is the identity matrix. This basis and the entire construction are similar to the construction of "ordinary" splines of interpolation described in Sections A.8.1 and A.8.2; compare Figures 4.13 and A.20. With these functions we can solve Hermite interpolation problems with two double knots, α_i and α_{i+1}. Namely, for any numbers $a_i, b_i, a_{i+1}, b_{i+1}$, the function

$$s(\varphi) = a_i H_{i,00}(\varphi) + b_i H_{i,01}(\varphi) + a_{i+1} H_{i,10}(\varphi) + b_{i+1} H_{i,11}(\varphi)$$

is an odd cubic trigonometric polynomial such that $s(\alpha_i) = a_i$, $s'(\alpha_i) = b_i$, $s(\alpha_{i+1}) = a_{i+1}$, $s'(\alpha_{i+1}) = b_{i+1}$. For arbitrary numbers $a_0, b_0, \ldots, a_{k-1}, b_{k-1}$ the function, which in each interval $[\alpha_i, \alpha_{i+1}]$ is a trigonometric polynomial obtained in this way,[9] is an element of the space $\mathcal{T}_\Delta^{(3,1)}$, i.e., an odd cubic trigonometric spline of class C^1. To obtain a function of class C^2, taking the prescribed values at the points $\alpha_0, \ldots, \alpha_{k-1}$, we need equations which express the continuity of the second-order derivatives at these points, with unknown values of the first-order derivatives

[9]All the time we treat the subscripts in a cyclic manner, taking $\alpha_k = \alpha_0 + 2\pi$ and $a_k = a_0$, $b_k = b_0$.

at these points. Below we derive these equations. Let

$$\begin{aligned} T_i(\varphi) &= [\cos(\varphi-\alpha_i), \sin(\varphi-\alpha_i), \cos 3(\varphi-\alpha_i), \sin 3(\varphi-\alpha_i)], \\ H_i(\varphi) &= [H_{i,00}(\varphi), H_{i,01}(\varphi), H_{i,10}(\varphi), H_{i,11}(\varphi)]. \end{aligned}$$

The two bases represented with the matrices above are related in the way described by the formula

$$T_i(\varphi) = H_i(\varphi)K_i,$$

where

$$K_i = \begin{bmatrix} 1 & 0 & 1 & 0 \\ 0 & 1 & 0 & 3 \\ \cos\delta_i & \sin\delta_i & \cos 3\delta_i & \sin 3\delta_i \\ -\sin\delta_i & \cos\delta_i & -3\sin 3\delta_i & 3\cos 3\delta_i \end{bmatrix}.$$

The coefficients in the first column of the matrix K_i are taken from the equality

$$\begin{aligned} \cos(\varphi-\alpha_i) = {} & \cos(\alpha_i-\alpha_i)H_{i,00}(\varphi) + \cos'(\alpha_i-\alpha_i)H_{i,01}(\varphi) + \\ & \cos(\alpha_{i+1}-\alpha_i)H_{i,10}(\varphi) + \cos'(\alpha_{i+1}-\alpha_i)H_{i,10}(\varphi), \end{aligned}$$

and the other columns are obtained in a similar way. To find the basis $H_i(\varphi) = T_i(\varphi)K_i^{-1}$ we need the matrix K^{-1}. A rather long and tedious symbolic calculation gives us the formula

$$K_i^{-1} = \begin{bmatrix} \frac{3}{4}(1-\frac{\cos^2\delta_i}{\sin^2\delta_i}) & \frac{\cos\delta_i}{2\sin\delta_i} & \frac{3\cos\delta_i}{4\sin^2\delta_i} & -\frac{1}{4\sin\delta_i} \\ \frac{3}{2}\frac{\cos^3\delta_i}{\sin^3\delta_i} & \frac{3}{4\sin^2\delta_i}-\frac{1}{2} & \frac{9\sin^2\delta_i-6}{4\sin^3\delta_i} & \frac{3\cos\delta_i}{4\sin^2\delta_i} \\ \frac{1}{4}(1+\frac{3\cos^2\delta_i}{\sin^2\delta_i}) & \frac{\cos\delta_i}{2\sin\delta_i} & -\frac{3\cos\delta_i}{4\sin^2\delta_i} & \frac{1}{4\sin\delta_i} \\ -\frac{\cos^3\delta_i}{2\sin^3\delta_i} & \frac{1}{2}-\frac{1}{4\sin^2\delta_i} & \frac{2-3\sin^2\delta_i}{4\sin^3\delta_i} & -\frac{\cos\delta_i}{4\sin^2\delta_i} \end{bmatrix}.$$

The matrix K^{-1} exists if and only if $\sin\delta_i \neq 0$, i.e., if the length δ_i of the interval $[\alpha_i, \alpha_{i+1}]$ is not an integer multiple of π. Let us recall that in the context of geometric continuity of surfaces there is $0 < \delta_i < \pi$ for $i = 0, \dots, k-1$, and $\delta_0 + \dots + \delta_{k-1} = 2\pi$. Using the formula for K_i^{-1}, we can find (after more tedious calculations) the second-order derivatives of our "local" basis functions constructed for the intervals $[\alpha_{i-1}, \alpha_i]$ and $[\alpha_i, \alpha_{i+1}]$:

$$\begin{aligned} H''_{i,00}(u_i) &= -3-6\frac{\cos^2\delta_i}{\sin^2\delta_i}, & H''_{i-1,00}(u_i) &= 6\frac{\cos\delta_{i-1}}{\sin^2\delta_{i-1}}, \\ H''_{i,01}(u_i) &= -4\frac{\cos\delta_i}{\sin^2\delta_i}, & H''_{i-1,01}(u_i) &= \frac{2}{\sin\delta_{i-1}}, \\ H''_{i,10}(u_i) &= 6\frac{\cos\delta_i}{\sin^2\delta_i}, & H''_{i-1,10}(u_i) &= -3-6\frac{\cos^2\delta_{i-1}}{\sin^2\delta_{i-1}}, \\ H''_{i,11}(u_i) &= -\frac{2}{\sin\delta_i}, & H''_{i-1,11}(u_i) &= 4\frac{\cos\delta_{i-1}}{\sin^2\delta_{i-1}}. \end{aligned}$$

Let p_{i-1} and p_i be odd cubic trigonometric polynomials, which describe a trigonometric spline function s in the intervals $[\alpha_{i-1}, \alpha_i]$ and $[\alpha_i, \alpha_{i+1}]$. We can write these polynomials in the form

$$\begin{aligned} p_{i-1}(\varphi) &= a_{i-1} H_{i-1,00}(\varphi) + b_{i-1} H_{i-1,01}(\varphi) + a_i H_{i-1,10}(\varphi) + b_i H_{i-1,11}(\varphi), \\ p_i(\varphi) &= a_i H_{i,00}(\varphi) + b_i H_{i,01}(\varphi) + a_{i+1} H_{i,10}(\varphi) + b_{i+1} H_{i,11}(\varphi). \end{aligned}$$

By substituting the expressions for the second-order derivatives of our basis functions to the equation $p''_{i-1}(\alpha_i) = p''_i(\alpha_i)$, after some rearrangement, we obtain the following equation:

$$\begin{aligned} &\sin\delta_i\, b_{i-1} + 2\sin(\delta_{i-1}+\delta_i)\, b_i + \sin\delta_{i-1}\, b_{i+1} = \\ &3\left(\frac{\cos\delta_{i-1}\sin\delta_i}{\sin\delta_{i-1}} a_{i-1} + \frac{\cos 2\delta_{i-1} - \cos 2\delta_i}{2\sin\delta_{i-1}\sin\delta_i} a_i + \frac{\cos\delta_i \sin\delta_{i-1}}{\sin\delta_i} a_{i+1}\right). \end{aligned} \tag{4.33}$$

The approach used here to obtain the equation of continuity of the second-order derivative at the knot of interpolation is fully analogous to the one used in Section A.8.2, and the resulting equation (4.33) is strikingly similar to (A.42).

Equations (4.33) form the system

$$A_3 \boldsymbol{b}_3 = B_3 \boldsymbol{a}_3, \tag{4.34}$$

where

$$\begin{aligned} \boldsymbol{a}_3 &= [a_0, \dots, a_{k-1}]^T = [s_3(\alpha_0), \dots, s_3(\alpha_{k-1})]^T, \\ \boldsymbol{b}_3 &= [b_0, \dots, b_{k-1}]^T = [s'_3(\alpha_0), \dots, s'_3(\alpha_{k-1})]^T, \end{aligned}$$

and the matrices A_3 and B_3 have the cyclic tridiagonal structure:

$$A_3 = \begin{bmatrix} 2\sin(\delta_{k-1}+\delta_0) & \sin\delta_{k-1} & & \sin\delta_0 \\ \sin\delta_1 & 2\sin(\delta_0+\delta_1) & \ddots & \\ & \ddots & \ddots & \sin\delta_{k-3} \\ \sin\delta_{k-2} & & \sin\delta_{k-1} & 2\sin(\delta_{k-2}+\delta_{k-1}) \end{bmatrix},$$

$$B_3 = 3\begin{bmatrix} \frac{\cos 2\delta_{k-1} - \cos 2\delta_0}{2\sin\delta_{k-1}\sin\delta_0} & \frac{\cos\delta_0\sin\delta_{k-1}}{\sin\delta_0} & & \frac{-\cos\delta_{k-1}\sin\delta_0}{\sin\delta_{k-1}} \\ \frac{-\cos\delta_0\sin\delta_1}{\sin\delta_0} & \frac{\cos 2\delta_0 - \cos 2\delta_1}{2\sin\delta_0\sin\delta_1} & \ddots & \\ & \ddots & \ddots & \frac{\cos\delta_{k-2}\sin\delta_{k-3}}{\sin\delta_{k-2}} \\ \frac{\cos\delta_{k-1}\sin\delta_{k-2}}{\sin\delta_{k-1}} & & \frac{-\cos\delta_{k-2}\sin\delta_{k-1}}{\sin\delta_{k-2}} & \frac{\cos 2\delta_{k-2} - \cos 2\delta_{k-1}}{2\sin\delta_{k-2}\sin\delta_{k-1}} \end{bmatrix}.$$

The consistency of the interpolation problem in the space $\mathcal{T}_\Delta^{(3,2)}$, in particular conditions which must be satisfied by the function values given, depend on the partition Δ in a rather complicated way. In many practical constructions one can take a partition such that $\delta_0 = \cdots = \delta_{k-1} =$

$2\pi/k$. As we shall see, in this case there are no restrictions on the function values, which means that there are no restrictions on the third-order derivatives of the common boundary curves of the patches surrounding the common point. However, for $k \in \{3, 4, 6\}$, the interpolation problem has infinitely many solutions. In special applications one might choose a special partition (see e.g. Figure 4.5), which is why the analysis below covers more general cases. The cases of $k = 3$, $k = 4$, $k = 5$, $k = 6$ and $k > 6$ are discussed separately.

Proposition 4.8 *Let $k = 3$. For any given function values at $\alpha_0, \alpha_1, \alpha_2$, the interpolation problem in the space $\mathcal{T}_\Delta^{(3,2)}$ has infinitely many solutions which are odd cubic trigonometric polynomials.*

Proof. By Corollary 4.5, for $k = 3$ and $h = 0$, there is $\dim \mathcal{T}_\Delta^{(3,2)} = 4$. The solution of the interpolation problem may be written explicitly:

$$s_3(\varphi) = \sum_{i=0}^{2} a_i \frac{\sin(\varphi - \alpha_{i-1}) \sin^2(\varphi - \alpha_i)}{\sin \delta_{i-1} \sin^2 \delta_i} + c \prod_{i=0}^{2} \sin(\varphi - \alpha_i),$$

with an arbitrary constant c. □

Proposition 4.9 *Let $k = 4$.*

1. *If $\alpha_2 \neq \alpha_0 + \pi$ and $\alpha_3 \neq \alpha_1 + \pi$ (i.e., $h = 0$), then the interpolation problem in $\mathcal{T}_\Delta^{(3,2)}$ has a unique solution which is an odd cubic trigonometric polynomial.*

2. *Suppose that either $\alpha_2 = \alpha_0 + \pi$, or $\alpha_3 = \alpha_1 + \pi$, i.e., $h = 1$. The interpolation problem in $\mathcal{T}_\Delta^{(3,2)}$ has solutions if the function values given satisfy the condition $a_2 = -a_0$ or $a_3 = -a_1$ respectively.*

3. *If $\alpha_2 = \alpha_0 + \pi$ and $\alpha_3 = \alpha_1 + \pi$ ($h = 2$), then the interpolation problem has solutions.*

If $h > 0$ and the set of solutions of the interpolation problem in the space $\mathcal{T}_\Delta^{(3,2)}$ is non-empty, then it is a two-dimensional coset of this space.

Proof. 1. By Corollary 4.5, $\dim \mathcal{T}_\Delta^{(3,2)} = 4$. The trigonometric polynomial

$$s_3(\varphi) = \sum_{i=0}^{3} a_i \frac{\sin(\varphi - \alpha_{i-1}) \sin^2(\varphi - \alpha_i)}{\sin \delta_{i-1} \sin^2 \delta_i}$$

is a solution of the interpolation problem—unique because the number of interpolation conditions is equal to the dimension of the space.

2. It suffices to study the case $\alpha_2 = \alpha_0 + \pi$, $\alpha_3 \neq \alpha_1 + \pi$. By Corollary 4.5, there is $\dim \mathcal{T}_\Delta^{(3,2)} = 5$. A basis of this space is made of four odd trigonometric polynomials and the function[10]

$$\sin_+^3(\varphi - \alpha_0) \stackrel{\text{def}}{=} \begin{cases} \sin^3(\varphi - \alpha_0) & \text{if } \sin(\varphi - \alpha_0) \geq 0, \\ 0 & \text{else.} \end{cases}$$

All elements of this basis, and consequently all their linear combinations, take values with the same absolute value and opposite signs at α_0 and α_2. Therefore, the equality $a_2 = -a_0$ is a consistency condition for the interpolation conditions. By Proposition 4.8, for arbitrary numbers a_0, a_1, a_3 there exist odd cubic trigonometric polynomials taking these values at α_0, α_1 and α_3. The trigonometric polynomials of interpolation form a one-dimensional coset. The presence of the function $\sin_+^3(\varphi - \alpha_0)$ extends the set of solutions of the interpolation problem in $\mathcal{T}_\Delta^{(3,2)}$ to a two-dimensional coset.

3. If $k = 4$, $h = 2$, then, by Corollary 4.5, $\dim \mathcal{T}_\Delta^{(3,2)} = 6$. Solutions of the interpolation problem may be described by the formula

$$s_3(\varphi) = \sum_{i=0}^{3} a_i \frac{\sin_+^3(\varphi - \alpha_i)}{\sin^3 \delta_i} + \\ c_0 \sin^2(\varphi - \alpha_0)\sin(\varphi - \alpha_1) + c_1 \sin(\varphi - \alpha_0)\sin^2(\varphi - \alpha_1),$$

with arbitrary constants c_0 and c_1. □

Proposition 4.10 *Let $k = 5$.*

1. *If $h = 0$, i.e., $\alpha_i \neq \alpha_j \pm \pi$ for all $i, j \in \{0, \ldots, k-1\}$, then the interpolation problem in $\mathcal{T}_\Delta^{(3,2)}$ has a unique solution.*

2. *If $h = 1$, i.e., $\alpha_i = \alpha_j + \pi$ for one pair of numbers $\{i, j\}$, then the interpolation problem in $\mathcal{T}_\Delta^{(3,2)}$ has solutions if $a_i = -a_j$, and these solutions form a one-dimensional coset.*

3. *If $h = 2$, then any interpolation problem in $\mathcal{T}_\Delta^{(3,2)}$ has solutions, which form a one-dimensional coset.*

Proof. 1. By Corollary 4.5, $\dim \mathcal{T}_\Delta^{(3,2)} = 5$. As the space dimension is equal to the number of interpolation conditions, it suffices to prove that the matrix A_3 in (4.34) is non-singular. A long and tedious symbolic calculation, omitted here, proves that $\det A_3 \neq 0$.

[10]Obviously, the function $\sin_+^3(\varphi - \alpha_i) \in \mathcal{T}_\Delta^{(3,2)}$ corresponds to the function $y_{i,+}^3 \in \mathcal{H}_\Delta^{(3,2)}$; in general, the functions $x_i^m y_{i,+}^{l-m}$, considered in the proof of Theorem 4.1, correspond to the trigonometric spline functions $\cos^m(\varphi - \alpha_i)\sin_+^{l-m}(\varphi - \alpha_i)$.

2. If $h = 1$, then, by Corollary 4.5, $\dim \mathcal{T}_{\Delta}^{(3,2)} = 5$. Without loss of generality, we can assume that $\alpha_2 = \alpha_0 + \pi$ and then we can find the same basis of the space $\mathcal{T}_{\Delta}^{(3,2)}$ as in the proof of point 2 of Proposition 4.9; in other words, if $\Delta_0 = \Delta \setminus \{\alpha_3\}$ or $\Delta_0 = \Delta \setminus \{\alpha_4\}$, then the spaces $\mathcal{T}_{\Delta}^{(3,2)}$ and $\mathcal{T}_{\Delta_0}^{(3,2)}$ are identical. Therefore, there must be $a_2 = -a_0$. If this condition is satisfied, then by Proposition 4.9 we can solve the interpolation problem in the subspace $\mathcal{T}_{\Delta_1}^{(3,2)}$ corresponding to the partition $\Delta_1 = \Delta \setminus \{\alpha_2\}$. The solution in the subspace is unique and, as its dimension is equal to $\dim \mathcal{T}_{\Delta}^{(3,2)} - 1$, the set of solutions in the entire space $\mathcal{T}_{\Delta}^{(3,2)}$ is a one-dimensional coset.

3. If $h = 2$, then, by Corollary 4.5, $\dim \mathcal{T}_{\Delta}^{(3,2)} = 6$. We can assume that $\alpha_3 = \alpha_1 + \pi$ and $\alpha_4 = \alpha_2 + \pi$. To solve the interpolation problem in $\mathcal{T}_{\Delta}^{(3,2)}$, we can take any odd cubic trigonometric polynomial $p(\varphi)$ such that $p(\alpha_0) = a_0$, $p(\alpha_1) = a_1$ and $p(\alpha_4) = a_4$ (it exists by Proposition 4.8), and then take

$$s_3(\varphi) = p(\varphi) + (a_2 - p(\alpha_2))\frac{\sin_+^3(\varphi - \alpha_1)}{\sin^3 \delta_2} + (a_3 - p(\alpha_3))\frac{\sin_+^3(\varphi - \alpha_2)}{\sin^3 \delta_2}.$$

The dimension of the coset made of these solutions is equal to the dimension of the coset of solutions of the interpolation problem with $k = 3$. □

The analysis of the interpolation problems in the spaces $\mathcal{T}_{\Delta}^{(3,2)}$ becomes harder with the growth of the number of elements of the partition Δ (i.e., of the junction halflines) and, to my knowledge, nice and elegant conditions for the existence of solutions of interpolation problems (expressed in terms of $\delta_0, \ldots, \delta_{k-1}$) are not known yet. Nevertheless, the results of the analysis below cover many practical cases. Given a partition for which this analysis does not give conclusive answers, we can use numerical computations (or symbolic calculations, if we can) to find out whether the matrix A_3 is singular and if so, then find coefficients of the linear equations to be satisfied by the numbers $a_0, \ldots, a_{k-1}$. However, "wild" partitions dividing the full angle in a quite uneven manner are unlikely to appear in practical applications.

The determinant of the matrix A_3 is a function of the numbers $\delta_0, \ldots, \delta_{k-2}$ (which determine $\delta_{k-1} = 2\pi - \sum_{i=0}^{k-2} \delta_i$). The domain of this function is the simplex $\{ (\delta_0, \ldots, \delta_{k-2})\colon \delta_0, \ldots, \delta_{k-2} > 0,\ \sum_{i=0}^{k-2} \delta_i < 2\pi \}$. Due to Lemmas 4.6 and 4.7, we can arbitrarily fix the numbers δ_{k-3} and δ_{k-4} so that their sum be less than π (we discuss it in the proof of Proposition 4.11). In this way we obtain a further reduction of the number of variables in our problem. A good illustration of its complexity is in Figure 4.14, which shows the set of zeros of $\det A_3$ for partitions with $k = 6$ elements and such that $\delta_4 = \delta_5 = \frac{\pi}{3}$. The set of zeros of the determinant is a surface inscribed in the simplex $\{ (\delta_0, \delta_1, \delta_2)\colon \delta_0, \delta_1, \delta_2 > 0,\ \delta_0 + \delta_1 + \delta_2 < \frac{4\pi}{3} \}$.

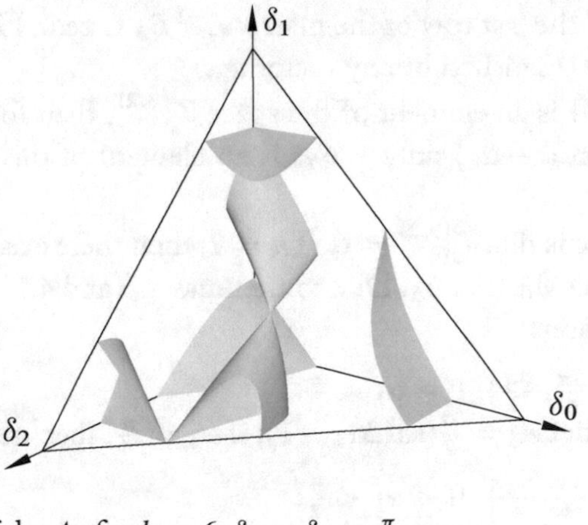

Figure 4.14: Set of zeros of $\det A_3$ for $k = 6, \delta_4 = \delta_5 = \frac{\pi}{3}$

Proposition 4.11 *Let $k = 6$.*

1. *If $h = 0$ or $h = 1$, then there may be $\det A_3 \neq 0$ or $\det A_3 = 0$; in the latter case, solutions of the system (4.34) exist (and form a one-dimensional coset) if the numbers $a_0, \ldots, a_{k-1}$ satisfy a single homogeneous linear equation.*

2. *If $h = 2$, then the matrix A_3 is non-singular; there is no restriction on $a_0, \ldots, a_{k-1}$ and the solution $b_0, \ldots, b_{k-1}$ of the system (4.34) is unique.*

3. *If $h = 3$, then $\det A_3 = 0$; the system (4.34) is consistent for any numbers $a_0, \ldots, a_{k-1}$ and the set of solutions is a one-dimensional coset of the space $\mathbb{R}^6$.*

Proof. 3. If $h = 3$, then, by Corollary 4.5, $\dim \mathcal{T}_\Delta^{(3,2)} = 7$. The halflines $\hbar_0, \ldots, \hbar_5$ are parts of three lines, which divide the plane to six convex cones. Due to Lemmas 4.6 and 4.7, we can reduce the analysis to the case $\alpha_0 = 0, \alpha_1 = \frac{\pi}{3}$ and $\alpha_2 = \frac{2\pi}{3}$. This is because any restriction on the values of the trigonometric spline s_3 (an element of the space $\mathcal{T}_\Delta^{(3,2)}$) at the points α_i is exactly the same restriction as for the values of the spline $\tilde{s}_3$, being an element of the space $\mathcal{T}_{\tilde{\Delta}}^{(3,2)}$ defined over the partition $\tilde{\Delta}$ related to Δ in the way described in Lemma 4.6.

Due to the above, the analysis of the case $k = 6, h = 3$ may be done for the partition such that $\alpha_i = \frac{i\pi}{3}$ for $i = 0, \ldots, 5$. There is $\delta_0 = \cdots = \delta_5 = \frac{\pi}{3}$. We can use the Gaussian elimination (without pivoting) to obtain matrices L and U such that $A_3 = LU$ and the matrix L is lower triangular, with all diagonal coefficients equal to 1. The matrix U is upper triangular and all coefficients in its last row are zeros; hence, the matrix A_3 is singular and its rank is 5.

System (4.34) is equivalent to $U\boldsymbol{b}_3 = L^{-1}B_3\boldsymbol{a}_3$; as the last row of the matrix U is zero, the last coordinate of the right-hand side vector $L^{-1}B_3\boldsymbol{a}_3$, must be zero; this is the consistency condition. With a calculation, omitted here, we can check that the last row of the matrix L^{-1} is

$[-1, 1, -1, 1, -1, 1]$ and the last row of the matrix $L^{-1}B_3$ is zero. Hence, our consistency condition is the equation $0 = 0$ satisfied by any vector $\boldsymbol{a}_3$.

If a function $s_3(\varphi)$ is an element of the space $\mathcal{T}_\Delta^{(3,2)}$, then for any number c the function $s_3(\varphi) + c\sin(\varphi - \alpha_0)\sin(\varphi - \alpha_1)\sin(\varphi - \alpha_2)$ is an element of this space and it takes the same values at the points α_i.

2. For $h < 3$, there is $\dim \mathcal{T}_\Delta^{(3,2)} = 6$. If $h = 2$, then there exist four lines in $\mathbb{R}^2$ which contain the junction halflines $h_0, \dots, h_5$. Due to Lemmas 4.6 and 4.7, the problem may be reduced to the following three cases:

- $\Delta = \{0, \alpha_1, \frac{\pi}{2}, \pi, \frac{5\pi}{4}, \frac{3\pi}{2}\}$, $0 < \alpha_1 < \frac{\pi}{2}$:
  we calculate $\det A_3 = \frac{9}{2}(\sin 2\alpha_1 - 1)$; if $\alpha_1 \neq \frac{\pi}{4}$, then $\det A_3 \neq 0$.
- $\Delta = \{0, \alpha_1, \frac{\pi}{2}, \pi, \frac{3\pi}{2}, \frac{7\pi}{4}\}$, $0 < \alpha_1 < \frac{\pi}{2}$:
  we obtain $\det A_3 = -\frac{9}{2}(1 + \sin 2\alpha_1) \neq 0$.
- $\Delta = \{0, \alpha_1, \frac{\pi}{4}, \frac{\pi}{2}, \pi, \frac{3\pi}{2}\}$, $0 < \alpha_1 < \frac{\pi}{4}$:
  we find $\det A_3 = -\frac{9}{2}\cos^2\alpha_1 \neq 0$.

1. Suppose that $h < 2$ and the matrix A_3 is singular. Then there exists a non-zero function $s(\varphi)$ such that $s(\alpha_i) = 0$ for $i = 0, \dots, 5$. If the partition Δ contains at most one pair $\{\alpha_j, \alpha_j + \pi\}$, then by removing α_j we obtain a partition Δ_0 such that its five junction halflines are located on five different lines. By Proposition 4.10, any interpolation problem in the space $\mathcal{T}_{\Delta_0}^{(3,2)}$ has a unique solution, say, $z(\varphi)$. The space $\mathcal{T}_\Delta^{(3,2)}$ is the direct algebraic sum of the space $\mathcal{T}_{\Delta_0}^{(3,2)}$ and the one-dimensional space spanned by $\{s\}$. Hence, any solution s_3 of the interpolation problem in the space $\mathcal{T}_\Delta^{(3,2)}$ may be written as

$$s_3(\varphi) = z(\varphi) + cs(\varphi),$$

with an arbitrary number c; as $s(\alpha_i) = 0$, there must be $a_i = z(\alpha_i)$. Due to the uniqueness of the function z, the number a_i is determined by the numbers a_j, where $j \neq i$. □

Proposition 4.12 *Let $k \geq 7$. If $\delta_{i-1} + \delta_i < \frac{2\pi}{3}$ for all i, then the interpolation problem in the space $\mathcal{T}_\Delta^{(3,2)}$ for any numbers $a_0, \dots, a_{k-1}$ has a unique solution.*

Proof. If $\delta_{i-1} + \delta_i = \alpha_{i+1} - \alpha_{i-1} < \pi$, then $\sin\delta_{i-1} > 0$, $\sin\delta_i > 0$ and $\sin(\delta_{i-1} + \delta_i) > 0$. Let

$$\begin{aligned} f_i(\delta_{i-1}, \delta_i) &\stackrel{\text{def}}{=} |2\sin(\delta_{i-1} + \delta_i)| - |\sin\delta_{i-1}| - |\sin\delta_i| = \\ &2\sin(\delta_{i-1} + \delta_i) - \sin\delta_{i-1} - \sin\delta_i = 2\sin x(2\cos x - \cos y), \end{aligned}$$

where $x = \frac{1}{2}(\delta_{i-1} + \delta_i)$, $y = \frac{1}{2}(\delta_{i-1} - \delta_i)$. For $0 < 2x = \delta_{i-1} + \delta_0 < \frac{2\pi}{3}$, there is $2\cos x > 2\cos\frac{\pi}{3} = 1 \geq \cos y$ and $\sin x > 0$, hence $f_i(\delta_{i-1}, \delta_i) > 0$. If this inequality is satisfied for all i, then the matrix A_3 is diagonally dominant; by Gershgorin's theorem, known from the linear algebra, such a matrix is non-singular. □

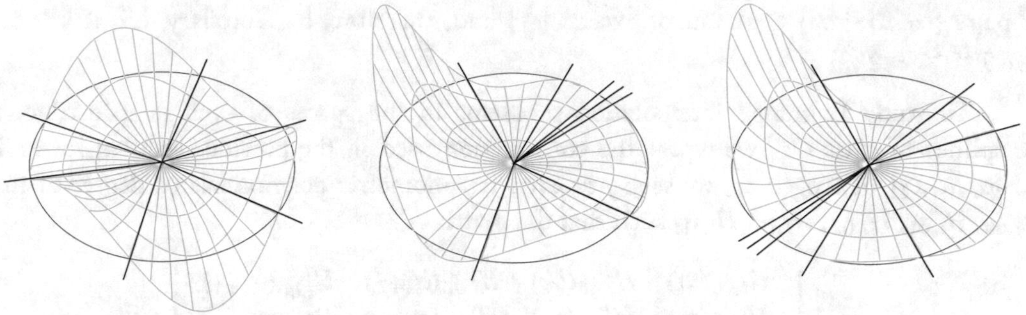

Figure 4.15: Piecewise cubic monkey saddles of class C^2 over partitions with $k = 6, 7, 8$

Due to Proposition 4.12, if $k \geq 7$ and the full angle is partitioned evenly, then there are no restrictions on the numbers $a_0, \ldots, a_{k-1}$. Then, the third-order derivatives of the common curves of patches surrounding a common point may be arbitrary; in this case, they determine the mixed third-order derivatives at the common point. To conclude this long and perhaps exhausting section, in Figure 4.15 we show examples of partitions with the corresponding matrices A_3 singular. These partitions were found in numerical experiments. The pictures show graphs of bivariate cubic splines of class $C^2(\mathbb{R}^2)$ (elements of the spaces $\mathcal{H}_\Delta^{(3,2)}$) determined by non-zero solutions of System (4.34) with $a_0 = \cdots = a_{k-1} = 0$. The existence of such a solution spoils the uniqueness of the interpolation problem and, if $k \geq 7$, it also imposes a restriction (having the form of a homogeneous linear equation) on the numbers $a_0, \ldots, a_{k-1}$, because in that case $\dim \mathcal{T}_\Delta^{(3,2)} = k$.

G^2 compatibility conditions and even quartic trigonometric splines

The G^2 compatibility conditions do not impose any restriction on the fourth-order derivatives of the common curves of patches; this was shown (using the direct method) in Ye and Nowacki [1996]. To reveal *all* degrees of freedom in the construction of the fourth-order derivatives of the patches surrounding the common point, we use the same method as for the third-order derivatives, making use of the connection between these derivatives and even quartic trigonometric splines over a partition Δ.

Let $\Delta = \{\alpha_0, \ldots, \alpha_{k-1}\}$ be a (fixed) partition of the full angle and let s_4 be a function of class C^2 which in each interval $[\alpha_i, \alpha_{i+1}]$ is a linear combination of the functions

$$1,\ \cos 2\varphi,\ \sin 2\varphi,\ \cos 4\varphi,\ \sin 4\varphi.$$

The values and derivatives of the function s_4 are related to the fourth-order (partial and mixed) derivatives of the patches surrounding a common corner in the way described in Section 4.3.3. The total number of degrees of freedom in choosing the fourth-order derivatives of the patches at their common corner is the dimension of the linear vector space $\mathcal{T}_\Delta^{(4,2)}$. Note that the number h

of pairs $\{\alpha_i, \alpha_i + \pi\} \subset \Delta$ cannot exceed $\lfloor \frac{k}{2} \rfloor$ and, therefore, by Corollary 4.5, if $k \geq 5$, then $\dim \mathcal{T}_\Delta^{(4,2)} = 2k$.

To study Lagrange interpolation problems in the spaces of even quartic trigonometric splines of class C^2, we repeat the construction used in the previous section; for an interval $[\alpha_i, \alpha_{i+1}]$ of length δ_i, we seek *five* even trigonometric polynomials of degree at most 4, $H_{i,00}, H_{i,01}, H_{i,02}, H_{i,10}, H_{i,11}$, such that the matrix

$$\begin{bmatrix} H_{i,00}(\alpha_i) & H'_{i,00}(\alpha_i) & H_{i,00}(\alpha_{i+1}) & H'_{i,00}(\alpha_{i+1}) \\ H_{i,01}(\alpha_i) & H'_{i,01}(\alpha_i) & H_{i,01}(\alpha_{i+1}) & H'_{i,01}(\alpha_{i+1}) \\ H_{i,10}(\alpha_i) & H'_{i,10}(\alpha_i) & H_{i,10}(\alpha_{i+1}) & H'_{i,10}(\alpha_{i+1}) \\ H_{i,11}(\alpha_i) & H'_{i,11}(\alpha_i) & H_{i,11}(\alpha_{i+1}) & H'_{i,11}(\alpha_{i+1}) \end{bmatrix}$$

is the identity matrix and, in addition, $H''_{i,01}(\alpha_{i+1}) = H''_{i,11}(\alpha_i) = 0$, $H_{i,02}(\alpha_i) = H'_{i,02}(\alpha_i) = H_{i,02}(\alpha_{i+1}) = H'_{i,02}(\alpha_{i+1}) = 0$, and $H''_{i,02}(\alpha_i) = 1$. The matrices

$$T_i(\varphi) = [1, \cos 2(\varphi - \alpha_i), \sin 2(\varphi - \alpha_i), \cos 4(\varphi - \alpha_i), \sin 4(\varphi - \alpha_i)]$$

and

$$H_i(\varphi) = [H_{i,00}(\varphi), H_{i,01}(\varphi), H_{i,02}(\varphi), H_{i,10}(\varphi), H_{i,11}(\varphi)]$$

represent two bases; the former is our starting point and the latter is what we want to find. The relation between the two bases is described by the formula

$$T_i(\varphi) = K_i H_i(\varphi)$$

where the 5×5 matrix K_i is

$$K_i = \begin{bmatrix} 1 & 1 & 0 & 1 & 0 \\ 0 & 0 & 2 & 0 & 4 \\ 0 & 4(1 + \cos 2\delta_i) & 4 \sin 2\delta_i & 8(3 + \cos 2\delta_i) \cos 2\delta_i & 8(3 + \cos 2\delta_i) \sin 2\delta_i \\ 1 & \cos 2\delta_i & \sin 2\delta_i & \cos 4\delta_i & \sin 4\delta_i \\ 0 & -2 \sin 2\delta_i & 2 \cos 2\delta_i & -4 \sin 4\delta_i & 4 \cos 4\delta_i \end{bmatrix}.$$

A symbolic calculation gives us the formula

$$K_i^{-1} = \begin{bmatrix} \frac{1}{2} & -\frac{3}{4} \frac{\sin 2\delta_i}{(\cos 2\delta_i - 1)^2} & \frac{\cos 2\delta_i + 2}{8(1 - \cos 2\delta_i)} & \frac{1}{2} & \frac{3}{4} \frac{\sin 2\delta_i}{(\cos 2\delta_i - 1)^2} \\ \frac{\cos 2\delta_i}{\cos 2\delta_i - 1} & \frac{(\cos 2\delta_i + 1) \sin 2\delta_i}{2(\cos 2\delta_i - 1)^2} & \frac{\cos 2\delta_i + 1}{4(\cos 2\delta_i - 1)} & \frac{\cos 2\delta_i}{1 - \cos 2\delta_i} & -\frac{\sin 2\delta_i}{(\cos 2\delta_i - 1)^2} \\ \frac{\sin 2\delta_i \cos 2\delta_i}{(\cos 2\delta_i - 1)^2} & \frac{\cos 2\delta_i + 2}{2(1 - \cos 2\delta_i)} & \frac{\sin 2\delta_i}{4(\cos 2\delta_i - 1)} & -\frac{\cos 2\delta_i \sin 2\delta_i}{(\cos 2\delta_i - 1)^2} & \frac{1}{2(\cos 2\delta_i - 1)} \\ \frac{1 + \cos 2\delta_i}{2(1 - \cos 2\delta_i)} & \frac{(1 - 2\cos 2\delta_i) \sin 2\delta_i}{4(\cos 2\delta_i - 1)^2} & \frac{\cos 2\delta_i}{8(1 - \cos 2\delta_i)} & \frac{\cos 2\delta_i + 1}{2(\cos 2\delta_i - 1)} & \frac{\sin 2\delta_i}{4(\cos 2\delta_i - 1)^2} \\ -\frac{\sin 2\delta_i \cos 2\delta_i}{2(\cos 2\delta_i - 1)^2} & \frac{2\cos 2\delta_i + 1}{4(\cos 2\delta_i - 1)} & \frac{\sin 2\delta_i}{8(1 - \cos 2\delta_i)} & \frac{\cos 2\delta_i \sin 2\delta_i}{2(\cos 2\delta_i - 1)^2} & \frac{1}{4(1 - \cos 2\delta_i)} \end{bmatrix},$$

which allows us to obtain $H_i(\varphi) = K_i^{-1}T_i(\varphi)$. As we can see, the matrix K^{-1} exists if $\cos 2\delta_i \neq 1$, i.e., $\delta_i \neq \pi$. Further calculations give us

$$\begin{aligned}
H''_{i,00}(\alpha_i) &= -H''_{i,10}(\alpha_i) = -H''_{i,00}(\alpha_{i+1}) = H''_{i,10}(\alpha_{i+1}) = \frac{4(\cos 2\delta_i + 2)}{\cos 2\delta_i - 1}, \\
H''_{i,01}(\alpha_i) &= -H''_{i,11}(\alpha_{i+1}) = \frac{6\sin 2\delta_i}{\cos 2\delta_i - 1}, \\
H''_{i,02}(\alpha_i) &= H''_{i,02}(\alpha_{i+1}) = 1.
\end{aligned}$$

Examples of functions found in this way are shown in Figure 4.16.

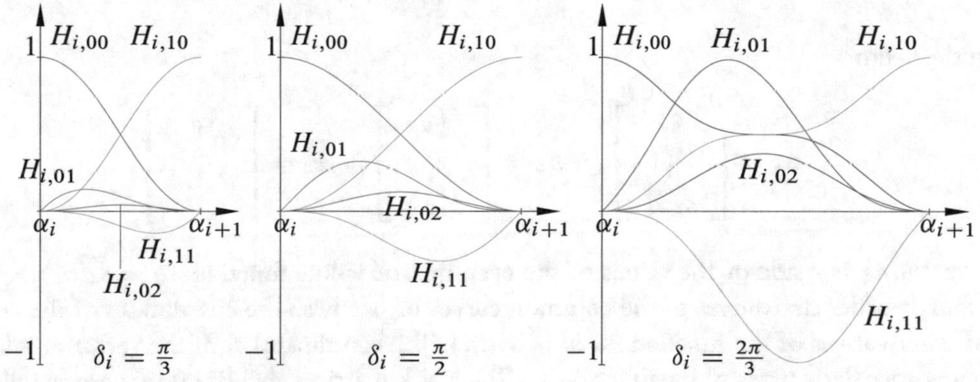

Figure 4.16: "Local" bases of the space of even quartic trigonometric polynomials

Let p_{i-1} and p_i be even quartic trigonometric polynomials which describe a trigonometric spline function in the intervals $[\alpha_{i-1}, \alpha_i]$ and $[\alpha_i, \alpha_{i+1}]$. We can write them in the form

$$\begin{aligned}
p_{i-1}(\varphi) &= a_{i-1}H_{i-1,00}(\varphi) + b_{i-1}H_{i-1,01}(\varphi) + c_{i-1}H_{i-1,02}(\varphi) + \\
&\quad a_i H_{i-1,10}(\varphi) + b_i H_{i-1,11}(\varphi), \\
p_i(\varphi) &= a_i H_{i,00}(\varphi) + b_i H_{i,01}(\varphi) + c_i H_{i,02}(\varphi) + \\
&\quad a_{i+1}H_{i,10}(\varphi) + b_{i+1}H_{i,11}(\varphi).
\end{aligned}$$

No matter how the coefficients $a_{i-1}, b_{i-1}, c_{i-1}, a_i, b_i, c_i, a_{i+1}, b_{i+1}$ are chosen, these two functions at α_i have the same value, $p_{i-1}(\alpha_i) = p_i(\alpha_i) = a_i$, and derivative, $p'_{i-1}(\alpha_i) = p'_i(\alpha_i) = b_i$. The formulae for the second-order derivatives of the basis functions, shown previously, make it possible to write the sides of the equation $p''_{i-1}(\alpha_i) = p''_i(\alpha_i)$ explicitly:

$$\begin{aligned}
-\frac{4(\cos 2\delta_{i-1} + 2)}{\cos 2\delta_{i-1} - 1}a_{i-1} + c_{i-1} + \frac{4(\cos 2\delta_{i-1} + 2)}{\cos 2\delta_{i-1} - 1}a_i - \frac{6\sin 2\delta_{i-1}}{\cos 2\delta_{i-1} - 1}b_i = \\
\frac{4(\cos 2\delta_i + 2)}{\cos 2\delta_i - 1}a_i + \frac{6\sin 2\delta_i}{\cos 2\delta_i - 1}b_i + c_i - \frac{4(\cos 2\delta_i + 2)}{\cos 2\delta_i - 1}a_{i+1}.
\end{aligned}$$

After reordering this equation we obtain the following:

$$\begin{aligned}(\cos 2\delta_{i-1} - 1)(\cos 2\delta_i - 1)(c_i - c_{i-1}) +&\\ 6(\sin 2(\delta_{i-1} + \delta_i) - \sin 2\delta_{i-1} - \sin 2\delta_i)\, b_i =&\\ 4(\cos 2\delta_{i-1} + 2)(\cos 2\delta_i - 1)(a_i - a_{i-1}) +&\\ 4(\cos 2\delta_i + 2)(\cos 2\delta_{i-1} - 1)(a_{i+1} - a_i).&\end{aligned} \tag{4.35}$$

For a partition Δ with k elements, there are k equations (4.35), which may be written together in matrix form:

$$A_4 \boldsymbol{c}_4 + B_4 \boldsymbol{b}_4 = C_4 \boldsymbol{a}_4 \tag{4.36}$$

with the vectors

$$\boldsymbol{a}_4 = \begin{bmatrix} a_0 \\ \vdots \\ a_{k-1} \end{bmatrix}, \quad \boldsymbol{b}_4 = \begin{bmatrix} b_0 \\ \vdots \\ b_{k-1} \end{bmatrix}, \quad \boldsymbol{c}_4 = \begin{bmatrix} c_0 \\ \vdots \\ c_{k-1} \end{bmatrix}.$$

The vector $\boldsymbol{a}_4$ is made of the values of the even quartic spline function, $a_i = s_4(\alpha_i)$, related to the fourth-order derivatives of the common curves of patches. The coordinates of the vector $\boldsymbol{b}_4$ are the derivatives of the function s_4, $b_i = s_4'(\alpha_i)$. The coordinates c_i of the vector $\boldsymbol{c}_4$ add to the second-order derivatives of the function s_4. The $k \times k$ matrices A_4, B_4 and C_4 are as follows:

$$\begin{aligned} A_4 &= \operatorname{diag}\bigl((\cos 2\delta_{k-1} - 1)(\cos 2\delta_0 - 1), \ldots, \\ &\qquad (\cos 2\delta_{k-2} - 1)(\cos 2\delta k - 1 - 1)\bigr) \cdot D_-, \\ B_4 &= 6 \operatorname{diag}\bigl(\sin 2(\delta_{k-1} + \delta_0) - \sin 2\delta_{k-1} - \sin 2\delta_0, \ldots, \\ &\qquad \sin 2(\delta_{k-2} + \delta_{k-1}) - \sin 2\delta_{k-2} - \sin 2\delta_{k-1}\bigr), \\ C_4 &= 4 \operatorname{diag}\bigl((\cos 2\delta_{k-1} + 2)(\cos 2\delta_0 - 1), \ldots, \\ &\qquad (\cos 2\delta_{k-2} + 2)(\cos 2\delta_{k-1} - 1)\bigr) \cdot D_- + \\ &\quad 4 \operatorname{diag}\bigl((\cos 2\delta_0 + 2)(\cos 2\delta_{k-1} - 1), \ldots, \\ &\qquad (\cos 2\delta_{k-1} + 2)(\cos 2\delta_{k-2} - 1)\bigr) \cdot D_+. \end{aligned}$$

Here diag(...) denotes a diagonal matrix with the diagonal coefficients listed in the parentheses, and the $k \times k$ matrices D_- and D_+ are cyclic bidiagonal:

$$D_- = \begin{bmatrix} 1 & & & -1 \\ -1 & 1 & & \\ & \ddots & \ddots & \\ & & -1 & 1 \end{bmatrix}, \quad D_+ = \begin{bmatrix} -1 & 1 & & \\ & \ddots & \ddots & \\ & & -1 & 1 \\ 1 & & & -1 \end{bmatrix}.$$

Proposition 4.13

1. *If* $\delta_{i-1} + \delta_i \neq \pi$ *(i.e.,* $\alpha_{i+1} \neq \alpha_{i-1} + \pi$*) for all* i*, then System* (4.36) *with arbitrarily fixed vectors* $\boldsymbol{a}_4$ *and* $\boldsymbol{c}_4$ *has a unique solution* $\boldsymbol{b}_4$.
2. *The homogeneous system* $A_4\boldsymbol{c}_4 = \boldsymbol{0}$ *has solutions of the form* $c[1,\ldots,1]^T$ *for all* $c \in \mathbb{R}$*. System* (4.36) *with the vectors* $\boldsymbol{a}_4$ *and* $\boldsymbol{b}_4$ *fixed has solutions* $\boldsymbol{c}_4$ *if and only if*

$$\sum_{i=0}^{k-1} \frac{\sin 2(\delta_{i-1}+\delta_i) - \sin 2\delta_{i-1} - \sin 2\delta_i}{(\cos 2\delta_{i-1} - 1)(\cos 2\delta_i - 1)} b_i = 4\sum_{i=0}^{k-1} \frac{\cos 2\delta_{i-1} - \cos 2\delta_i}{(\cos 2\delta_{i-1} - 1)(\cos 2\delta_i - 1)} a_i. \tag{4.37}$$

3. *If* $k = 4$ *and* $\alpha_2 = \alpha_0 + \pi$, $\alpha_3 = \alpha_1 + \pi$*, then System* (4.36) *with the vectors* $\boldsymbol{a}_4$ *and* $\boldsymbol{b}_4$ *arbitrarily fixed has a non-empty set of solutions* $\boldsymbol{c}_4$.
4. *In other cases System* (4.36) *with unknown vectors* $\boldsymbol{b}_4$ *and* $\boldsymbol{c}_4$ *is consistent and the dimension of the space of solutions of the homogeneous system* $A_4\boldsymbol{c} + B_4\boldsymbol{b} = \boldsymbol{0}$ *is equal to* k.

Proof. 1. Let $f_i(\delta_{i-1},\delta_i) \stackrel{\text{def}}{=} \sin 2(\delta_{i-1}+\delta_i) - \sin 2\delta_{i-1} - \sin 2\delta_i$. We can check that $f_i(\delta_{i-1},\delta_i) = 2\sin x(\cos x - \cos y)$, where $x = \delta_{i-1}+\delta_i$ and $y = \delta_{i-1}-\delta_i$. Assuming that $\delta_{i-1} > 0, \delta_i > 0, \delta_{i-1}+\delta_i < 2\pi, \delta_{i-1} \neq \pi$ and $\delta_i \neq \pi$, there is $f_i(\delta_{i-1},\delta_i) = 0$ if and only if $\delta_{i-1}+\delta_i = \pi$. If the condition $\delta_{i-1}+\delta_i \neq \pi$ is satisfied for all i, then all diagonal coefficients of the matrix B_4 are non-zero, and this matrix is non-singular.

2. The rank of the matrix A_4 is equal to the rank of the matrix D_-, which is $k-1$. The systems of equations $A_4\boldsymbol{c}_4 = \boldsymbol{0}$ and $D_-\boldsymbol{c}_4 = \boldsymbol{0}$ are equivalent. Equation (4.37) is obtained by adding all equations (4.35) and rearranging the sum.

3. If $k = 4$ and $\alpha_2 = \alpha_0 + \pi, \alpha_3 = \alpha_1 + \pi$, then B_4 is the zero matrix and Equation (4.37), with any $\boldsymbol{a}_4$ and $\boldsymbol{b}_4$, is $0 = 0$.

4. It is obvious that apart from the case considered just above the matrix B_4 is non-zero and the rank of the $k \times 2k$ matrix $[A_4, B_4]$ is equal to k. A system of linear equations with such a matrix is consistent and the space of solutions of the homogeneous system with this matrix is equal to k. □

The analysis of G^2 compatibility conditions using trigonometric splines confirmed the absence of restrictions imposed on the fourth-order derivatives of the common curves of patches; moreover, it revealed *all* the degrees of freedom in constructing trigonometric splines of interpolation. The properties of the trigonometric splines may be easily translated to compatibility conditions for patches surrounding a common point using the formulae in Section 4.3.3. However, our knowledge of the restrictions on the derivatives of the common curves does not translate easily to practical construction of smooth surfaces. The next section contains remarks about possible methods of using the theory developed in this chapter in practical constructions.

4.4 BEYOND THE CURVATURE CONTINUITY AND TOWARDS PRACTICE

The higher the order of geometric continuity, the more restrictive and harder to analyse the compatibility conditions are. For surfaces of class G^3, they involve derivatives up to the order 6. The idea of choosing the 6-th order derivatives of the common curves using a polynomial function of degree 6, which is in accordance with Hahn's approach, for practical purposes seems hardly acceptable—as have seen, the G^2 compatibility conditions do not impose any restriction at all on the fourth-order derivatives of the common curves, and even the third-order derivatives in many cases (depending on the partition of the full angle determined by the tangent halflines of the curves) may be arbitrary. Similar discoveries may be expected for higher order compatibility conditions.

The possibility of constructing function spaces $\mathcal{H}_\Delta^{(l,n)}$, with the help of corresponding trigonometric splines, widely extends the possible choice of shapes of smooth surfaces. To construct a surface of class G^n, instead of choosing just *one* polynomial function of degree $2n$ and then using it to fix the derivatives of the common curves and the patches at the central point (which ensures satisfying the compatibility conditions), we can fix a partition Δ and then construct a basis of the linear vector space whose elements are bivariate splines of degree $2n$, of class C^n over that partition of the full angle. Having the basis (which is the union of bases of the spaces $\mathcal{H}_\Delta^{(l,n)}$, for $l = 0, \dots, 2n$), we can choose three elements of this space in order to obtain a *piecewise polynomial* vector function, which represents a surface in the three-dimensional space. The derivatives of this function determine the derivatives of the patches (and their common boundary curves) at the common point; as all elements (bivariate splines) of our space satisfy the compatibility conditions, so do the chosen ones.

The approach suggested above raises another problem: *how* to choose these bivariate splines. The trouble is that it is necessary to choose elements of a space of a rather high dimension; for example, five patches surrounding a common point in a construction of a surface of class G^1 give us $3 \cdot (1 + 2 + 5) = 24$ degrees of freedom, while in case of G^2, assuming $h = 0$ (see Theorem 4.1), there are $3 \cdot (1 + 2 + 3 + 5 + 10) = 63$ numbers to be chosen. For surfaces of class G^1 and G^2, Hahn's solution gives us respectively $3 \cdot (1 + 2 + 3) = 18$ and $3 \cdot (1 + 2 + 3 + 4 + 5) = 45$ parameters to choose. Designers should be able to avoid such an *embarras de richesses*; instead, a procedure built into a design system ought to make an automatic choice (perhaps allowing the user to make corrections). Usually patches surrounding a common point are necessary to fill a polygonal hole in a given smooth surface. In that case, it is often desirable to make the surface filling the hole indistinguishable from that surface. The choice of derivatives of the patches filling the hole at the common point should be made so as to achieve this goal. Some methods for this are described in the next chapter.

CHAPTER 5

Filling polygonal holes

As we saw in the previous chapter, a common corner of $k \neq 4$ patches forming a smooth surface may cause trouble in constructions. Often in practice we cannot eliminate such points completely, but we can try to minimise their number. In this chapter, we assume that a surface is represented by a mesh made of quadrangular facets, as described in Section A.6. We assume that the mesh has extraordinary elements, distant enough from each other. Thus, the majority of the surface is made of B-spline patches and it has polygonal holes which have to be filled in an extraordinary way.

The patches filling the holes will differ from the other patches of the surface; in particular, their degree will be higher. In general, low degree of surface patches is desirable and researchers put a lot of effort to develop constructions yielding patches of low degree. An example is the construction given by Peters [2002], who obtained surfaces of class G^2 made of patches of degree $(3, 3)$ filling the hole and patches of degree $(3, 5)$ surrounding it. However, focusing on the degree above everything else may compromise the quality of the surface shape. To obtain a surface filling the hole, visually indistinguishable from the rest of the surface, one may need patches of a higher degree.

Usually a surface filling the hole is not determined uniquely by interpolation conditions (i.e., by the patches surrounding the hole), and there are too many degrees of freedom for "manual" tuning. To obtain the best results, a functional measuring the "badness" of the surface is defined and an optimisation algorithm is applied to find the surface with the minimal badness. Recently Karčiauskas and Peters [2015] showed how to fill holes with biquintic patches, which minimise such functionals and result in a surface of class G^2. In this chapter, patches of a higher degree are the result but the construction is more flexible. In particular, it makes use of *all* the degrees of freedom allowed by compatibility conditions.

Finding a minimal point of a functional is often equivalent to solving numerically a partial differential equation, and the most popular method of doing this is the finite element method. After constructing a suitable finite-dimensional function space, a system of algebraic equations is solved to choose an element of this space. Constructions of surfaces by solving partial differential equations were first described by Bloor and Wilson [1989]. The reader may refer to textbooks, e.g. Ciarlet [1978], to learn about the finite element method, whose (very) special case is the construction described in this chapter.

We are going to fill holes in bicubic surfaces represented by a mesh (see Section A.6) with all facets quadrangular. Each submesh made of 16 vertices, 24 edges and 9 facets, whose vertices and

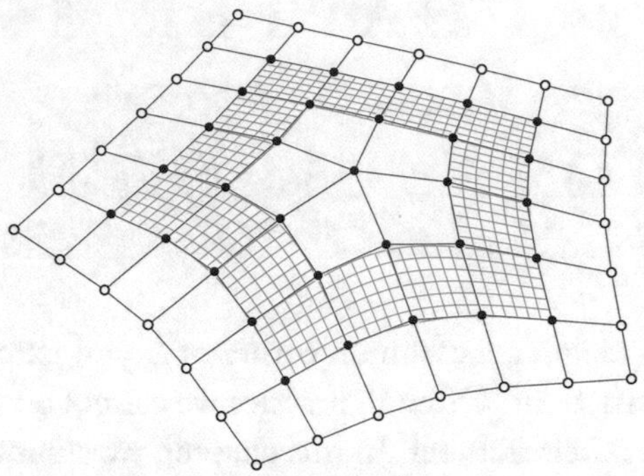

Figure 5.1: A Sabin net and the polygonal hole in a surface made of bicubic patches

edges form four rows and four columns, represents a bicubic polynomial patch (see Figure A.14). If these patches are regular, then they form a surface of class G^2 with holes. We assume that each extraordinary element is an inner vertex incident with $k \neq 4$ edges and is the central point of a Sabin net of radius 3 being a part of the mesh (see Section A.6.5). In this way, every k-gonal hole is surrounded by $3k$ bicubic patches, and its boundary is made of $2k$ cubic curves (Fig. 5.1).

Many constructions of patches filling polygonal holes have an element deserving special attention: an explicitly defined domain of a parametrisation of the part of the surface, made of the (given) patches surrounding the hole and the patches filling it. This domain, denoted below by A, is a planar area and its part, denoted by Ω, is the domain of the patches filling the hole. The parametrisation of the part of the surface over A must be of class C^n and regular, thus fulfilling the definition of G^n continuity of the surface.

Below, we assume that A and Ω are curvilinear polygons. These two areas are represented by a *planar* Sabin net. The planar bicubic patches represented by regular submeshes of the Sabin net cover the area $A \setminus \Omega$. The area A will be divided to $4k$ curvilinear quadrangles; $3k$ of them are domains of the reparametrised bicubic patches surrounding the hole and the other k (obtained by dividing Ω and denoted by $\Omega_0, \dots, \Omega_{k-1}$) are the domains of some parametrisations of the patches filling the hole (Fig. 5.2). Our goal is to obtain a parametrisation of class $C^1(A)$ or $C^2(A)$.[1]

5.1 THEORETICAL BACKGROUND

In this section, equations of geometric continuity studied in Chapter 3 are revisited. This time, the goal is not to obtain surface patches in $\mathbb{R}^3$, but to construct a function space for the finite

[1]The curvature continuity is the best we can achieve with a piecewise bicubic surface. To obtain a higher order geometric continuity, we need surfaces of a higher degree. Note that a k-gonal hole in a surface of degree (n, n) represented by an appropriate Sabin net is surrounded by nk polynomial patches and its boundary consists of $(n-1)k$ polynomial arcs of degree n.

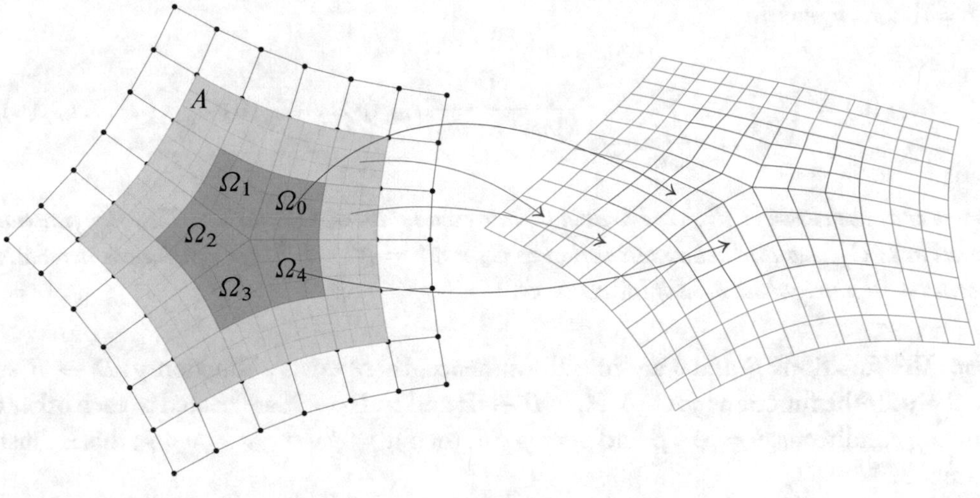

Figure 5.2: Parametrisation of a part of the surface represented by a Sabin net

element method; the functions from this space are defined piecewise, with pairs of pieces having smooth junctions.

Let $B, C, D \subset \mathbb{R}^2$ be open areas. Let s, t be local coordinates in the area B and let u, v be local coordinates in the area D. Let $\Phi = \{ (s,t) : 0 < s < 1,\ t = t_0 \}$ and $\Psi = \{ (u,v) : 0 < u < 1,\ v = v_0 \}$ be line segments contained in the areas B and D respectively. We consider regular one-to-one mappings (bijections) $\boldsymbol{\beta}: B \to C$ and $\boldsymbol{\delta}: D \to C$ of class C^n; such mappings are called diffeomorphisms of class C^n. The image of the line segment Φ under the mapping $\boldsymbol{\beta}$ is a curve Θ. We assume that the curve Θ is also the image of the line segment Ψ under the mapping $\boldsymbol{\delta}$, and if $s = u \in (0, 1)$, then $\boldsymbol{\beta}(s, t_0) = \boldsymbol{\delta}(u, v_0)$. In other words, for all $u \in (0, 1)$, both expressions, $\boldsymbol{\beta}(u, t_0)$ and $\boldsymbol{\delta}(u, v_0)$, describe the same point on the curve $\Theta \subset C$ (Fig. 5.3).

Theorem 5.1 *Let $\phi: C \to \mathbb{R}$ be a function of class C^n and let $\nu = \phi \circ \boldsymbol{\beta}$. Let $\mu^*: D \to \mathbb{R}$ be a function of class C^n. Let*

$$\boldsymbol{\delta}(u, v_0) = \boldsymbol{\beta}(u, t_0), \tag{5.1}$$

$$\mu^*(u, v_0) = \nu(u, t_0), \tag{5.2}$$

for all $u \in (0, 1)$. Let there exist functions $s_1, t_1, \ldots, s_n, t_n$ of class C^n such that

$$\frac{\partial^j}{\partial v^j}\boldsymbol{\delta}(u, v)\Big|_{v=v_0} = \sum_{k=1}^{j}\sum_{h=0}^{k} a_{jkh}(u)\frac{\partial^k}{\partial s^h t^{k-h}}\boldsymbol{\beta}(u, t)\Big|_{t=t_0}, \tag{5.3}$$

$$\frac{\partial^j}{\partial v^j}\mu^*(u, v)\Big|_{v=v_0} = \sum_{k=1}^{j}\sum_{h=0}^{k} a_{jkh}(u)\frac{\partial^k}{\partial s^h t^{k-h}}\nu(u, t)\Big|_{t=t_0} \tag{5.4}$$

for $j = 1, \dots, n$, *where*

$$a_{jkh}(u) = \binom{k}{h} \sum_{\substack{m_1+\dots+m_k=j \\ m_1,\dots,m_k>0}} \frac{j!}{k! m_1! \dots m_k!} s_{m_1}(u) \dots s_{m_h}(u) t_{m_{h+1}}(u) \dots t_{m_k}(u).$$

Suppose that the curve Θ *divides the area* C *into two subareas,* C_1 *and* C_2*. Then, the function, whose restriction to* C_1 *is* ϕ *and whose restriction to* C_2 *is* $\phi^* = \mu^* \circ \boldsymbol{\delta}^{-1}$*, has continuous derivatives up to the order* n *in a neighbourhood of the curve* Θ*.*

Proof. The functions $\boldsymbol{\beta}$ and $\boldsymbol{\delta}$ are bijections; hence, there exists a function $\boldsymbol{\gamma}\colon D \to B$ such that $\boldsymbol{\delta} = \boldsymbol{\beta} \circ \boldsymbol{\gamma}$. If the functions $\phi\colon C \to \mathbb{R}$, $\nu\colon B \to \mathbb{R}$ and $\mu\colon D \to \mathbb{R}$ are related to each other with the following conditions: $\nu = \phi \circ \boldsymbol{\beta}$ and $\mu = \phi \circ \boldsymbol{\delta}$, then $\mu = \phi \circ \boldsymbol{\beta} \circ \boldsymbol{\gamma} = \nu \circ \boldsymbol{\gamma}$; this is illustrated in Figure 5.3.

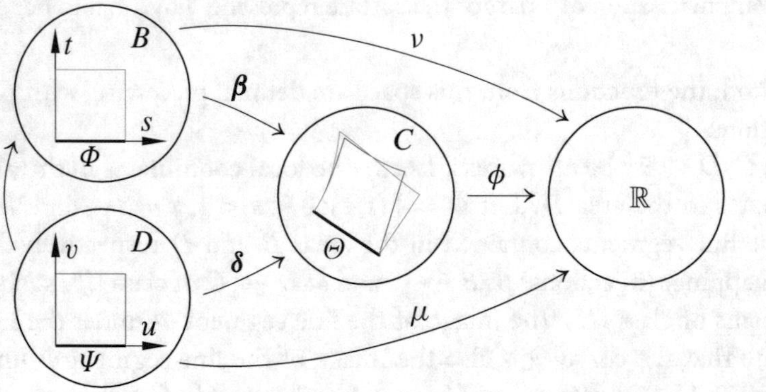

Figure 5.3: Areas and mappings in Theorem 5.1

The line segment Ψ consists of points, whose v-coordinate has a constant value v_0. Similarly, all points of the line segment $\Phi = \boldsymbol{\gamma}(\Psi)$ have the t coordinate equal to a fixed t_0. The function $\boldsymbol{\gamma}$, being the composition of $\boldsymbol{\beta}^{-1}$ and $\boldsymbol{\delta}$, is of class C^n. Let the two scalar functions which describe the coordinates of $\boldsymbol{\gamma}$ be denoted by $s(u, v)$ and $t(u, v)$. By (5.1), there is $s(u, v_0) = u$, $t(u, v_0) = t_0$ for all $u \in (0, 1)$. The image of the line segment Ψ under the mapping $\boldsymbol{\gamma}$ is the line segment Φ. We may assume that the functions which describe the mapping $\boldsymbol{\gamma}$ are such that

$$s(u, v) = u + \sum_{i=1}^{n} s_i(u) \frac{(v - v_0)^i}{i!} + o\big((v - v_0)^n\big),$$
$$t(u, v) = t_0 + \sum_{i=1}^{n} t_i(u) \frac{(v - v_0)^i}{i!} + o\big((v - v_0)^n\big),$$

where the functions $s_1, t_1, \ldots, s_n, t_n$ are of class C^n. The calculation of the partial derivatives of orders $1, \ldots, n$ of the functions $\boldsymbol{\delta} = \boldsymbol{\beta} \circ \boldsymbol{\gamma}$ and $\mu = \nu \circ \boldsymbol{\gamma}$ using the generalised Faà di Bruno's formula (A.55) gives us the equalities (5.3) and (5.4), with μ instead of μ^*.

The mapping $\mu = \phi \circ \boldsymbol{\delta}$ may be replaced by a mapping $\mu^*\colon D \to \mathbb{R}$ of class C^n having the same values and derivatives up to the order n that μ has at all points of the line segment Ψ. The function μ^* is a composition of some function ϕ^* with $\boldsymbol{\delta}$. The functions ϕ and ϕ^* have the same values and derivatives up to the order n at all points of the curve Θ, which completes the proof. □

It is worth comparing Formulae (5.3) and (5.4) with (3.5); all three formulae are in fact the same. The most important consequence of Theorem 5.1 is the fact that the junction functions $s_1, t_1, \ldots, s_n, t_n$ are *the same* in the description of the junction of the mapping $\boldsymbol{\beta}$ with $\boldsymbol{\delta}$ and the junction of the function ν with μ^*.

Theorem 5.1 may be used as follows: having a planar parametric patch $\boldsymbol{\beta}$, we choose junction functions $s_1, t_1, \ldots, s_n, t_n$ and then construct a mapping $\boldsymbol{\delta}$ that satisfies Equations (5.1) and (5.3). The mapping $\boldsymbol{\beta}$ is a parametrisation of a planar bicubic patch, represented by a regular subnet of the planar Sabin net, and whose boundary curve (Θ) delimits the area Ω. Then, we construct a **domain patch**, which corresponds to $\boldsymbol{\delta}$ in our scheme of mappings. Given a function ν (determined by a bicubic patch adjacent to the hole in the surface), we can construct the function μ using the junction functions fixed earlier. The function ϕ may describe (a coordinate of) a part of a parametrisation of class C^n of the surface, and then the function μ describes a parametrisation of this part over the rectangular domain;[2] μ will be a polynomial mapping describing a coordinate of a (Coons) patch filling a part of the hole.

A pair of functions whose domains have a common curve inside Ω necessitates using Theorem 5.1 in a little more complicated way. Just as in Section 3.5, we introduce auxiliary patches. This allows us to deal with the final patches symmetrically: instead of fitting one of them to the other, we shall fit both final patches to an auxiliary patch. For each pair of the final patches having a common curve, we introduce what we call an **auxiliary domain patch**, described by the mapping $\boldsymbol{\beta}$ in our scheme. The auxiliary domain patch will determine a curve Θ inside Ω. If Ω is a domain of patches to fill a k-sided hole, then these curves will divide Ω to k curvilinear quadrilaterals. For an auxiliary domain patch we construct *two* sets of junction functions, which we use to construct cross-boundary derivatives of two domain patches. Note that the tangent halflines of k common curves of domain patches at their common point determine a partition of the full angle, which in turn determines compatibility conditions for the auxiliary patches.

Having junction functions for all polynomial curves delimiting and dividing the area Ω, we may define the auxiliary patches (which correspond to the mapping ν), taking care of the compatibility conditions. The last two steps are the constructions of the cross-boundary derivatives of the final patches filling the hole and then the final patches themselves.

[2]In the finite element method, such a rectangle is called a standard element.

The final patches are described by triples of scalar functions; but instead of constructing just the patches, we construct an entire *space* of scalar functions of class C^n over the area A. By definition, if a parametrisation made of a triple of such functions is regular, then it describes a surface of class G^n. It is possible to optimise the shape of the surface; the optimisation chooses the best triple of functions from this space. To define optimisation criteria (which we do in Sections 5.3 and 5.4), usually one needs an explicit parametrisation of the area Ω, described piecewise by the mappings $\boldsymbol{\delta}$ from our scheme.

5.2 CONSTRUCTING FUNCTION SPACES

We consider a part of a bicubic surface represented by a Sabin net of radius 3, with the central vertex incident with k edges. Such a net is the control net of our part of the surface, which consists of $3k$ bicubic patches surrounding a k-sided hole. We also consider a planar Sabin net of radius 3, which we call a **domain net.** The domain net determines $3k$ planar bicubic patches surrounding a k-sided curvilinear polygon Ω; each of the k patches has only one common corner with Ω, whose boundary is made of cubic boundary curves of the other $2k$ patches.

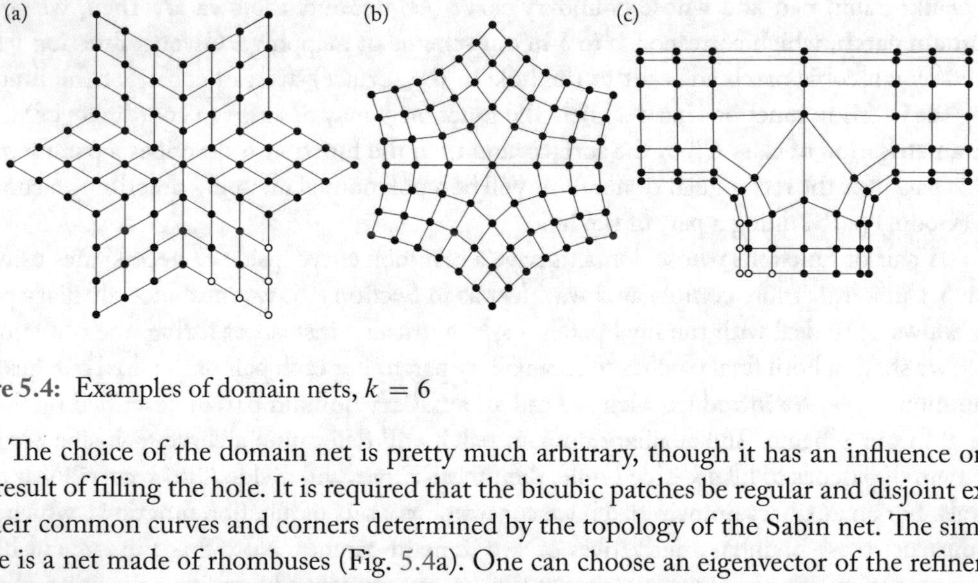

Figure 5.4: Examples of domain nets, $k = 6$

The choice of the domain net is pretty much arbitrary, though it has an influence on the final result of filling the hole. It is required that the bicubic patches be regular and disjoint except for their common curves and corners determined by the topology of the Sabin net. The simplest choice is a net made of rhombuses (Fig. 5.4a). One can choose an eigenvector of the refinement operator discussed in Section A.6.5 (Fig. 5.4b). Another possibility is to project the control net on the best-fitted plane in the space,[3] which may result in a net shown in Figure 5.4c.

The vertices of the Sabin subnet of radius 2, marked by black dots in Figure 5.4, are relevant for the construction. The other vertices are needed to describe the patches surrounding the hole

[3]The fitting may be done by minimising the sum of squares of distances of the vertices of the Sabin net in the three-dimensional space from the plane. The best-fitted plane passes through the gravity centre of the vertices, say, $\boldsymbol{c}$. Let $A = \sum_i (\boldsymbol{v}_i - \boldsymbol{c})(\boldsymbol{v}_i - \boldsymbol{c})^T$, where $\boldsymbol{v}_i$ are the vertices of the net. The normal vector of the best-fitted plane is the eigenvector of the matrix A, corresponding to its smallest eigenvalue.

in Figures 5.1 and 5.4, but they have no influence on the boundary of the hole and the cross-boundary derivatives of the first and second order.

The elements of the space to construct are functions of class C^1 or C^2 in the area A, which is the union of Ω and the surrounding area covered by the $3k$ bicubic planar patches (see Figure 5.2). The restrictions of these functions to the latter area may describe reparametrised bicubic patches around the hole. The area Ω will be divided into k curvilinear quadrangles, $\Omega_0, \dots, \Omega_{k-1}$, which will be the domains of reparametrised patches filling the hole in the surface. Any symmetry of the domain net is worth preserving in this division, as it makes it possible to obtain symmetric surfaces filling the holes in the surfaces that have symmetric control nets.

The division is done by k curves, which meet at the **domain central point** inside Ω. Their tangent halflines determine a partition of the full angle, whose role for the compatibility conditions was studied in Chapter 4.

The boundary of Ω consists of $2k$ cubic polynomial curves. Each curvilinear quadrangle Ω_i has the boundary made of two of them and two curves inside Ω; we have a total number of $3k$ polynomial curves, which make the boundaries of these quadrangles. The parametrisation of each curve (Θ in Theorem 5.1) is extended to define a mapping $\boldsymbol{\beta}$, representing an auxiliary domain patch. For the boundary curves of Ω, the auxiliary domain patches are the bicubic patches represented by the regular submeshes of the domain net. For each curve dividing Ω to the quadrangles Ω_i, the auxiliary domain patch is represented by this curve and one or two cross-boundary derivatives.

The construction of the function space is done in the steps outlined below:

1. Find Bézier representations of the $2k$ planar bicubic patches represented by the regular subnets of the domain net and surrounding the area Ω.
2. Choose the domain central point.
3. Construct curves dividing Ω into the curvilinear quadrangles $\Omega_0, \dots, \Omega_{k-1}$, and cross-boundary derivatives along these curves—of order 1 if G^1 continuity is to be obtained, or 1 and 2 in the construction of a surface of class G^2.
4. Construct the junction functions for all four curves bounding each area Ω_i. The set of junction functions for each side represents the mapping $\boldsymbol{\gamma}$ in Theorem 5.1. The two sets of junction functions for the curves meeting at each corner of Ω_i must satisfy the Hermite interpolation conditions determined by the compatibility conditions.
5. For $i = 0, \dots, k-1$, construct the cross-boundary derivatives of domain patch (the mapping $\boldsymbol{\delta}$ in Theorem 5.1), and then the domain patch, being a parametrisation of Ω_i. The domain patch is a Coons patch (represented in Bézier form—see Section A.9) that is bicubically or biquintically blended in the construction of a surface of class G^1 or G^2 respectively. Its domain is the unit square.
6. Construct functions of class C^1 or C^2 in the area A covered by the bicubic patches found in Step 1 and by the domain patches obtained in Step 5. To define any of them in the

area Ω_i, it is necessary to choose a mapping ν for each side of Ω_i. This mapping, called an **auxiliary basis function patch**, is determined by the values at the line segment Φ and cross-boundary derivatives. Using the junction functions found in Step 4, cross-boundary derivatives of basis function patches (corresponding to μ^*) may be constructed for each side of the square being the domain of the domain patch. The **basis function patch** is obtained as a bicubically or biquintically blended Coons patch. In Ω_i the function being the result of the entire construction is the composition of the mapping inverse to the domain patch and the basis function patch.

Finding bicubic patches surrounding Ω

This step is easy: from the domain net we reject k facets at the "corners", and from what is left we extract k subnets, each consisting of 12 facets, 31 edges and 20 vertices, which may be stored in arrays with four rows and five columns. The subnet is the control net of a bicubic B-spline patch with two uniform knot sequences, which may be taken as $(u_0, \ldots, u_8) = (-4, -3, -2, -1, 0, 1, 2, 3, 4)$ and $(v_0, \ldots, v_7) = (-3, -2, -1, 0, 1, 2, 3, 4)$; the patch consists of two polynomial pieces. Knot insertion (see Sections A.4 and A.5) may be used to obtain the B-spline representation with the additional knots $-1, -1, 0, 0, 1, 1$ in the "u" knot sequence and $0, 0, 1, 1$ in the "v" knot sequence. As the appropriate knots in the new representation have the multiplicity 3, the B-spline control net contains Bézier control nets of the bicubic patches adjacent to the area Ω (see Figure 5.5).

The parameters of the $2k$ Bézier patches surrounding the area Ω, ranging from 0 to 1, will be denoted by the letters u and s or v and t (Fig. 5.6). We assume that the boundary curves which form the boundary of Ω correspond to $s = 0$ or $t = 0$. Having the Bézier control nets of the patches, we can easily find the control points of the boundary curves and the cross-boundary derivatives, i.e., derivatives of the patches with respect to s or t.

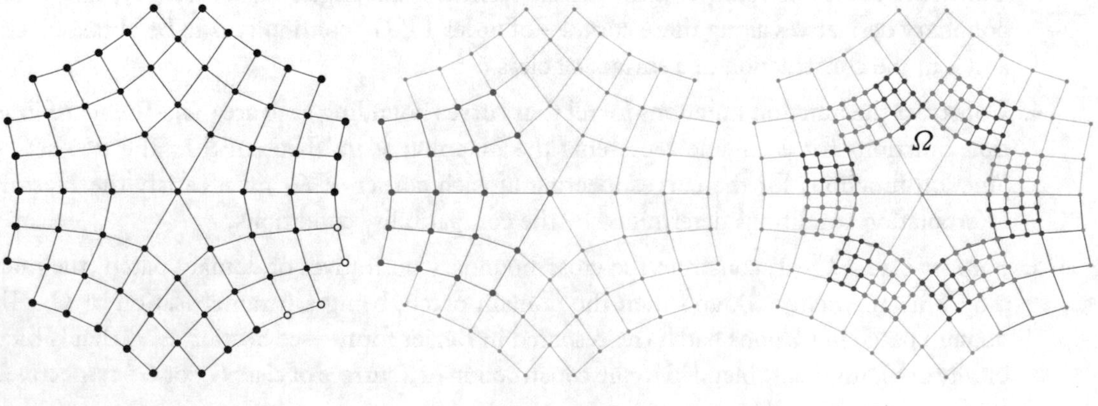

Figure 5.5: Finding Bézier control nets of patches around Ω

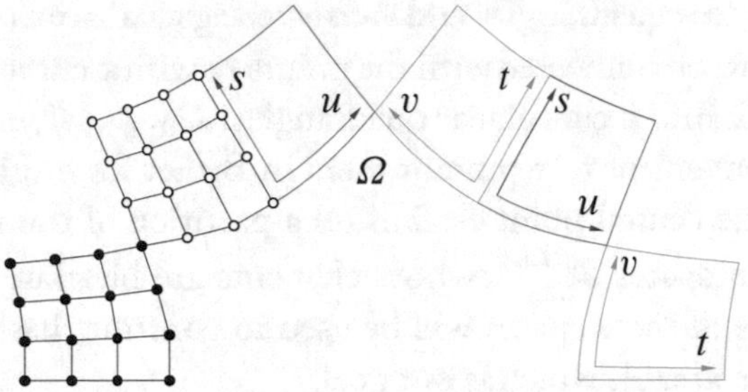

Figure 5.6: Parametrisations of bicubic patches surrounding Ω

Constructing the central point

Though the choice of the central point is pretty much arbitrary, it is desirable to choose it in a way that preserves any symmetry of the domain net. Each side curve of the area Ω consists of two cubic polynomial arcs meeting at a point denoted here by $\boldsymbol{r}_i$ (the subscript corresponds to the side). The simplest method of choosing the central point $\boldsymbol{c}$ is to take it at the gravity centre of the points $\boldsymbol{r}_i$. These points are vertices of the Bézier control nets found in the previous step—see Figures 5.5 and 5.7.

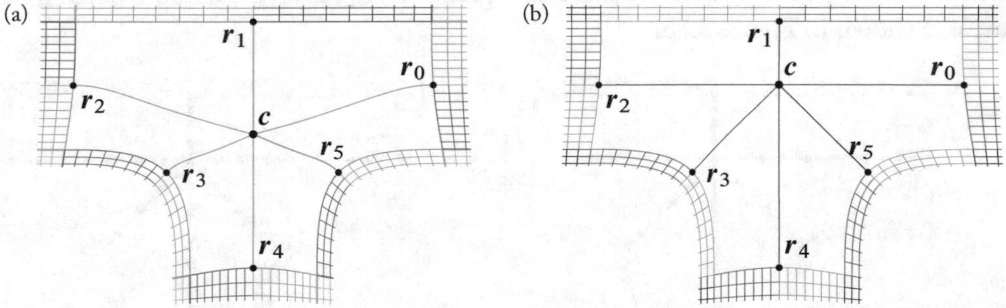

Figure 5.7: The central points chosen with different methods: (a) by taking the gravity centre, (b) by solving a linear least-squares problem

The simplest method is not always the best one; for special domain nets, like the one shown in Figure 5.4c, which determines the area Ω shown in Figure 5.7, better results may be obtained by choosing the central point as follows: we construct k lines passing through the points $\boldsymbol{r}_i$ and having directions of the cross-boundary derivatives of the bicubic patches surrounding Ω (i.e., lines tangent to the common curves of pairs of patches surrounding Ω at $\boldsymbol{r}_i$). In general, these lines need not have a common point. However, we can choose the point $\boldsymbol{c}$ such that the sum of squares of its distances from all these lines is minimal. A system of linear equations (usually inconsistent) with unknown $\boldsymbol{c}$ may be written and solved as a linear least-squares problem. Its derivation is left as an exercise.

Dividing Ω into curvilinear quadrangles and constructing auxiliary domain patches

The central point $\boldsymbol{c}$ has to be connected with the points $\boldsymbol{r}_i$ with k curves, denoted by $\underline{\boldsymbol{q}}_0, \dots, \underline{\boldsymbol{q}}_{k-1}$, which divide the area Ω into k curvilinear quadrangles, $\Omega_0, \dots, \Omega_{k-1}$. These should be polynomial curves and it is convenient to represent them in Bézier form. Fixing the first-order derivatives of the curves at the central point establishes a partition of the full angle Δ, which in turn determines the function spaces $\mathcal{H}_\Delta^{(l,n)}$, whose elements are bivariate splines considered in Sections 4.3.2–4.3.5. Bases of these spaces will be used to construct bases of the space of functions of class C^1 or C^2 in the area A, which is our goal.

The G^2 compatibility conditions at the central point involve derivatives of the curves up to the order 4, while at the other end points, $\boldsymbol{r}_i$, only the first- and second-order derivatives are relevant (see Equations (4.7′)–(4.14′)). As we focus on one curve $\underline{\boldsymbol{q}}_i$, below we drop the subscript. The end points of the curve $\underline{\boldsymbol{q}}$, corresponding to 0 and 1, are $\boldsymbol{c}$ and $\boldsymbol{r}_i$. It is simplest to take $\underline{\boldsymbol{q}}'(0) = \boldsymbol{r}_i - \boldsymbol{c}$. If the angle between two junction halflines obtained in this way is very close, but not equal, to π, a modification is necessary, producing a pair of junction halflines on one line. Such a modification improves the stability of the construction.

For reasons explained later, we take $\underline{\boldsymbol{q}}''(0) = \underline{\boldsymbol{q}}'''(0) = \underline{\boldsymbol{q}}''''(0) = \boldsymbol{0}$. The vectors $\underline{\boldsymbol{q}}'(1)$ and $\underline{\boldsymbol{q}}''(1)$ should be equal to the first- and second-order cross-boundary derivatives of the patches adjacent to Ω at their common point $\boldsymbol{r}_i$. After fixing the derivatives at 0 and 1, it is easy to construct the control points of a Bézier curve of degree 7 using the formulae from Section A.2.2. An example is shown in Figure 5.8b.

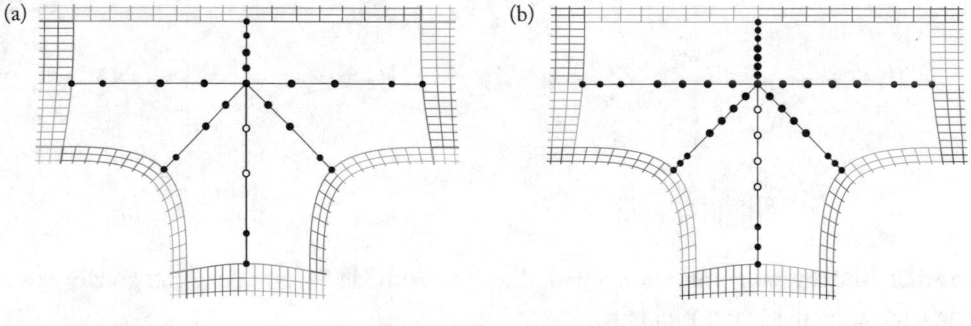

Figure 5.8: Bézier curves dividing Ω into curvilinear quadrangles: (a) of degree 4 in the G^1 construction, (b) of degree 7 in the G^2 construction

In the construction of a surface of class G^1 we need less: only the first- and second-order derivative at the central point and the first-order derivative at the point on the boundary of Ω. Thus, we have five interpolation conditions (in particular, we take $\underline{\boldsymbol{q}}''(0) = \boldsymbol{0}$), and the Bézier curve which satisfies this Hermite interpolation problem may be quartic (Fig. 5.8a).

The curves dividing Ω correspond to Θ in Theorem 5.1 (see Figure 5.3). They are parts of the definition of the auxiliary domain patches $\boldsymbol{q}_0, \dots, \boldsymbol{q}_{k-1}$, which correspond to the mapping $\boldsymbol{\beta}$. We denote $\underline{\boldsymbol{q}}(s) = \boldsymbol{q}(s, 0)$. In the construction, we need only the curve $\underline{\boldsymbol{q}}$ and one or two cross-boundary derivatives of the patch $\boldsymbol{q}$. At the central point we may choose the vector $\boldsymbol{q}_t(0,0)$ with the same length as $\boldsymbol{q}'(0)$ and orthogonal to this vector, and $\boldsymbol{q}_{st}(0,0) = \mathbf{0}$. In the G^2 construction, we also need the vectors $\boldsymbol{q}_{sst}(0,0)$, $\boldsymbol{q}_{ssst}(0,0)$, $\boldsymbol{q}_{tt}(0,0)$, $\boldsymbol{q}_{stt}(0,0)$ and $\boldsymbol{q}_{sstt}(0,0)$, which may be equal to $\mathbf{0}$. At the boundary of Ω we need $\boldsymbol{q}_{st}(1,0)$, and in the G^2 construction also the vectors $\boldsymbol{q}_{stt}(1,0)$, $\boldsymbol{q}_{sst}(1,0)$ and $\boldsymbol{q}_{sstt}(1,0)$. We choose the appropriate derivative vectors of the bicubic patches surrounding Ω.

In the G^2 construction, we have seven interpolation conditions for the first, and six conditions for the second-order cross-boundary derivative. Hence, the cross-boundary derivatives of the auxiliary patches may be polynomial curves of degrees 6 and 5 respectively. The construction of a surface of class G^1 requires only the first-order cross-boundary derivative of the auxiliary domain patch with four interpolation conditions; it may be a cubic curve. Constructing Bézier representations of the cross-boundary derivatives may be done in a way similar to that of the curves $\underline{\boldsymbol{q}}$. Figure 5.9 shows examples—here explicit Bézier representations of the auxiliary domain patches were found and used to draw some constant parameter lines.

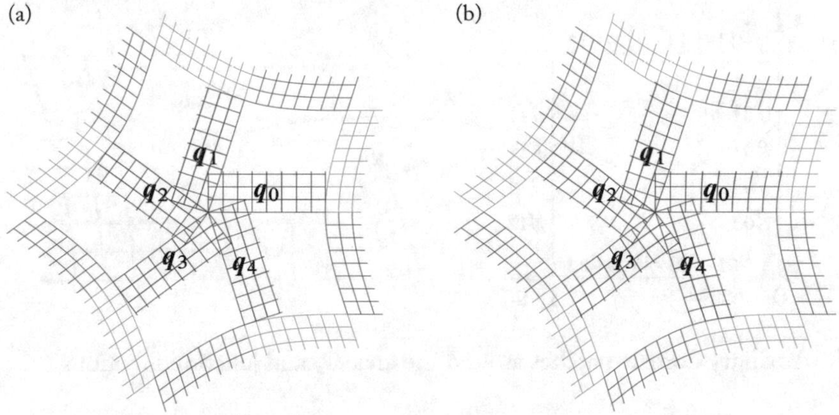

Figure 5.9: Auxiliary domain patches in the construction of a function space for a filling surface (a) of class G^1, (b) of class G^2

A number of arbitrary decisions described above may be changed in order to improve the result of the construction, and considering alternatives is left as an exercise (or rather a subject for further studies). One of the goals might be obtaining surfaces made of patches of lower degree. While it is easy to decrease the degrees of the boundary curves of the auxiliary patches (i.e., the curves $\underline{\boldsymbol{q}}$) and their cross-boundary derivatives, doing so does not translate directly to the degree of the final patches filling the hole in the surface. This is because this degree is determined by the degrees of the polynomials used as junction functions and auxiliary basis function patches, which we deal with below. Modifications resulting in a lower degree are possible, but non-trivial, and they are not necessarily advantageous for the shape of the final surface filling the hole.

Constructing junction functions and domain patches

We focus on one curvilinear quadrangle Ω_i, and to make things easier we use new symbols. A part of the mapping scheme from Figure 5.3 is repeated four times in Figure 5.10, where all four copies share the same mapping $\boldsymbol{\delta}_i$ ($\boldsymbol{\delta}$ in Theorem 5.1), which we named the domain patch. We are going to construct it now. Each side of the unit square on the left side corresponds to one line segment Ψ, which is mapped by $\boldsymbol{\delta}_i$ to the corresponding boundary curve of Ω_i. In the previous steps, we constructed a mapping $\boldsymbol{\beta}$ for each side of the square; two of those mappings are represented by the bicubic patches determined by the domain net, while the other two are the auxiliary domain patches represented by the curves dividing Ω and by the cross-boundary derivatives. The auxiliary domain patches corresponding to the line segments $u = 0$ and $v = 0$ will be denoted here by $\boldsymbol{q}_0$ and $\boldsymbol{r}_0$, while the symbols $\boldsymbol{q}_1$ and $\boldsymbol{r}_1$ will be used for the other two, corresponding to $u = 1$ and $v = 1$.

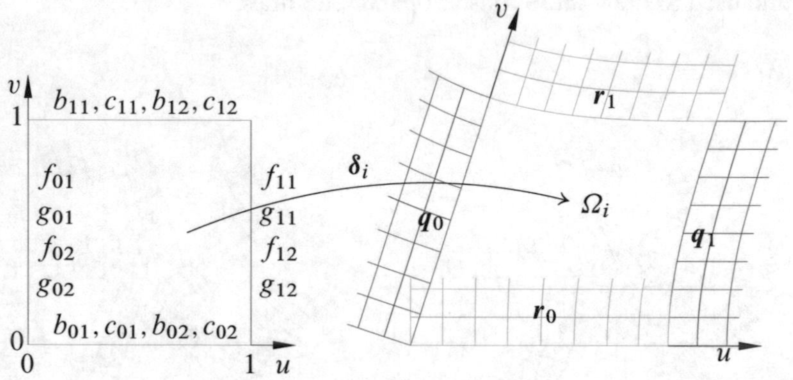

Figure 5.10: Auxiliary domain patches around the area Ω_i and junction functions

The domain patch will be defined as a planar Coons patch, defined by four boundary curves, which we already have, and the cross-boundary derivatives of the first-order, or of the first- and second-order. To construct the cross-boundary derivatives for each side of the unit square, we need the mapping $\boldsymbol{\gamma}$, or rather its partial derivatives described by the junction functions denoted by b_{lm}, c_{lm}, f_{lm} and g_{lm}, associated with the sides of the square as shown in Figure 5.10. To construct a surface of class G^1, we need only two junction functions (b_{l1} and c_{l1} or f_{l1} and g_{l1}) for each side of the domain of $\boldsymbol{\delta}_i$; all four junction functions for each side are necessary in the G^2 construction.

The junction functions and their derivatives have to satisfy equations given in Section 4.2, which we write for four vertices of the unit square, and in which we substitute the derivatives of the appropriate pair of the patches, $\boldsymbol{q}_l$ and $\boldsymbol{r}_j$. As the patches are regular and planar, Equations (4.7)–(4.14), with unknown values and derivatives of the junction functions, are consistent; each of them is a system of *two* scalar equations. By solving these equations, we obtain Hermite interpolation conditions (i.e., the function values and derivatives at 0 and 1), and then we obtain the junction functions as polynomials of interpolation. Some equations are indefinite. By choosing particular solutions we are going to obtain polynomials of low degrees, thus preventing excessive growth of degree of the final patches filling the hole.

Possible orderings of the equations to solve, for the construction of a surface of class G^1 and G^2, are shown in Tables 5.1 and 5.2. The last column indicates either the equation to solve in the current step, or the degree restriction for a particular junction function. For example, in the G^1 construction, first we compute the values of all junction functions at 0 and 1. Then, we choose the functions c_{01} and g_{01} as the unique polynomials of degree at most 1, which take the computed values at 0 and 1. In Step 2 we compute the derivatives of c_{01} and g_{01}, and we substitute them to the equations solved next. In both constructions (G^1 and G^2) the equations solved in subsequent steps are linear with respect to the yet unknown function values and derivatives. The polynomials should be represented in Bernstein bases; finding their coefficients so as to satisfy the interpolation conditions at 0 and 1 and computing their derivatives is easy (if in doubt, see Section A.2.2).

Having the junction functions, we can find the cross-boundary derivatives of the mapping $\boldsymbol{\delta}_i$ for all sides of the unit square using Formulae (4.3) and (4.4). In the G^2 construction we need in addition to compute the second-order cross-boundary derivatives using Formulae (4.5) and (4.6). We do this by applying the algorithms of multiplication of polynomials in Bernstein bases described in Section A.2.3.

In the G^1 construction the degree of the resulting cross-boundary derivatives is equal to 5. We can, therefore, raise the degree of the boundary curves of Ω_i (i.e., of the patch $\boldsymbol{\delta}_i$) to 5 and then find the Bézier representation of the biquintic patch $\boldsymbol{\delta}_i$, defined as the bicubically blended Coons patch (see Section A.9). The second-order cross-boundary derivatives obtained in the G^2 construction are of degree 9. The domain patch in this case is a biquintically blended Coons patch whose degree with respect to both parameters is 9. Examples of the domain patches are shown

Table 5.1: Computing values and derivatives of the junction functions in a G^1 construction

step	compute	using
1	$b_{01}(0)$, $c_{01}(0)$, $f_{01}(0)$, $g_{01}(0)$, $b_{01}(1)$, $c_{01}(1)$, $f_{11}(0)$, $g_{11}(0)$, $b_{11}(0)$, $c_{11}(0)$, $f_{01}(1)$, $g_{01}(1)$, $b_{11}(1)$, $c_{11}(1)$, $f_{11}(1)$, $g_{11}(1)$	Equations (4.7), (4.8)
2	$c'_{01}(0)$, $c'_{01}(1)$, $g'_{01}(0)$, $g'_{01}(1)$	$\deg c_{01} = \deg g_{01} = 1$
3	$b'_{01}(0)$, $f'_{01}(0)$	Equation (4.9)
4	$b'_{01}(1)$, $f'_{01}(1)$,	$\deg b_{01} = \deg f_{01} = 2$
5	$b'_{11}(0)$, $c'_{11}(0)$, $f'_{11}(0)$, $g'_{11}(0)$	Equation (4.9)
6	$c'_{11}(1)$, $g'_{11}(1)$	$\deg c_{11} = \deg g_{11} = 2$
7	$b'_{11}(1)$, $f'_{11}(1)$	Equation (4.9)

Table 5.2: Computing values and derivatives of the junction functions in a G^2 construction

step	compute	using
1	$b_{01}(0), c_{01}(0), f_{01}(0), g_{01}(0),$ $b_{01}(1), c_{01}(1), f_{11}(0), g_{11}(0),$ $b_{11}(0), c_{11}(0), f_{01}(1), g_{01}(1),$ $b_{11}(1), c_{11}(1), f_{11}(1), g_{11}(1)$	Equations (4.7), (4.8)
2	$b_{02}(0), c_{02}(0), f_{02}(0), g_{02}(0),$ $b_{02}(1), c_{02}(1), f_{12}(0), g_{12}(0),$ $b_{12}(0), c_{12}(0), f_{02}(1), g_{02}(1),$ $b_{12}(1), c_{12}(1), f_{12}(1), g_{12}(1)$	Equations (4.10), (4.11)
3	$c'_{01}(0), c'_{01}(1), c''_{01}(0), c''_{01}(1),$ $g'_{01}(0), g'_{01}(1), g''_{01}(0), g''_{01}(1)$	$\deg c_{01} = \deg g_{01} = 1$
4	$b'_{01}(0), f'_{01}(0)$	Equation (4.9)
5	$b'_{01}(1), b''_{01}(0), b''_{01}(1), f'_{01}(1), f''_{01}(0), f''_{01}(1)$	$\deg b_{01} = \deg f_{01} = 2$
6	$b'_{11}(0), c'_{11}(0), f'_{11}(0), g'_{11}(0)$	Equation (4.9)
7	$b'_{02}(0), c'_{02}(0), f'_{02}(0), g'_{02}(0),$ $b'_{12}(0), c'_{12}(0), f'_{12}(0), g'_{12}(0)$	Equations (4.12), (4.13)
8	$c''_{02}(0), c'_{02}(1), c''_{02}(1), g''_{02}(0), g'_{02}(1), g''_{02}(1)$	$\deg c_{02} = \deg g_{02} = 2$
9	$b''_{02}(0), f''_{02}(0)$	Equation (4.14)
10	$b'_{02}(1), b''_{02}(1), f'_{02}(1), f''_{02}(1)$	$\deg b_{02} = \deg f_{02} = 3$
11	$f''_{11}(0), g''_{11}(0), b''_{11}(0), c''_{11}(0)$	Equations (4.12), (4.13)
12	$f''_{12}(0), g''_{12}(0), b''_{12}(0), c''_{12}(0)$	Equation (4.14)
13	$c'_{11}(1), c''_{11}(1), g'_{11}(1), g''_{11}(1)$	$\deg c_{11} = \deg g_{11} = 3$
14	$b'_{11}(1), f'_{11}(1)$	Equation (4.9)
15	$b''_{11}(1), f''_{11}(1)$	$\deg b_{11} = \deg f_{11} = 4$
16	$f'_{12}(1), g'_{12}(1), b'_{12}(1), c'_{12}(1)$	Equations (4.12), (4.13)
17	$c''_{12}(1), g''_{12}(1)$	$\deg c_{12} = \deg g_{12} = 4$
18	$b''_{12}(1), f''_{12}(1)$	Equation (4.14)

(using lines of constant parameters) in Figure 5.11. The difference between the patches for the G^1 and G^2 construction is small; perhaps it is best visible near the corners of the area Ω.

Note that if the domain net has a k-fold rotational symmetry, which is preserved also by the auxiliary domain patches, then the junction functions are the same for all areas Ω_i, and also the set of domain patches is symmetric.

Constructing basis function patches

Consider a Sabin net in $\mathbb{R}^3$ whose projection on $\mathbb{R}^2$, obtained by rejecting the z-coordinate of all vertices, is the domain net. The surface made of bicubic patches represented by this Sabin net

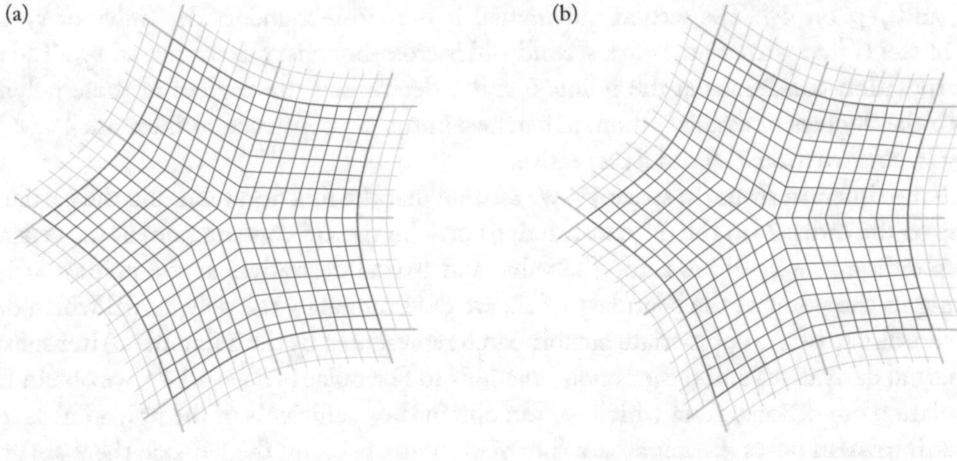

Figure 5.11: Domain patches in the construction of a surface (a) of class G^1 (b) of class G^2

is the graph of a scalar function of class C^2, defined in the area $A \setminus \Omega$. We are going to extend any such function to obtain a function ϕ_j of class C^1 or C^2 in the entire area A. To do this, for each curvilinear quadrangle Ω_i which is an image of the unit square under the mapping $\boldsymbol{\delta}_i$, we construct a scalar function (a bivariate polynomial, denoted by μ in Figure 5.3, and by μ_{ij} if an indication of the area Ω_i is needed) whose domain is the unit square. The extension of the function from $A \setminus \Omega$ to the entire area A is in Ω_i the composition $\mu_{ij} \circ \boldsymbol{\delta}_i^{-1}$.

Actually, we are going to find bases of *two* linear vector spaces whose elements are functions in A. The elements $\hat{\phi}_j$ of a basis of the first space, denoted by V_1, are functions taking non-zero values at the boundary of the area Ω; any such function is related to a Sabin net of radius 2 having only one vertex with the z-coordinate not equal to 0. The Sabin net of radius 2 with the extraordinary vertex incident with k edges (corresponding to a k-sided hole in the surface) has $6k + 1$ vertices. Therefore, we need $6k + 1$ functions, which form a basis of the space V_1; each of them corresponds to a Sabin net having one vertex with the coordinate $z = 1$ and the other vertices in the xy plane. The orthogonal projection of all these Sabin nets on this plane is the domain net.

The second space, V_0, is made of functions taking non-zero values only in the area Ω. This space is needed to construct regular final patches and to optimise their shape. Its dimension depends on the partition of the full angle determined by the halflines tangent to the curves dividing Ω at the central point.

Each basis function ϕ_j, which we are going to construct, is described with k polynomials called **basis function patches,** $\mu_{j0}, \ldots, \mu_{j,k-1}$. In the construction, we need k **auxiliary basis function patches,** $\nu_{j0}, \ldots, \nu_{j,k-1}$, one for each curve dividing the area Ω into the quadrangles Ω_i. The auxiliary basis function patch is represented by two or three polynomials of one variable. The first polynomial determines the values of the function ν_{ji} on the line segment Φ (and the functions

$\mu_{j,i-1}$ and μ_{ji} on Ψ). The second polynomial is the cross-boundary derivative of ν_{ji} and the third, in the G^2 construction, is the second-order cross-boundary derivative of ν_{ji}. The number of interpolation conditions at the points 0 and 1 determines the degrees of these polynomials, equal to the degrees of auxiliary domain patches: in the G^1 construction these are 4 and 3, while in the G^2 construction 7, 6 and 5 respectively.

For a function from the space V_1, we assume that the function value and the partial derivatives up to the order 2 (in the G^1 construction) or 4 (in the G^2 construction) at the central point (corresponding to $s = 0$) are zero. The value and partial derivatives at the point $s = 1$, corresponding to the point at the boundary of Ω, are determined by the polynomial which describes the z-coordinate of the appropriate bicubic patch represented by the Sabin net. After substituting their partial derivatives and the junction functions to Formulae (4.7′)–(4.14′), we obtain Hermite interpolation conditions, from which we can obtain the coefficients of the polynomials in tensor product Bernstein bases. Examples are shown in Figure 5.12; on the left side there are graphs of the functions $\hat{\nu}_{ji} \circ \boldsymbol{\beta}_i^{-1}$ for each side of the areas Ω_i. On the right side there are graphs of the corresponding basis functions of class $C^2(A)$.

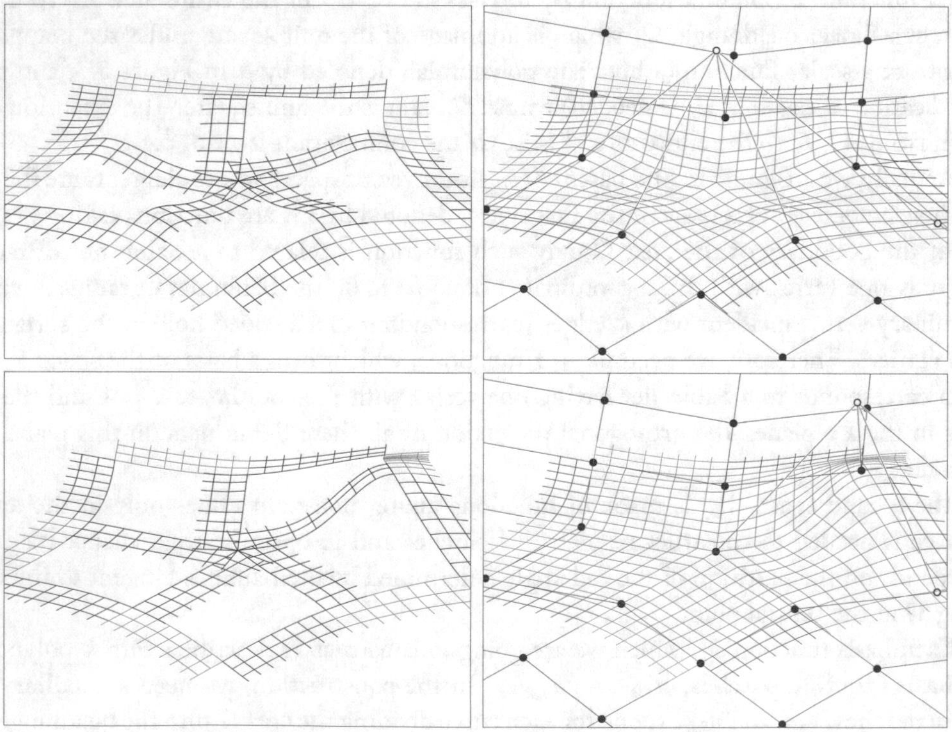

Figure 5.12: Auxiliary basis function patches and corresponding functions from the space V_1

Functions from the space V_0 are equal to 0 at the entire boundary of the area Ω and outside it. The polynomials which describe the values and cross-boundary derivatives of ν_i at $s = 1$ are equal to 0, together with their derivatives of the first (and, in the G^2 construction, second) order. Let Δ be the partition of the full angle determined by the halflines tangent to the curves dividing Ω at the central point (junction halflines, see Section 4.3.2). The basis functions correspond to the elements of bases of the spaces $\mathcal{H}_\Delta^{(0,1)}, \mathcal{H}_\Delta^{(1,1)}, \mathcal{H}_\Delta^{(2,1)}$ in the G^1 construction, or $\mathcal{H}_\Delta^{(0,2)}, \ldots, \mathcal{H}_\Delta^{(4,2)}$ in the G^2 construction.

Elements of bases of these spaces may be constructed in the way described in Section 4.3.2. Recall that each element of these bases is either a homogeneous polynomial of two variables, a truncated homogeneous polynomial (equal to 0 in a halfplane), or it is described by distinct homogeneous polynomials in the cones bounded by the junction halflines. For such an element, say, f_j, we construct k functions ν_{ji} in such a way that the value and partial derivatives of the composition $\nu_{ji} \circ \boldsymbol{\beta}_i^{-1}$ agree with the value and partial derivatives of the bivariate polynomials which describe the function f_j in the two cones adjacent to the i-th junction halfline. Note that to compute the values and derivatives of the polynomials, which represent our auxiliary basis function patch, we used a linear reparametrisation of the function f_j; this is why the curves dividing Ω have the derivatives of order 2 (in the G^1 construction) or of order 2, 3, 4 (in the G^2 construction) equal to $\mathbf{0}$.

Examples of auxiliary basis function patches and the corresponding basis functions of class $C^2(A)$ are shown in Figure 5.13. Example (a) shows a function corresponding to the constant polynomial, equal to 1 in $\mathbb{R}^2$. Example (b) is a function corresponding to $p(x, y) = x$, and (c) corresponds to a piecewise cubic polynomial, constructed using a cubic B-spline function on the line ℓ extended on $\mathbb{R}^2$ and truncated cubic polynomials, as described in Section 4.3.2. As this construction is somewhat troublesome, it is possible and much simpler to construct only the basis functions corresponding to bivariate polynomials, like in the original scheme given by Hahn (Section 4.1).

Having the auxiliary basis function patches and junction functions, we can construct the cross-boundary derivatives of the basis function patches in just the same way as the cross-boundary derivatives of domain patches (except that now we are dealing with functions taking scalar values). The construction ensures that the compatibility conditions required to construct bicubically or biquintically blended Coons patches (see Section A.9) are satisfied. The basis function patches are polynomials of two variables, defined as bicubically or biquintically blended Coons patches. Their degrees are the same as the degrees of the domain patches, i.e., $(5, 5)$ in the G^1 construction and $(9, 9)$ in the G^2 construction.

The degree of basis function patches μ_{ji} determines the degree of the polynomial patches filling the hole, being the result of the entire construction, i.e., $(5, 5)$ or $(9, 9)$. We need to obtain

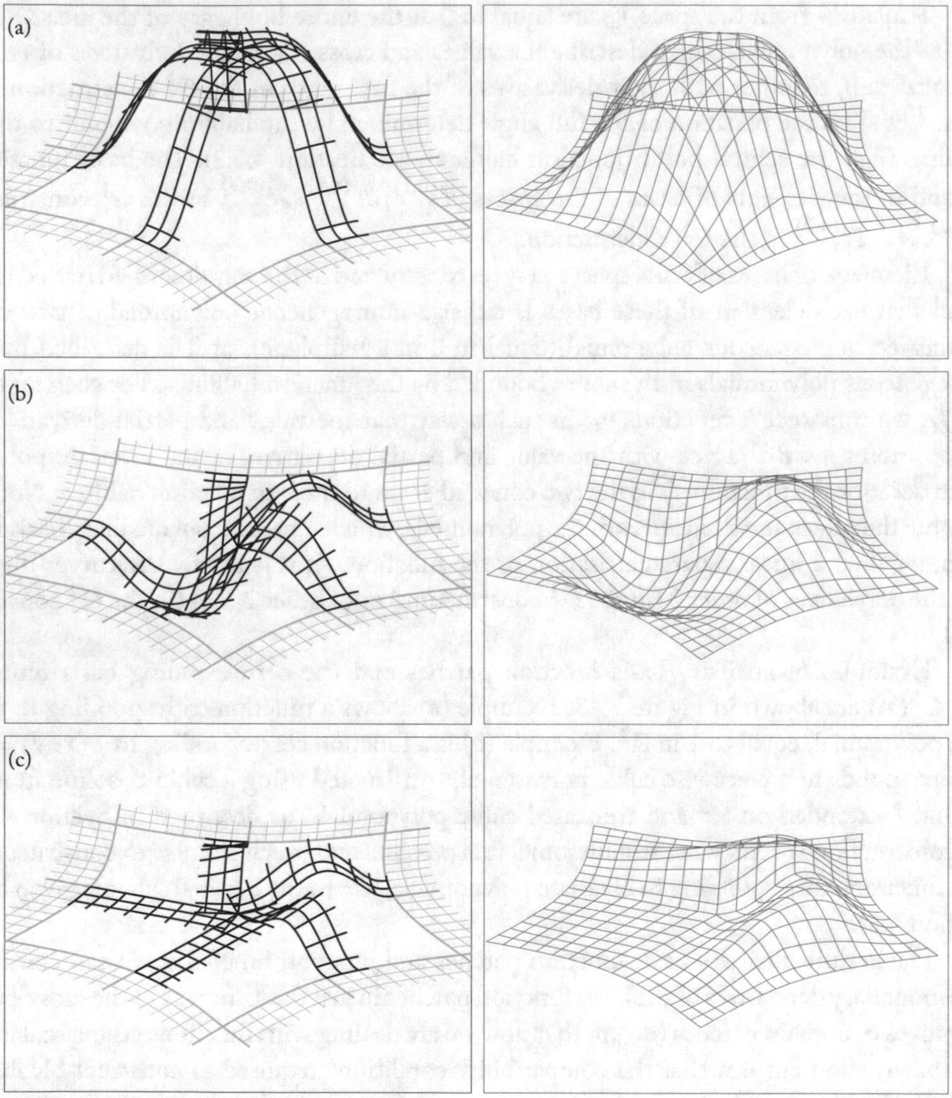

Figure 5.13: Auxiliary basis function patches and corresponding functions from the space V_0

them in Bézier form, i.e., to find the coefficients $b_{ji,pq}$ such that

$$\mu_{ji}(u,v) = \sum_{p=0}^{9}\sum_{q=0}^{9} b_{ji,pq} B_p^9(u) B_q^9(v).$$

A Bézier patch of degree (9, 9) is represented with 100 control points $\boldsymbol{p}_{pq}$, where $p, q \in \{0, \ldots, 9\}$. If $p, q \in \{3, \ldots, 6\}$, then the point $\boldsymbol{p}_{pq}$ does not influence the boundary or the first- and second-order cross-boundary derivative at any side of the unit square (the domain of the Bézier patch). Therefore, it is possible to extend the space V_0, by adding more elements to the basis found so far. The new elements are functions of class $C^2(A)$ equal to zero except for one area Ω_i. A function ϕ_j non-zero in this area is the composition: $\phi_j = \mu_{ji} \circ \boldsymbol{\delta}_i^{-1}$, where μ_{ji} (with the index j depending on i, p and q) is the tensor product of the Bernstein polynomials:

$$\mu_{ji}(u, v) = B_p^9(u) B_q^9(v), \quad p, q \in \{3, \ldots, 6\}.$$

For a k-sided area Ω, there are $16k$ functions of this form. Examples are shown in Figure 5.14.

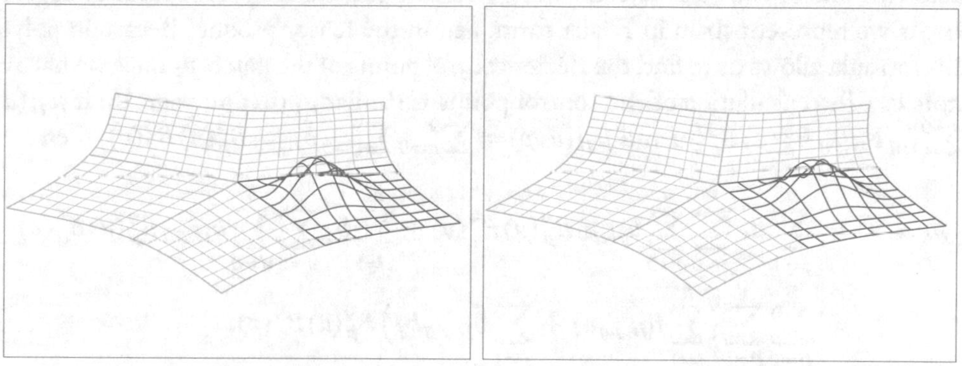

Figure 5.14: Additional basis functions of class C^2

Similarly, considering a Bézier patch of degree (5, 5) we can see that 32 of its 36 control points determine the boundary and the first-order cross-boundary derivatives. The other four control points, $\boldsymbol{p}_{pq}$, where $p, q \in \{2, 3\}$, may be choosen freely. Therefore, in the G^1 construction, we can extend the space V_0 by adding $4k$ elements to the basis; each of them is a function of class $C^1(A)$ non-zero in only one area Ω_i, whose basis function patch μ_{ji} is the tensor product of Bernstein polynomials of degree 5.

Now we have bases of two linear vector spaces, V_1 and V_0, whose elements are functions of class $C^1(A)$ or $C^2(A)$. The dimension of the space V_1 is $m = 6k + 1$; its basis functions, denoted by $\hat{\phi}_1, \ldots, \hat{\phi}_m$, are associated with the vertices $\boldsymbol{b}_1, \ldots, \boldsymbol{b}_m$ of the Sabin net black in Figure 5.1. The dimension of the space $V_0 = \mathrm{lin}\{\phi_1, \ldots, \phi_n\}$ is determined by the desired order of geometric continuity (G^1 or G^2) of the surface filling the hole, by the partition of the full angle introduced with the division of the area Ω and by the inclusion or omission of functions different from 0 in only one subarea Ω_i. The formula

$$\boldsymbol{s} = \sum_{j=1}^{n} \boldsymbol{a}_j \phi_j + \sum_{j=1}^{m} \boldsymbol{b}_j \hat{\phi}_j, \tag{5.5}$$

where $\boldsymbol{a}_j$ are arbitrary vectors in $\mathbb{R}^3$, describes a parametrisation of class $C^1(\Omega)$ or $C^2(\Omega)$ of a surface filling the polygonal hole in the surface. The problem is to choose the vectors $\boldsymbol{a}_j$ so as to obtain a surface of a good shape.

Suggestions what is a good shape are given in the next two sections; now consider the formula

$$\boldsymbol{p}_i(u,v) = \sum_{j=1}^{n} \boldsymbol{a}_j \mu_{ji}(u,v) + \sum_{j=1}^{m} \boldsymbol{b}_j \hat{\mu}_{ji}(u,v), \tag{5.6}$$

obtained by replacing each basis function ϕ_j or $\hat{\phi}_j$ in (5.5) by its basis function patch (μ_{ji} or $\hat{\mu}_{ji}$) associated with one of the areas Ω_i. The basis function patches are polynomials of degree (5, 5) or (9, 9). As we represent them in Bézier form, i.e., in the tensor product Bernstein polynomial basis, this formula allows us to find the Bézier control points of the patch $\boldsymbol{p}_i$ once we have chosen the vectors $\boldsymbol{a}_j$. The calculation of the control points is similar to that on page 18: if $\mu_{ji}(u,v) = \sum_{p=0}^{9}\sum_{q=0}^{9} b_{ji,pq} B_p^9(u) B_q^9(v)$ and $\hat{\mu}_{ji}(u,v) = \sum_{p=0}^{9}\sum_{q=0}^{9} \hat{b}_{ji,pq} B_p^9(u) B_q^9(v)$, then

$$\begin{aligned}\boldsymbol{p}_i(u,v) &= \sum_{j=1}^{n} \boldsymbol{a}_j \sum_{p=0}^{9}\sum_{q=0}^{9} b_{ji,pq} B_p^9(u) B_q^9(v) + \sum_{j=1}^{m} \boldsymbol{b}_j \sum_{p=0}^{9}\sum_{q=0}^{9} \hat{b}_{ji,pq} B_p^9(u) B_q^9(v)\\ &= \sum_{p=0}^{9}\sum_{q=0}^{9}\Bigl(\sum_{j=1}^{n} b_{ji,pq}\boldsymbol{a}_j + \sum_{j=1}^{m} \hat{b}_{ji,pq}\boldsymbol{b}_j\Bigr) B_p^9(u) B_q^9(v);\end{aligned}$$

hence, the control points of the patch $\boldsymbol{p}_i$ are $\boldsymbol{p}_{i,pq} = \sum_{j=1}^{n} b_{ji,pq}\boldsymbol{a}_j + \sum_{j=1}^{m} \hat{b}_{ji,pq}\boldsymbol{b}_j$.

In principle, the vectors $\boldsymbol{a}_j$ may be chosen by "manual" manipulation. However, their number, equal to the dimension of the space V_0, is too big. For example, if $k = 5$ and $h = 0$ (see Theorem 4.1), in the G^1 construction we have

$$\dim V_0 = \dim \mathcal{H}_\Delta^{(0,1)} + \dim \mathcal{H}_\Delta^{(1,1)} + \dim \mathcal{H}_\Delta^{(2,1)} + 4k = 1 + 2 + 5 + 20 = 28,$$

and in the G^2 construction there is

$$\dim V_0 = \dim \mathcal{H}_\Delta^{(0,2)} + \cdots + \dim \mathcal{H}_\Delta^{(4,2)} + 16k = 1 + 2 + 3 + 5 + 10 + 80 = 101.$$

Therefore, the vectors $\boldsymbol{a}_j$ ought to be chosen by an automatic procedure. The task for the computer is to choose the vectors which describe the "optimal" surface (in the set of surfaces represented with the given vectors $\boldsymbol{b}_j$) and then to find the control points of the Bézier patches filling the hole.

A surface filling the hole may be obtained as the minimum of a functional measuring the "badness" of surfaces. There are two possible targets: a good parametrisation or a good shape of the surface. A good parametrisation is described by three scalar functions whose graphs have small undulations; note that it does not guarantee a good shape of the surface (nor even the assumed order of geometric continuity), because such a parametrisation *may* have singularities. On the other hand, in many practical cases this approach yields satisfactory results and is simpler than the optimisation of functionals being explicit measures of the shape badness, increasing with undulations of the surface. Examples of the two approaches are discussed in the next two sections.

Functionals measuring undulations are most often integrals, with integrands defined using partial derivatives of a surface parametrisation. The minima of such functionals satisfy some differential equations called the Euler–Lagrange equations. Here we discuss them only as much as necessary to understand the optimisation problem and the algorithm. Interested readers may refer to textbooks of variational calculus, e.g. Giaquinta and Hildebrandt [1996], to widen and deepen their knowledge of this subject.

5.3 MINIMISATION OF QUADRATIC FORMS

The functionals considered in this section are quadratic forms, and the corresponding differential equations are linear. The optimal surface filling the hole may, therefore, be found by solving a system of algebraic linear equations, which is relatively easy.

Let us recall the notion of a quadratic form in a real linear vector space V. We begin with a **bilinear form**, which is a functional φ whose arguments are two vectors in V, linear with respect to both arguments. A **quadratic form**, say, Φ, is obtained by substituting the same vector for both arguments of a bilinear mapping: $\Phi(f) = \varphi(f, f)$. It is easy to prove that any quadratic form is associated with a unique *symmetric* bilinear form such that $\varphi(f, g) = \varphi(g, f)$ for all $f, g \in V$. A quadratic form Φ is said to be **positive-definite** in the space V, if for all elements f of this space other than the zero vector there is $\Phi(f) > 0$.[4]

We consider a quadratic form Φ (associated with a symmetric bilinear form φ) in the finite-dimensional space V being the direct algebraic sum of the spaces V_0 and V_1. We need to find the vector s (being a function in Ω taking scalar, i.e., real values), minimising the form Φ in the set of functions having the form $s = \sum_{j=1}^{n} a_j\phi_j + \sum_{j=1}^{m} b_j\hat{\phi}_j$ (see Formula 5.5), where the coefficients b_j are given. Let

$$F(a_1, \dots, a_n) \stackrel{\text{def}}{=} \Phi\Bigg(\sum_{j=1}^{n} a_j\phi_j + \sum_{j=1}^{m} b_j\hat{\phi}_j\Bigg).$$

Using the symmetric bilinear form φ, we can rewrite the formula defining F as follows:

$$F(a_1, \dots, a_n) = \sum_{i=1}^{n}\sum_{j=1}^{n} \varphi(\phi_i, \phi_j)a_i a_j + 2\sum_{i=1}^{n}\sum_{j=1}^{m} \varphi(\phi_i, \hat{\phi}_j)a_i b_j + \sum_{i=1}^{m}\sum_{j=1}^{m} \varphi(\hat{\phi}_i, \hat{\phi}_j)b_i b_j,$$

[4]The bilinear form φ is then a scalar product in the space V.

which shows that the function F is a quadratic polynomial of the variables $a_1, \ldots, a_n$. It is easy to find its partial derivatives:

$$\frac{\partial}{\partial a_i} F(a_1, \ldots, a_n) = 2\sum_{j=1}^{n} \varphi(\phi_i, \phi_j)a_j + 2\sum_{j=1}^{m} \varphi(\phi_i, \hat{\phi}_j)b_j.$$

All partial derivatives at the minimal point of F must be zero, which gives us the following system of linear equations:

$$\sum_{j=1}^{n} \varphi(\phi_i, \phi_j)a_j = -\sum_{j=1}^{m} \varphi(\phi_i, \hat{\phi}_j)b_j, \quad i = 1, \ldots, n.$$

We can rewrite it in matrix form

$$A\boldsymbol{a} = -B\boldsymbol{b}, \tag{5.7}$$

where A and B are matrices of dimensions $n \times n$ and $n \times m$, with coefficients $a_{ij} = \varphi(\phi_i, \phi_j)$ and $b_{ij} = \varphi(\phi_i, \hat{\phi}_j)$ respectively, the vector $\boldsymbol{b} \in \mathbb{R}^m$ is made of the coefficients $b_1, \ldots, b_m$, and the vector $\boldsymbol{a} = [a_1, \ldots, a_n]^T$ is unknown.

The symmetric matrix A is positive-definite if and only if the quadratic form Φ is positive-definite in the space V_0 spanned by the basis functions $\phi_1, \ldots, \phi_n$. Then, for any choice of the coefficients b_j, which determine the right-hand side, System (5.7) has a unique solution. It represents the minimum of the function F in $\mathbb{R}^n$, i.e., the minimum of the functional Φ in the set of functions s defined with these coefficients b_j.

Three scalar functions, which together make a parametrisation of the surface filling the hole, may be minimal functions of a quadratic form; each of them may be found independently by solving System (5.7) with a different vector $\boldsymbol{b}$.[5] Now, the question is *how* to choose the quadratic form and how to evaluate the corresponding bilinear form in order to obtain the coefficients of the matrices A and B.

The coefficients b_j determine interpolation conditions for the function s at the boundary of Ω; such interpolation conditions are called **boundary conditions**. The boundary of Ω is a closed curve, which we denote by the symbol Γ. This curve is smooth except for k points—the corners of Ω. The actual form of boundary conditions depends on the desired order of geometric continuity of the surface. A surface filling the hole always has a prescribed boundary; therefore, our boundary conditions always determine the value of the function s at each point of the curve Γ. To obtain a surface of class G^1, we need in addition a tangent plane at each point of the boundary of the hole. It may be determined by the first-order cross-boundary derivative of the parametrisation of the surface around the hole, and then the boundary condition for the function s fixes the

[5]In the construction of a surface in $\mathbb{R}^3$, the quadratic form is minimised three times—the numbers a_j and b_j are respectively the x-, y- and z-coordinates of the vectors $\boldsymbol{a}_j$ and $\boldsymbol{b}_j$ in Formula (5.5).

first-order normal derivative, i.e., the derivative in the direction perpendicular to Γ. Taking into account also the second-order cross-boundary derivative allows us to obtain a surface of class G^2, etc.

The quadratic forms which we are going to use are integral functionals which may be minimised by solving numerically their associated Euler–Lagrange equations. To choose a quadratic form appropriate for the construction of a surface of class G^r it is necessary to consider various elliptic partial differential equations—candidates for the Euler–Lagrange equation—and choose an equation which together with the boundary conditions forms a well-posed boundary problem, i.e., a problem whose solution exists and is unique. The search for an appropriate positive-definite quadratic form is, thus, done by considering boundary problems for partial differential equations. The type of boundary conditions which specify the values of a function and its derivatives up to the order r at Γ has the name of Dirichlet. In general, such boundary conditions may define a well-posed boundary problem for an elliptic equation of order $2r+2$.

Note that the function $s = \sum_{j=1}^{n} a_j\phi_j + \sum_{j=1}^{m} b_j\hat{\phi}_j$ at all points of the curve Γ is equal to the function $\hat{s} = \sum_{j=1}^{m} b_j\hat{\phi}_j$, as all basis functions ϕ_j constructed in Section 5.2 are equal to zero on Γ. Also, the first-order partial derivatives of the functions ϕ_j vanish on Γ, and, therefore, the first-order partial derivatives of the functions s and $\hat{s}$ on the curve Γ, regardless of the coefficients $a_1, \dots, a_n$, are the same. The functions ϕ_j of class $C^2(A)$, constructed in order to obtain surfaces of class G^2, have in addition the second-order derivatives equal to zero at all points of the curve Γ.

First we deal with the G^1 construction. Each of the three scalar functions, which describe the coordinates of a parametrisation $\boldsymbol{s}$ of the surface filling the hole, may be obtained by solving the following boundary problem[6] (with the function p unknown):

$$\begin{cases} \Delta^2 p(x,y) = 0 & \text{for } (x,y) \in \Omega, \\ p(x,y) = \hat{s}(x,y) & \text{for } (x,y) \in \Gamma, \\ \dfrac{\partial p}{\partial \boldsymbol{n}}(x,y) = \dfrac{\partial \hat{s}}{\partial \boldsymbol{n}}(x,y) & \text{for } (x,y) \in \Gamma. \end{cases} \tag{5.8}$$

The following symbols are used above: Ω is the (open) domain, Γ denotes its boundary, and $\boldsymbol{n}$ denotes the unit normal vector of the curve Γ at each of its points except the k corners of Ω, where it is undefined. The derivative in the direction of the vector $\boldsymbol{n}$, $\frac{\partial p}{\partial \boldsymbol{n}}$, is a cross-boundary derivative. The function $\hat{s} = \sum_{j=1}^{m} b_j\hat{\phi}_j$, restricted to the area $A \setminus \Omega$, is a coordinate of a parametrisation of the surface made of the bicubic patches surrounding the hole. The first boundary condition is introduced to obtain the positional continuity, while the second boundary condition ensures the interpolation of the cross-boundary derivative. If a parametrisation $\boldsymbol{s}$ made of functions satisfying these boundary conditions is regular, then it represents a surface filling the hole with G^1 continuity.

[6]Here the symbols x, y denote coordinates in the plane containing Ω, *not* in the three-dimensional space, in which the surface resides.

The symbol Δ is Laplace's operator (often called Laplacian) whose value for a function f of class C^2 is

$$\Delta f(x,y) = \frac{\partial^2 f}{\partial x^2}(x,y) + \frac{\partial^2 f}{\partial y^2}(x,y),$$

and, therefore,

$$\Delta^2 f(x,y) = \frac{\partial^4 f}{\partial x^4}(x,y) + 2\frac{\partial^4 f}{\partial x^2 \partial y^2}(x,y) + \frac{\partial^4 f}{\partial y^4}(x,y)$$

for a function f of class C^4. The equation $\Delta^2 p = 0$, called the **biharmonic equation**, is linear; it is the simplest elliptic partial differential equation of the fourth order and the Euler–Lagrange equation of a functional, which we may want to minimise.

This functional is the quadratic form

$$\Phi(f) \stackrel{\text{def}}{=} \int_\Omega (\Delta f)^2 \,\mathrm{d}\Omega, \tag{5.9}$$

associated with the bilinear form

$$\varphi(f,g) \stackrel{\text{def}}{=} \int_\Omega \Delta f \cdot \Delta g \,\mathrm{d}\Omega. \tag{5.10}$$

To prove it, let a function p^* be the minimal point of the functional Φ in the set of functions satisfying the boundary conditions, and let q be another function whose values and first-order derivatives at Γ are zero. At first we assume that both functions have infinitely many continuous derivatives. Then, the function $F(t) \stackrel{\text{def}}{=} \Phi(p^* + tq)$ has the derivative:

$$\begin{aligned}\frac{\mathrm{d}}{\mathrm{d}t}F(t) &= \frac{\mathrm{d}}{\mathrm{d}t}\int_\Omega \big(\Delta(p^*+tq)\big)^2 \,\mathrm{d}\Omega = \int_\Omega \frac{\mathrm{d}}{\mathrm{d}t}\big((\Delta p^*)^2 + 2t\Delta p^* \cdot \Delta q + t^2(\Delta q)^2\big)\,\mathrm{d}\Omega \\ &= 2\int_\Omega \big(\Delta p^* \cdot \Delta q + t(\Delta q)^2\big)\,\mathrm{d}\Omega = 2\varphi(p^*,q) + 2t\Phi(q).\end{aligned}$$

As the function p^* is the minimal point of the functional Φ, the derivative of F at $t = 0$ must be 0; hence,

$$\varphi(p^*,q) = \int_\Omega \Delta p^* \cdot \Delta q \,\mathrm{d}\Omega = 0. \tag{5.11}$$

Using Green's theorem (see e.g. Spivak [1965]), one can prove the formula[7]

$$\int_\Omega \Delta f \cdot g \,\mathrm{d}\Omega = -\int_\Omega \langle \nabla f, \nabla g\rangle \,\mathrm{d}\Omega + \int_\Gamma \langle \nabla f, \boldsymbol{n}\rangle \,\mathrm{d}\Gamma.$$

[7] Note that $\langle \nabla f, \boldsymbol{n}\rangle = \frac{\partial f}{\partial \boldsymbol{n}}$.

Using it twice to the integral in (5.11), we can obtain the following equation:

$$\int_{\Omega} \Delta^2 p^* \cdot q \,\mathrm{d}\Omega = 0,$$

whose derivation is an exercise for the reader. Note that the integrals over the curve Γ in this calculation are equal to 0 for functions q, whose values and derivatives at Γ are zero. As the last equation must be satisfied for *all* such functions q, there must be $\Delta^2 p^* = 0$, i.e., the function p^* must be a solution of the biharmonic equation in the area Ω.

A reader interested only in the algorithm, or unfamiliar with the functional analysis, may skip this and the next two paragraphs or postpone reading them. They describe, in the most concise way, the concept of a weak solution of our boundary problem, which is approximated by the result of the numerical computations. On the other hand, interested readers may refer to the literature about differential equations, e.g. Evans [1998], or about the numerical methods of solving them, e.g. Ciarlet [1978].

Equation (5.11) makes sense also for functions p^* and q whose first-order partial derivatives are continuous in Ω and whose second-order derivatives are limited and *piecewise* continuous, with the points of discontinuity located on a finite number of smooth curves in Ω. Note that the functions of class $C^1(\Omega)$, whose construction is described in Section 5.2, have this property. Going one step further, we can consider an infinite-dimensional linear vector space whose elements are *all* functions with this property. We denote it by the symbol $\tilde{H}^2(\Omega)$ and its subspace, whose elements are functions vanishing together with the first-order derivatives at the boundary curve Γ, will be denoted by $\tilde{H}_0^2(\Omega)$. The bilinear form φ given by (5.10) is a scalar product in this subspace. Thus, $\tilde{H}_0^2(\Omega)$ is a linear vector space equipped with the norm

$$\|f\|_{\tilde{H}_0^2} = \sqrt{\Phi(f)} = \left(\int_{\Omega} (\Delta f)^2 \,\mathrm{d}\Omega\right)^{1/2}.$$

This space is *not* closed, i.e., it does not contain limits of some Cauchy sequences of its elements. Its closure, denoted by $H_0^2(\Omega)$, is an example of a Sobolev space (see Evans [1998]).

The quadratic form Φ used to define the norm in $H_0^2(\Omega)$ is positive-definite in this space and, as a consequence, the form Φ has a unique minimal point in the subset of functions satisfying the boundary conditions being the starting point of these considerations. The minimal point is not necessarily a function of class $C^4(\Omega)$; note that the boundary conditions determined by the piecewise bicubic surface around the hole usually enforce discontinuities of the third-order derivatives in a neighbourhood of some points of the curve Γ. As the minimal point does not have to satisfy the biharmonic equation in the classical sense, it is called a **weak solution** of the boundary problem (5.8). The minimal element of the quadratic form Φ in the coset of the finite-dimensional space $V = V_0 \oplus V_1$, whose elements satisfy the boundary condition, is an approximation of this weak solution.

To interpret the functional Φ given by Formula (5.9) as a measure of the badness, assume that the surface with the hole is represented by a Sabin net whose projection on the xy plane in $\mathbb{R}^3$ is the domain net and Ω is the projection of the hole in the surface. Let the surface filling the hole be the graph of a scalar function $p(x, y)$. By taking $u = x$ and $v = y$, we obtain the obvious parametrisation of this surface, made of the functions $x(u, v) = u$, $y(u, v) = v$, $z(u, v) = p(u, v) = p(x, y)$. The coefficients of matrices of its first and second fundamental form (see Section A.10.2) are

$$g_{11} = 1 + p_x^2, \quad g_{12} = p_x p_y, \quad g_{22} = 1 + p_y^2$$
$$b_{11} = \frac{p_{xx}}{\sqrt{\det G}}, \quad b_{12} = \frac{p_{xy}}{\sqrt{\det G}}, \quad b_{22} = \frac{p_{yy}}{\sqrt{\det G}},$$

where $\det G = 1 + p_x^2 + p_y^2$. By substituting these coefficients to (A.52), we obtain the formula which describes the mean curvature of the graph of the function p:

$$H = \frac{p_{xx}(1 + p_y^2) + p_{yy}(1 + p_x^2) - 2p_x p_y p_{xy}}{2(1 + p_x^2 + p_y^2)^{3/2}}. \tag{5.12}$$

Looking at this formula, we notice that if at a certain point (x, y) there is $p_x = p_y = 0$, then the Laplacian of p is twice greater than the mean curvature of the surface at the point $(x, y, p(x, y))$: $\Delta p = p_{xx} + p_{yy} = 2H$. If the partial derivatives of the function p are close to zero, then the Laplacian of p approximates the mean curvature of its graph multiplied by 2. Thus, in principle, the functional Φ increases with the growth of the mean curvature of the graph. Its minima correspond to functions whose graphs have the mean curvature small, which is the case when the curvature is not changing rapidly because of undulations. Note also that a planar surface has the mean curvature equal to zero, and the functional Φ is equal to 0 for all functions whose graphs are planar.

Before solving the system of equations (5.7) one has to compute the coefficients of the matrices A and B. These coefficients are integrals,

$$a_{jl} = \varphi(\phi_j, \phi_l) = \int_\Omega \Delta\phi_j \cdot \Delta\phi_l \, \mathrm{d}\Omega, \quad b_{jl} = \varphi(\phi_j, \hat{\phi}_l) = \int_\Omega \Delta\phi_j \cdot \Delta\hat{\phi}_l \, \mathrm{d}\Omega,$$

and the only practical way of doing this step of the construction is to use a quadrature. Note that an integral over the area Ω is the sum of integrals over the curvilinear quadrangles $\Omega_0, \dots, \Omega_{k-1}$. The quadrangle Ω_i is the image of the unit square under the mapping $\boldsymbol{\delta}_i$, which we named the domain patch. Therefore, we can use the formula

$$\int_{\Omega_i} f(x, y) \, \mathrm{d}\Omega = \int_0^1 \int_0^1 f\big(\boldsymbol{\delta}_i(u, v)\big) J_i(u, v) \, \mathrm{d}u \, \mathrm{d}v,$$

where the integrand f is the product of $\Delta\phi_i$ and $\Delta\phi_j$ or $\Delta\hat{\phi}_j$, and $J_i(u, v)$ is the Jacobian of the mapping $\boldsymbol{\delta}_i$, i.e., the absolute value of the determinant of the 2×2 matrix $\big[\frac{\partial \boldsymbol{\delta}_i}{\partial u}, \frac{\partial \boldsymbol{\delta}_i}{\partial v}\big]$. The simplest

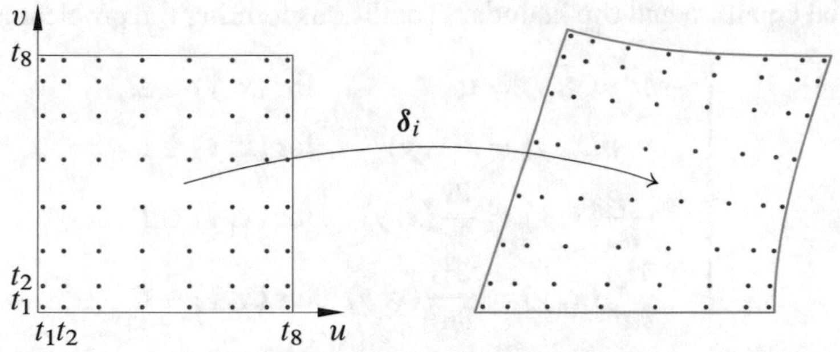

Figure 5.15: Knots of a quadrature to evaluate integrals in one of the areas Ω_i; $t_1, \dots, t_8$ are knots of the Gauss–Legendre quadrature of order 16 in the interval [0, 1]

way to evaluate the last integral is using a quadrature given by the formula[8]

$$\int_{\Omega_i} f(x, y)\,\mathrm{d}\Omega \approx \sum_{p=1}^{n} A_p \sum_{q=1}^{n} A_q f\big(\boldsymbol{\delta}_i(t_p, t_q)\big) J_i(t_p, t_q),$$

with the **quadrature knots** $t_1, \dots, t_n \in [0, 1]$ and **coefficients** $A_1, \dots, A_n$ chosen so as to obtain a sufficient accuracy. Figure 5.15 shows an example of a set of points (t_p, t_q) in the unit square and the points $\boldsymbol{\delta}_i(t_p, t_q)$ in the curvilinear quadrangle Ω_i, at which the integrand f is to be evaluated. More information on quadratures may be found in textbooks of numerical methods, e.g. Kincaid and Cheney [2002].

Note that the domain patch $\boldsymbol{\delta}_i$ is a polynomial mapping and basis function patches μ_{ji} and $\hat{\mu}_{ji}$ are polynomials. To evaluate the expressions $f\big(\boldsymbol{\delta}_i(t_p, t_q)\big)$, which are equal to $\Delta\phi_j \cdot \Delta\phi_l$ or $\Delta\phi_j \cdot \Delta\hat{\phi}_l$, we need the derivatives of the first and second order of the basis functions, e.g. $\phi_j = \mu_{ji} \circ \boldsymbol{\delta}_i^{-1}$. We can obtain them by computing the derivatives of the polynomials at (t_p, t_q) and then solving the systems of linear equations resulting from the generalised Fàa di Bruno's formula, as described in Section A.11.

A surface filling the hole with G^2 continuity may be constructed in a similar way, but this time we have an additional boundary condition: the filling surface must interpolate also the cross-boundary derivative of the second order. A well-posed problem with the additional boundary condition is obtained with a differential equation of order 6. As before, we need to construct three scalar functions which describe the coordinates of a parametrisation $\boldsymbol{s}$ of the filling surface. The

[8]By Fubini's theorem (see Spivak [1965]), an integral over a multidimensional domain may be expressed with one-dimensional integrals. The formula, called the "iterated integral", is converted directly into the "iterated quadrature" proposed here. If the dimension of the integrand domain is high, then this approach fails because of the computational complexity, but it works well for two-dimensional rectangles.

partial differential equation and the boundary conditions forming the problem are the following:

$$\begin{cases} -\Delta^3 p(x,y) = 0 & \text{for } (x,y) \in \Omega, \\ p(x,y) = \hat{s}(x,y) & \text{for } (x,y) \in \Gamma, \\ \dfrac{\partial p}{\partial \boldsymbol{n}}(x,y) = \dfrac{\partial \hat{s}}{\partial \boldsymbol{n}}(x,y) & \text{for } (x,y) \in \Gamma, \\ \dfrac{\partial^2 p}{\partial \boldsymbol{n}^2}(x,y) = \dfrac{\partial^2 \hat{s}}{\partial \boldsymbol{n}^2}(x,y) & \text{for } (x,y) \in \Gamma. \end{cases} \tag{5.13}$$

The triharmonic equation, $-\Delta^3 p = 0$, is the Euler–Lagrange equation of the functional (quadratic form) given by the formula

$$\Phi(f) \stackrel{\text{def}}{=} \int_\Omega \|\nabla\Delta f\|_2^2 \, \mathrm{d}\Omega, \tag{5.14}$$

where

$$\|\nabla\Delta f\|_2^2 = \left(\frac{\partial}{\partial x}\Delta f\right)^2 + \left(\frac{\partial}{\partial y}\Delta f\right)^2 = \left(\frac{\partial^3 f}{\partial x^3} + \frac{\partial^3 f}{\partial x \partial y^2}\right)^2 + \left(\frac{\partial^3 f}{\partial x^2 \partial y} + \frac{\partial^3 f}{\partial y^3}\right)^2.$$

The bilinear form φ such that $\Phi(f) = \varphi(f,f)$ is

$$\varphi(f,g) = \int_\Omega \langle \nabla\Delta f, \nabla\Delta g\rangle \, \mathrm{d}\Omega, \tag{5.15}$$

with the integrand $\langle \nabla\Delta f, \nabla\Delta g\rangle = (\frac{\partial}{\partial x}\Delta f \cdot \frac{\partial}{\partial x}\Delta g) + (\frac{\partial}{\partial y}\Delta f \cdot \frac{\partial}{\partial y}\Delta g)$. The connection between the boundary problem (5.13) and the quadratic form Φ given by (5.14) may be proved in a similar way as the connection between the boundary problem (5.8) and the quadratic form given by (5.9).

The evaluation of the bilinear form Φ given by (5.15) (or, rather, its approximation, with the integral replaced by a quadrature) may be done in a way similar to the one described with the G^1 construction. This time we have to evaluate the third-order derivatives of the compositions $\mu_{ij} \circ \boldsymbol{\delta}_i^{-1}$, which is also possible by solving systems of linear equations derived from the generalised Fàa di Bruno's formula.

An interpretation of the functional defined by Formula (5.14) as a measure of the surface badness is the following: as the Laplacian of a function approximates (up to a constant factor) the mean curvature of the graph of this function, its gradient grows with the rate of changes of the mean curvature. Thus, the minimisation of this functional gives us a function with a graph whose mean curvature is distributed evenly. Note that if the surface with a hole to be filled were a sphere or a cylinder, i.e., a surface with a constant mean curvature, then the perfect surface to fill the hole would be a part of this sphere or cylinder—a viewer would be completely unable to tell where the hole was. The minimisation of the quadratic form associated with the triharmonic equation gives us an approximation of such perfection.

Note that in both cases, G^1 and G^2, the result depends on the choice of the domain net considered in Section 5.2, and it may be poor if the domain net is poorly chosen. Nevertheless, filling surfaces obtained by the minimisation of quadratic forms are good enough in many cases, and computer procedures producing such surfaces (despite the complicated formulae inside) are very fast.

5.4 CONSTRUCTIONS WITH SHAPE OPTIMISATION

Solving numerically linear elliptic equations, e.g. biharmonic or triharmonic, produces a surface whose *parametrisation* has the minimal badness; if its shape is good (which is often the case), then it is only a side effect. It is, however, possible to search for minima of functionals being explicit measures of the *shape* badness. This approach is more troublesome, because to find a minimum one has to solve a system of nonlinear equations.

Functionals whose values measure the badness of surface shape have been studied since the 1990s; see e.g. Greiner [1994] or Sarraga [1998]. Such functionals are integrals with the integrands defined using the surface curvatures (principal, Gaussian or mean). A detailed study of shape optimisation exceeds the scope of this book, and only the two simplest functionals are briefly discussed here.

To construct a surface of class G^1, we can use the functional

$$\Psi(\mathcal{M}) = \int_{\mathcal{M}} H^2 \,\mathrm{d}\mathcal{M}. \tag{5.16}$$

The symbol H denotes the mean curvature of the surface $\mathcal{M}$. Note that if this surface is a graph of a function f, whose first-order partial derivatives are close to 0, then the functional Ψ is approximated by the functional Φ given by (5.9) multiplied by the constant factor 1/4. One can expect the mean curvature of surfaces obtained by the minimisation of the functional Ψ in the set of surfaces whose boundary is fixed together with the tangent plane at each point of the boundary to be rather small and evenly distributed.

Formula (5.16) has been obtained from (5.9) by replacing the Laplacian of a function f with the mean curvature of the surface, and the integral with respect to the (Lebesgue) measure in the domain Ω by the integral with respect to the measure on the surface $\mathcal{M}$. An analogous modification of Formula (5.14), needed in the construction of a surface of class G^2, requires finding a substitute of the "ordinary" gradient with an operator acting on functions defined on a surface.

What is the gradient of a function? The simplest answer, "the vector made of partial derivatives", is only a part of the truth. The gradient of a function f of n variables at a certain point of the function's domain is a linear functional whose argument is a vector in $\mathbb{R}^n$ and whose value is the directional derivative of f in the direction of that vector. Such a functional may indeed be represented by the vector whose coordinates are partial derivatives of f, and then the derivative of the function f in the direction of a vector $\boldsymbol{x} \in \mathbb{R}^n$ is the scalar product of the two vectors. However, it is more natural to represent the gradient by a row matrix, e.g. $\nabla f = [\frac{\partial f}{\partial x}, \frac{\partial f}{\partial y}]$, because the directional derivative may be obtained by matrix multiplication, $\frac{\partial f}{\partial \boldsymbol{x}} = (\nabla f)\boldsymbol{x}$. The second norm of the gradient (the square root of the sum of squares of the partial derivatives) is the steepest ascent (or descent), i.e., the maximal absolute value of the directional derivative of f in the direction of a unit vector.

To define the **gradient on a surface,** we consider a function $\boldsymbol{p}\colon \Omega \to \mathbb{R}^3$ of class C^1 which is a regular one-to-one parametrisation of the surface $\mathcal{M}$, another function $f\colon \Omega \to \mathbb{R}$ of class C^1, and the composition $h = f \circ \boldsymbol{p}^{-1}$, which is a function defined on the surface $\mathcal{M}$. Let $(u, v) \in \Omega$. The 3×2 matrix $[\boldsymbol{p}_u(u, v), \boldsymbol{p}_v(u, v)]$ represents the differential of the parametrisation $\boldsymbol{p}$ at (u, v), denoted by $\mathrm{D}\boldsymbol{p}|_{(u,v)}$ (for brevity below we write $[\boldsymbol{p}_u, \boldsymbol{p}_v]$ and $\mathrm{D}\boldsymbol{p}$). The differential of $\boldsymbol{p}$ is a linear mapping, which maps any vector in $\mathbb{R}^2$ to a vector in the tangent space of the surface $\mathcal{M}$, i.e., the two-dimensional subspace $T_{\mathcal{M}}$ of $\mathbb{R}^3$ parallel to the plane tangent to the surface $\mathcal{M}$ at the point $\boldsymbol{p}(u, v)$ (see Section A.10.2). By differentiating the equality $f = h \circ \boldsymbol{p}$, we obtain the formula

$$\nabla f = \nabla_{\mathcal{M}} h \circ \mathrm{D}\boldsymbol{p},$$

with a mapping $\nabla_{\mathcal{M}} h$. As the mappings ∇f and $\mathrm{D}\boldsymbol{p}$ are linear, so must be $\nabla_{\mathcal{M}} h$. The argument of the mapping $\nabla_{\mathcal{M}} h$ at a point $\boldsymbol{p}(u, v) \in \mathcal{M}$ is a vector from the subspace $T_{\mathcal{M}}$. The value of this mapping is the directional derivative of the function h. To represent $\nabla_{\mathcal{M}}$, we need a linear functional in $\mathbb{R}^3$, but there exist infinitely many linear functionals satisfying the formula above; this is because the matrix $[\boldsymbol{p}_u, \boldsymbol{p}_v]$ is not square, so it cannot be inverted. To find the proper linear functional, we can use the pseudoinverse,[9] traditionally denoted by $[\boldsymbol{p}_u, \boldsymbol{p}_v]^+$. Due to the regularity of $\boldsymbol{p}$, the matrix $[\boldsymbol{p}_u, \boldsymbol{p}_v]$ is of full rank. The 2×2 matrix $G = [\boldsymbol{p}_u, \boldsymbol{p}_v]^T[\boldsymbol{p}_u, \boldsymbol{p}_v]$ represents the first fundamental form of the parametrisation $\boldsymbol{p}$; it is non-singular, and the pseudoinverse of the matrix $[\boldsymbol{p}_u, \boldsymbol{p}_v]$ is

$$[\boldsymbol{p}_u, \boldsymbol{p}_v]^+ = \left([\boldsymbol{p}_u, \boldsymbol{p}_v]^T[\boldsymbol{p}_u, \boldsymbol{p}_v]\right)^{-1}[\boldsymbol{p}_u, \boldsymbol{p}_v]^T = G^{-1}[\boldsymbol{p}_u, \boldsymbol{p}_v]^T.$$

This allows us to write the formula

$$\nabla_{\mathcal{M}} h = (\nabla f)[\boldsymbol{p}_u, \boldsymbol{p}_v]^+ = (\nabla f)G^{-1}[\boldsymbol{p}_u, \boldsymbol{p}_v]^T,$$

which describes the 1×3 matrix $\nabla_{\mathcal{M}} h$ representing the gradient on the surface using the partial derivatives of the function f and the parametrisation $\boldsymbol{p}$. In fact, this matrix represents a linear functional in the space $\mathbb{R}^3$ whose value for any vector orthogonal to the subspace $T_{\mathcal{M}}$ is zero. Of all 1×3 matrices representing the linear functionals in $\mathbb{R}^3$ whose restrictions to $T_{\mathcal{M}}$ are the same,

[9]A pseudoinverse of a matrix A is a matrix A^+ such that $AA^+A = A$, $A^+AA^+ = A^+$, $A^+A = (A^+A)^T$ and $AA^+ = (AA^+)^T$. For any matrix $A \in \mathbb{R}^{m \times n}$, there exists a unique pseudoinverse matrix $A^+ \in \mathbb{R}^{n \times m}$. The formula $A^+ = (A^TA)^{-1}A^T$, used here, is appropriate for matrices A having linearly independent columns.

this matrix has the smallest second norm. Further (rather simple) calculation gives us the formula

$$\|\nabla_{\mathcal{M}} h\|_2^2 = (\nabla_{\mathcal{M}} h)(\nabla_{\mathcal{M}} h)^T = (\nabla f)G^{-1}(\nabla f)^T, \tag{5.17}$$

which we use below.

By replacing Δf in Formula (5.14) by the mean curvature H, the gradient in Ω by the gradient on the surface $\mathcal{M}$ and the integration over Ω by the integration over the surface $\mathcal{M}$ with respect to the measure on this surface, we obtain the functional

$$\Psi(\mathcal{M}) = \int_{\mathcal{M}} \|\nabla_{\mathcal{M}} H\|_2^2 \,\mathrm{d}\mathcal{M}, \tag{5.18}$$

which may be used as a measure of the badness of the shape in constructions of surfaces of class G^2, having a prescribed boundary and tangent plane and curvature at each point of the boundary. As the mean curvature depends on the derivatives of the parametrisation $\boldsymbol{s}$ of the first and second order, the integrand in (5.18) is an expression with derivatives of $\boldsymbol{s}$ up to the order three.

Though to evaluate the functionals defined with Formulae (5.16) and (5.18) we need a paramctrisation and its derivatives, the functionals are parametrisation-independent, i.e., using any (sufficiently smooth) parametrisation of a given surface $\mathcal{M}$ must result in obtaining the same functional value. It is a desirable property, as the functional value reflects the quality of the shape and nothing else. On the other hand, it is a source of trouble in numerical computations. Any surface has infinitely many parametrisations, and we are going to find a parametrisation of the optimal surface. Such a task is ill-posed. There are two practical ways to overcome this difficulty.

The first possibility is to modify the functional by adding a term measuring the badness of the parametrisation. The idea (not developed here) is to look for an optimal (in some sense) parametrisation of the surface with the optimal shape (i.e., one with the smallest shape badness measure). The badness of a parametrisation may be defined as a measure of undulations of its constant line parameters. This approach is not only useful in the problem of filling the hole, but it also works when the shape of the entire surface is to be optimised by repositioning vertices of the mesh representing the surface (see Kiciak [2013]).

The second approach is to assume that the surface filling the hole is the graph of a scalar function. The surface has to be represented in such a system of coordinates that a part of this surface around the hole is indeed the graph of a scalar function; this may be done by an appropriate rotation of the mesh representing the surface. Then, we can construct a surface filling the hole by minimising one of the functionals (quadratic forms) considered in the previous section. Having three scalar functions, $x(u,v)$, $y(u,v)$ and $z(u,v)$, which are minimal points of the quadratic form and describe a parametrisation $\boldsymbol{s}$, we fix the first two. The surface with the optimal shape is obtained by a modification of the function z. Let $\boldsymbol{q}$ be a function obtained from $\boldsymbol{s}$ by rejecting the third coordinate, i.e., described by the functions x, y. If the function $\boldsymbol{q}$ is one-to-one and regular, then we can define the function $p(x,y) = z\big(\boldsymbol{q}^{-1}(x,y)\big)$, whose graph fills the hole in our surface.

The domain of the function p is the area $\tilde{\Omega}$ obtained by projecting the hole on the plane xy; the function $\boldsymbol{q}$ maps the area Ω, the domain of the functions ϕ_j and $\hat{\phi}_j$ in Formula (5.5), to $\tilde{\Omega}$. The integral in Formula (5.16) may be transformed as follows:

$$\int_{\mathcal{M}} H^2 \,\mathrm{d}\mathcal{M} = \int_{\tilde{\Omega}} H^2 \sqrt{\det G}\,\mathrm{d}\tilde{\Omega}. \tag{5.19}$$

The mean curvature H of the graph of the function p, as a function of the variables x, y being arguments of p, is described by Formula (5.12). The factor $\sqrt{\det G}$, where G is the matrix of the first fundamental form, is the Jacobian of the transformation of variables of integration. Note that the symbol H on the opposite sides of the equality sign above denotes *two* functions, defined on the surface $\mathcal{M}$ and in the area $\tilde{\Omega}$ respectively; both functions take the same values being the mean curvature of the surface. I hope that this ambiguity is not confusing the readers who got so far.

A similar transformation may be applied to the integral in Formula (5.18), which defines a functional appropriate in constructions of surfaces of class G^2. Formula (5.17) allows us to express the gradient of the mean curvature on the surface using the ordinary gradient of the mean curvature seen as a function in $\tilde{\Omega}$. Using it, we obtain

$$\int_{\mathcal{M}} \|\nabla_{\mathcal{M}} H\|_2^2 \,\mathrm{d}\mathcal{M} = \int_{\tilde{\Omega}} (\nabla H)G^{-1}(\nabla H)^T \sqrt{\det G}\,\mathrm{d}\tilde{\Omega}. \tag{5.20}$$

How to numerically find minima of functionals such as the two above? The surface has a parametrisation described by Formula (5.5), with the vectors $\boldsymbol{b}_j$ (positions of vertices of the Sabin net in $\mathbb{R}^3$) given; the problem is to find the vectors $\boldsymbol{a}_j$. The minimisation of the quadratic form defined by Formula (5.9) or (5.14) gives us their x- and y-coordinates. Now we define a function F whose arguments are the coordinates $z_1, \dots, z_n$ of the vectors $\boldsymbol{a}_1, \dots, \boldsymbol{a}_n$. The value of this function is the value of the functional defined by Formula (5.16) or (5.18). The function F takes its (locally) minimal value at any point in $\mathbb{R}^n$ at which the gradient of F vanishes and the Hessian ($n \times n$ matrix of the second-order partial derivatives) is positive-definite.

The minimisation of a functional by searching a zero of the gradient of its discretisation, i.e., a function with a finite number of variables representing the argument of the functional, is known as the **Ritz method**. It is done iteratively, by constructing a sequence of points $\boldsymbol{x}^{(l)} \in \mathbb{R}^n$ which converge to the solution. If $\boldsymbol{g}^{(l)}$ is a vector in $\mathbb{R}^n$ whose coordinates are the first-order partial derivatives of the function F at the point $\boldsymbol{x}^{(l)}$, then the sequence of vectors $\boldsymbol{g}^{(l)}$ must converge to $\boldsymbol{0}$.

Partial derivatives of a function F given by

$$F(z_1, \dots, z_n) = \int_{\tilde{\Omega}} f(z_1, \dots, z_n)\,\mathrm{d}\tilde{\Omega}$$

are described by the formula

$$\frac{\partial F}{\partial z_j}(z_1, \dots, z_n) = \int_{\tilde{\Omega}} \frac{\partial f}{\partial z_j}(z_1, \dots, z_n)\,\mathrm{d}\tilde{\Omega}.$$

To evaluate the integrals we can use a quadrature in the way considered later in this section.

There are numerical methods of solving systems of nonlinear equations, which need to evaluate only the vectors $\boldsymbol{g}^{(l)}$ corresponding to the points $\boldsymbol{x}^{(l)}$. On the other hand, going one step further and computing the Hessian matrices $H^{(l)}$, whose coefficients are

$$\frac{\partial^2 F}{\partial z_i \partial z_j}(z_1, \ldots, z_n) = \int_{\tilde{\Omega}} \frac{\partial^2 f}{\partial z_i \partial z_j}(z_1, \ldots, z_n)\, \mathrm{d}\tilde{\Omega},$$

makes it possible to use Newton's method, which generates subsequent points as follows:

$$\boldsymbol{x}^{(l+1)} = \boldsymbol{x}^{(l)} + \boldsymbol{\delta}, \quad \text{where } H^{(l)}\boldsymbol{\delta} = -\boldsymbol{g}^{(l)}.$$

Each iteration involves solving a system of linear equations with the Hessian matrix $H^{(l)}$, which is symmetric and in a vicinity of the minimal point of F also positive-definite. For the coordinates of the initial point $\boldsymbol{x}^{(0)}$ usually one can take the z-coordinates of the vectors $\boldsymbol{a}_j$ found by the minimisation of the quadratic form. If one of the matrices $H^{(l)}$ is not positive-definite, then the initial point is insufficient for Newton's method. This may happen when the surface represented by this point has big undulations.

The integrals over $\tilde{\Omega}$ have to be approximated by quadratures similar to those described in Section 5.3. Note that the initial parametrisation $\boldsymbol{s}^{(0)}$ obtained by the minimisation of a quadratic form and the surfaces $\mathcal{M}^{(l)}$ resulting from consecutive iterations of the numerical minimisation method are made of k polynomial (Bézier) patches obtained using Formula (5.6). The patch $\boldsymbol{p}_i^{(l)}$ is associated with the area Ω_i being a part of the domain Ω of the parametrisations $\boldsymbol{s}^{(l)}$, which are described piecewise by the compositions $\boldsymbol{p}_i^{(l)} \circ \boldsymbol{\delta}_i^{-1}$. By rejecting the z-coordinate of the patch $\boldsymbol{p}_i^{(l)}$, we obtain a mapping $\tilde{\boldsymbol{\delta}}_i$, which is *the same* for all l; the domain of $\tilde{\boldsymbol{\delta}}_i$ is the unit square and the image is the curvilinear quadrangle $\tilde{\Omega}_i$, which is a part of the area $\tilde{\Omega}$. The surface $\mathcal{M}^{(l)}$ represented by the parametrisation $\boldsymbol{s}^{(l)}$ is the graph of a scalar function $p^{(l)}$. To evaluate the integrals considered here, we replace the areas Ω_i by $\tilde{\Omega}_i$ and the mappings $\boldsymbol{\delta}_i$ by $\tilde{\boldsymbol{\delta}}_i$, and we evaluate the partial derivatives of $p^{(l)}$, which appear in the expressions for the integrands. The quadrature coefficients and knots in the unit square may be chosen in the same way as in Section 5.3 (see Figures 5.15 and 5.16).

In a computer program we need at least a procedure evaluating the function F to minimise, as its partial derivatives may be approximated with divided differences. On the other hand, it is possible to evaluate the partial derivatives of the integrand in the definition of the function F. The simple formulae which define the functions to minimise hide quite complicated expressions. Even more complicated (and, therefore, omitted here) expressions describe their partial derivatives. Fortunately, computers deal perfectly with such expressions—one can derive the formulae using a symbolic package and then code them in procedures of evaluating the function F and its gradient and Hessian. The results are worth the effort.

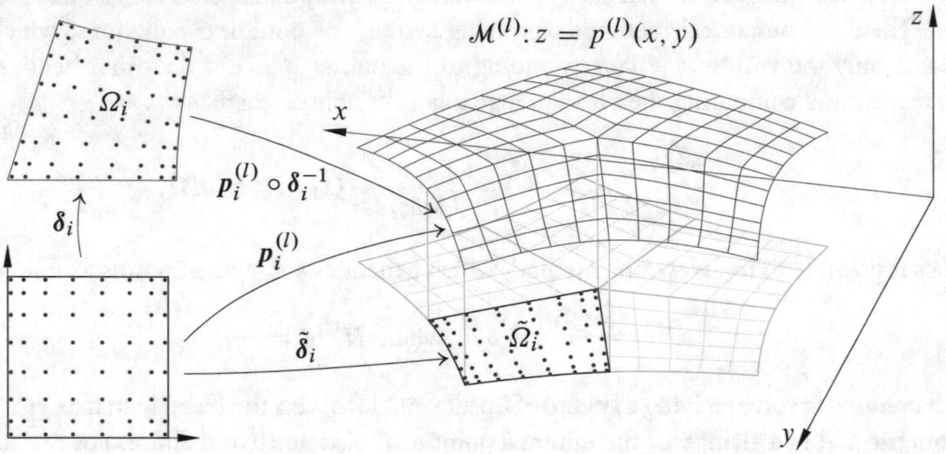

Figure 5.16: Quadrature knots for evaluation of the functionals measuring the shape quality

5.5 CONCLUSION

The example in Figure 5.17 shows a surface made of bicubic patches with four pentagonal holes, which have been filled with patches of degree (9, 9) so as to obtain a surface of class G^2. The patches filling the holes were obtained by the minimisation of the quadratic form given by Formula (5.14). It seems that distinguishing the surfaces filling the holes from the bicubic patches by looking at a rendered image is a very tough job. To find the holes one needs a more powerful visualisation tool.

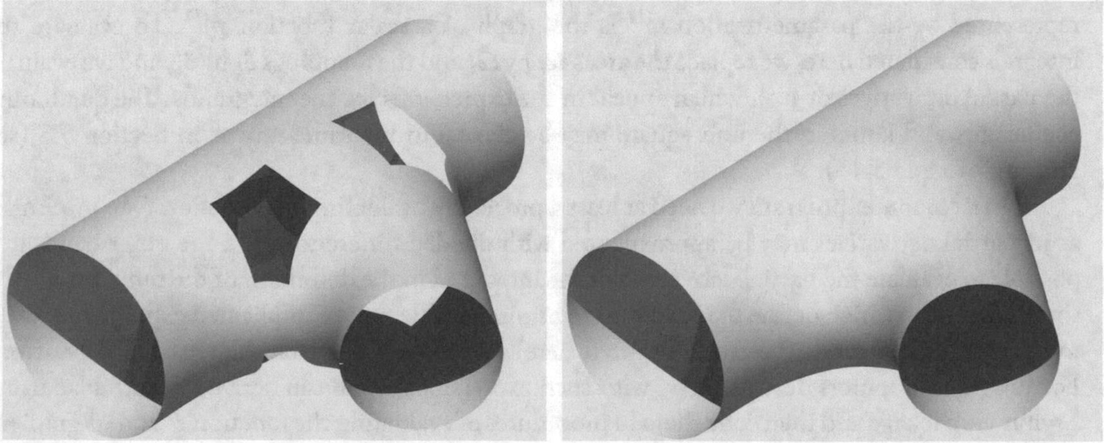

Figure 5.17: A surface with holes filled with G^2 continuity

To visualise the details of the shapes, the pictures illustrating the next example show the distribution of the mean curvature, which has been mapped to the colours of the surface points. Figure 5.18 shows four different surfaces filling a hole in the same surface; the pictures in the top row show the filling surfaces of class G^1, made of biquintic patches, while in the bottom row we can see surfaces of class G^2, made of patches of degree $(9, 9)$. The surfaces on the left side were obtained by the minimisation of quadratic forms, and the surfaces on the right side are minimal points of the functionals described in Section 5.4. It turns out that all surfaces are close to each other and if a part of the surface around the hole is flat enough, then the surface of class G^1 obtained by the minimisation of a quadratic form may be satisfactory in many practical designs. On the other hand, one may need a surface of class G^2 in a demanding application. The price to pay is the higher degree of the patches filling the surface.

The examples of constructions in this chapter did not exhaust the subject (perhaps as opposed to many readers). The functionals considered in the previous two sections are natural measures of the badness of the surfaces filling holes, but other functionals might be more appropriate in particular applications. It is possible to use them as optimisation criteria for B-spline surfaces made of many polynomial pieces or surfaces of arbitrary topology, represented by meshes. However, functionals, whose minimisation suppresses undulations of the surfaces filling (rather small) polygonal holes, may be inappropriate if a complicated surface is to be constructed by a numerical optimisation algorithm. This subject is beyond the scope of this book, but it is worth using this opportunity to take one more look at the functionals considered in this chapter.

One can notice that one of the elements of the space V being the algebraic sum of the spaces V_0 and V_1, whose bases were constructed in Section 5.2, is the constant function, equal to 1 in the area Ω. This function is obviously a solution of the biharmonic and triharmonic equation, and this observation may be used to prove that the relation between the Sabin net representing the part of the surface with a hole and the surface filling the hole, which has been obtained by the minimisation of the quadratic form given by (5.9) or (5.14), is an invariant of all affine transformations. The functionals defined in Section 5.4 using the mean curvature *do not* have this property. The relation between the surface with the hole and the optimal surface filling the hole is only an invariant of isometries and homotetiae. After applying a non-uniform scaling to the surface and filling the hole, we obtain a different result than the image (under the same scaling) of the surface filling the hole before the transformation.

The constructions based on the minimisation of the quadratic forms are also more robust; note that even when all vectors $\boldsymbol{b}_j$ in Formula (5.5) are the same, i.e., the relevant part of the mesh representing the surface is degenerated to a point, the system of linear equations (5.7) still has a unique solution, which admittedly represents a degenerated filling surface, being the best possible result in this case. The functionals depending on the surface shape are, then, undefined. A conclusion may be drawn that explicit shape optimisation is worth the effort if the surface with the hole is beautiful and we want to fill the hole perfectly.

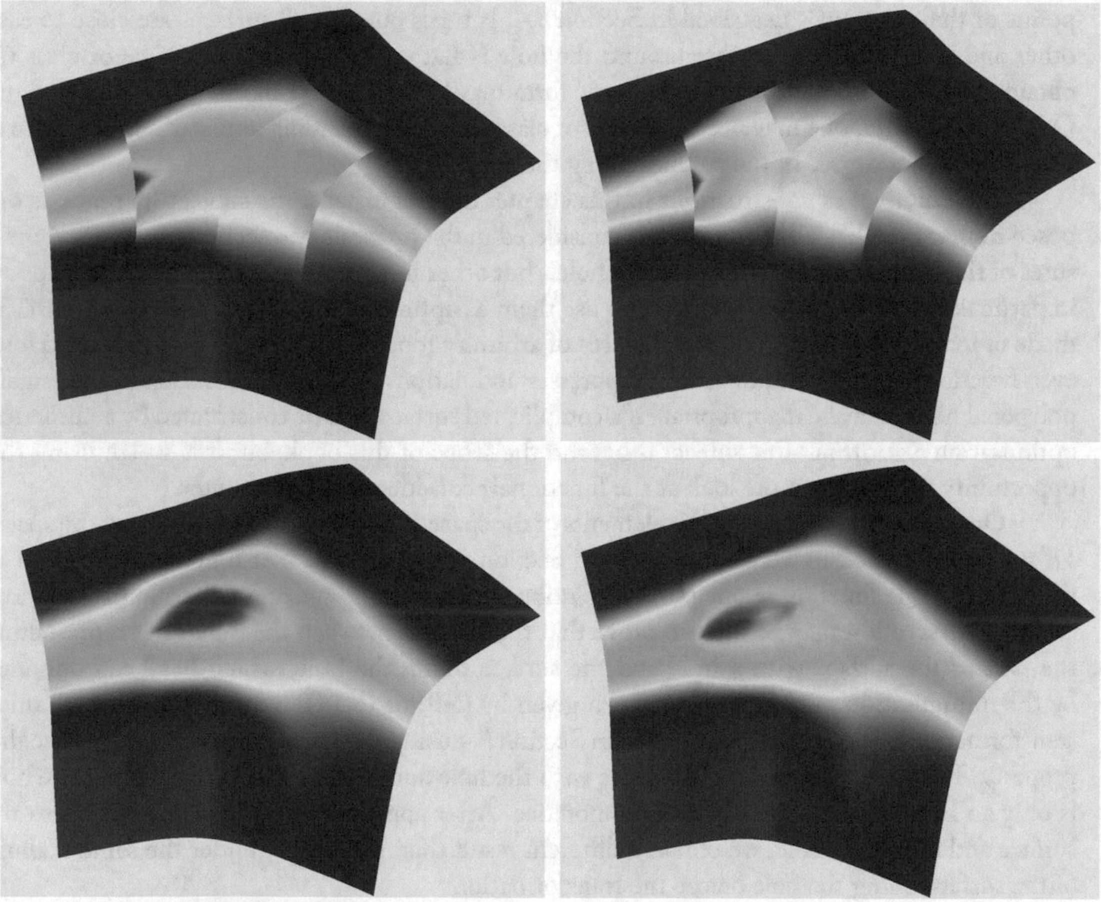

Figure 5.18: Surfaces filling a hole with G^1 and G^2 continuity (the colours correspond to the mean curvature)

CHAPTER 6

Images of surface shape

Using a computer to make an image of a surface, one can be willing to achieve one of the following three goals:

- an image showing a smoother surface, or
- an image showing the surface exactly as it is, or
- an image magnifying the surface details making it non-smooth.

In computer graphics, the first two situations occur more often than the last one. Having a geometric model, we usually want an exact image, or we may want to make the surface look smoother than it is, when our geometric model consists of a number of planar polygons which approximate a smooth surface. The process of "smoothing" images of polygons is known as shading.

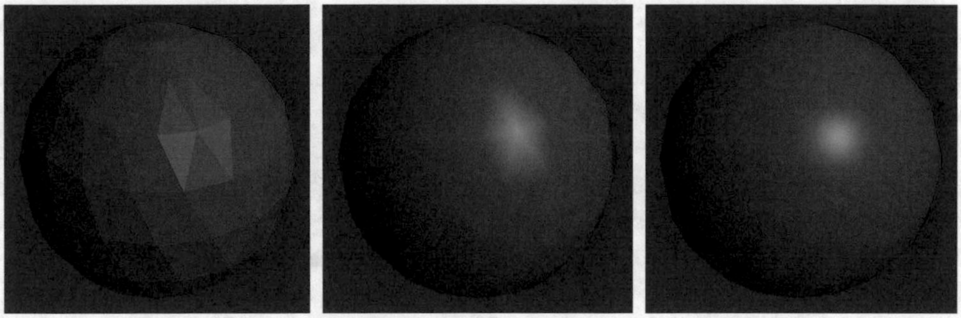

Figure 6.1: Faking a smooth surface: no shading, Gouraud shading and Phong shading

Figure 6.1 shows an example; the three pictures show the triangular facets of a polyhedron inscribed in a sphere, rendered with the intention of reproducing the sphere. Without any shading we can clearly see that the object is a polyhedron. As the lighting model describes a glossy surface, the effect of the Gouraud shading (a linear interpolation of the colour among the vertices of each triangle) is rather unnatural, giving a polygonal highlight. However, that the image obtained with the Phong shading (linear interpolation of the normal vector and then using a lighting model to compute the colours of pixels) shows a polyhedron, one can tell only by looking at the silhouette of the object; the silhouette is a polygon.

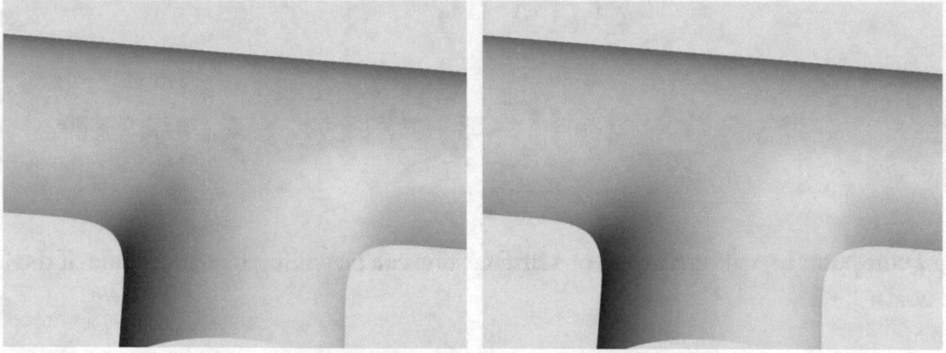

Figure 6.2: Surfaces of class G^1 and G^2 rendered with a lighting model

As human eyes are so easily deceived by shading, it turns out that realistic images, like the ones in Figure 6.2, are insufficient when one needs to see the actual shape of the surface. This is the third situation, when any details (or defects) of a surface which at first glance seems smooth have to be exposed. When the surface of an object must be subject to careful evaluation (e.g. before manufacturing the object in a million copies), one can use various visualisation tools available in CAD systems. In this chapter we discuss some of these tools.

6.1 CHARACTERISTIC LINES AND SHAPE FUNCTIONS

The simplest method of visualisation of a parametric surface is to draw a number of its constant parameter curves. However, from such an image one can get only a general idea of the shape of the surface; the image says more about the parametrisation than about the surface shape and after changing the parametrisation we can obtain an image of a completely different net of curves on the surface, whose shape has not changed at all. Moreover, when patches that are parts of a surface have a common boundary curve, their constant parameter curves may have non-smooth junctions, even if the junction of the patches is very smooth (see Figure 6.3). This is why when we are interested in showing the shape of a surface, we want parametrisation-independent images. Our goal is to obtain the same image of a given surface, regardless of the parametrisation used to represent it.

We are going to explore the idea that the shape of curves on the surface image may carry information about the shape of the surface. Suppose that we have a parametrisation $\boldsymbol{b}$ of a curve $\mathcal{B}$ in the domain A of a parametrisation $\boldsymbol{p}$ of our surface $\mathcal{M}$. The composition $\boldsymbol{c} = \boldsymbol{p} \circ \boldsymbol{b}$ is a parametrisation of a curve $\mathcal{C}$ located on the surface $\mathcal{M}$. If the parametrisations $\boldsymbol{p}$ and $\boldsymbol{b}$ were of class C^n and C^k respectively, and if the class of (analytic) continuity of the parametrisation $\boldsymbol{c}$ was visible on an image, then we could draw conclusions about the class of (analytic) continuity of the parametrisation $\boldsymbol{p}$. Analytic properties of parametrisations (except for obvious discontinuities) are of course invisible on the curve images, but we can see the shape of a curve and, in particular, we

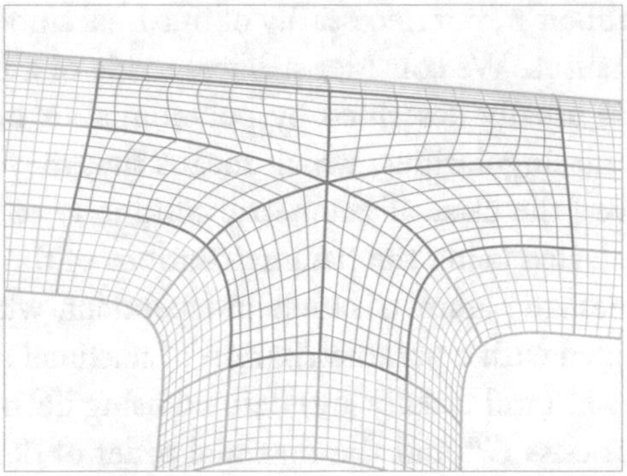

Figure 6.3: Constant parameter curves of patches making a smooth surface

can often recognise its class of geometric continuity (though sometimes it is not easy). To draw conclusions about the surface, we introduce the following definition:

Definition 6.1 *If a surface is of class G^n and not G^{n+1}, than any regular parametrisation of class C^n of (a part of) this surface will be called a* ***parametrisation of maximal smoothness****.*

Now suppose that the function $\boldsymbol{b}$ is a regular parametrisation of class C^k of a curve $\mathcal{B}$ in the domain of a parametrisation $\boldsymbol{p}$ of the surface $\mathcal{M}$ of class G^n, and $\boldsymbol{p}$ is a parametrisation of maximal smoothness. Then the composition $\boldsymbol{c} = \boldsymbol{p} \circ \boldsymbol{b}$ is a regular parametrisation of class $C^{\min\{n,k\}}$ of a curve $\mathcal{C}$ located on our surface. The class of geometric continuity of the curve $\mathcal{C}$ is $G^{\min\{n,k\}}$. If, seeing the image, we can tell that the order of geometric continuity of the curve $\mathcal{C}$ is $m < k$, then we conclude that the surface $\mathcal{M}$ is of class *at most* G^m. This possibility does not seem very attractive, because in general curves of class G^2 or even G^1 are seen as smooth. However, we can make much more of it, if the classes of geometric continuity of the surface $\mathcal{M}$ and the curve $\mathcal{B}$ are related due to the specific definition of the curve.

Instead of constructing parametric curves in the domain of a parametrisation of the surface, we are going to define them implicitly. Consider a scalar function $f \in C^k(A)$ (where $k \geq 1$), which takes a fixed value c at some points of A and such that its gradient in a neighbourhood of these points is non-zero. Let $f(u_0, v_0) = c$. Then, by the implicit function theorem, there exists an interval I and a parametrisation $\boldsymbol{b}\colon I \to A$ of class C^k of a curve $\mathcal{B}$ passing through the point (u_0, v_0), which is an isoline of the function f.

Definition 6.2 *Let $\boldsymbol{p}\colon A \to \mathbb{R}^3$ be a parametrisation of a surface $\mathcal{M}$. Let $f\colon A \to \mathbb{R}$ be a function such that for all $(u, v) \in A$ the value of f is determined by the shape of the surface patch $\boldsymbol{p}(\mathcal{N})$, where $\mathcal{N} \subset A$ is a neighbourhood of the point (u, v) (and, thus, the patch is a piece of the surface $\mathcal{M}$ with the point $\boldsymbol{p}(u, v)$). Any function having this property will be called a* ***shape function****. A curve obtained by mapping an isoline of a shape function in the domain A on the surface $\mathcal{M}$ by the parametrisation $\boldsymbol{p}$, will be called a* ***characteristic line*** *on the surface.*

Having a parametrisation $\boldsymbol{p}$, not necessarily of maximal smoothness, we can define various shape functions in its domain A. We consider surfaces made of a number of parametric patches whose parametrisations are usually described by polynomial or rational functions. Most often the parametrisations have no singularities, which means that they are of class C^∞. Our images are in fact supposed to reveal the class of geometric continuity of junctions of the patches. The shape functions will be evaluated using the parametrisations of the patches and their derivatives. However, characteristic lines are parametrisation-independent, which means that their shape is the same as if it were obtained with a parametrisation of maximal smoothness.

Shape functions *may* be (and usually are) defined using derivatives of the parametrisation. If the parametrisation is of class C^n and the maximal order of derivatives used is l, then (usually) the shape function will be of class C^{n-l} and, after mapping its isolines on the surface, we obtain characteristic lines of class G^{n-l}. Consequently, if we see that a characteristic line is only of class G^m (and *we have checked* that the gradient of the shape function is non-zero), then we can conclude that the surface is of class at most G^{m+l}. The number l will be called the **geometric discontinuity order magnification index** (or the magnification index for short) of the shape function.

The values of shape functions considered here are scalars, which ought to be mapped to the colours on the image. The mapping used to do it will be called a **palette**. For example, the value of a shape function might be mapped to the brightness, while the hue and saturation are fixed arbitrarily. Another possibility is to map the shape function value to the hue of a paint, and use a lighting model to obtain a "realistic" image of the painted surface. If the palette is a continuous function, then any discontinuities of the shape function should be visible on the image. On the other hand, a picture rendered with a discontinuous palette will show the characteristic lines on the surface. Palettes of both kinds were used to render the pictures in this chapter with the intention of showing qualitative images for the most popular shape functions. The continuous palette maps the entire range of the shape function, from the minimal to the maximal value, on the hue (going through the entire rainbow). The discontinuous palette is chosen to show the characteristic lines with a reasonable density for each picture.

Though, by definition, shape functions are intended to produce images independent of the way surfaces are parametrised, they may depend on "external" parameters, like those of the lighting model. The rest of this chapter discusses examples of the most popular shape functions.

6.2 PLANAR SECTIONS

The simplest shape function is defined using the formula

$$f(u, v) = \langle \boldsymbol{v}, \boldsymbol{p}(u, v) - \boldsymbol{o} \rangle.$$

The vector $\boldsymbol{v} \neq \boldsymbol{0}$ and the point $\boldsymbol{o}$, which may be arbitrarily chosen, represent a plane. The set of zeros of the function f, mapped by the parametrisation on the surface $\mathcal{M}$, is a planar section of this surface. Moreover, each isoline of the function f is a planar section of the surface. With a fixed vector $\boldsymbol{v}$ we obtain in this way the sections of the surface with a family of parallel planes.

Where the plane is not tangent to the surface $\mathcal{M}$ of class G^n, the common curve of the surface and the plane is of class G^n. The magnification index of the planar section curves is 0. In particular, the curves on the surfaces of class G^1 and G^2 in Figure 6.4 look almost the same. The discontinuities of the curvature of the curves on the right image are practically invisible, and to expose the curvature discontinuity of the surface we need a sharper tool.

Figure 6.4: Planar sections

6.3 ISOPHOTES

Let $\boldsymbol{n}\colon A \to S^2 \subset \mathbb{R}^3$ denote the Gauss map (see Section A.10.2) for the parametrisation $\boldsymbol{p}$, i.e., $\boldsymbol{n}(u,v)$ is a unit normal vector of the surface $\mathcal{M}$ at the point $\boldsymbol{p}(u,v)$, for all points (u,v) of the domain A. The unit normal vector may be obtained by computing the vector product of the partial derivatives of the parametrisation:

$$\boldsymbol{n}(u,v) = \frac{\boldsymbol{m}(u,v)}{\|\boldsymbol{m}(u,v)\|_2}, \quad \text{where} \quad \boldsymbol{m}(u,v) = \frac{\partial \boldsymbol{p}}{\partial u}(u,v) \wedge \frac{\partial \boldsymbol{p}}{\partial v}(u,v).$$

Let $\boldsymbol{l}$ be an arbitrary unit vector. The function given by the formula

$$f(u,v) = \langle \boldsymbol{l}, \boldsymbol{n}(u,v) \rangle$$

is a shape function. Its value at (u, v) is the cosine of the angle between the vector $\boldsymbol{l}$ and the normal vector of the surface at $\boldsymbol{p}(u, v)$. If the surface is perfectly matt and it is illuminated with the light coming only from the direction $\boldsymbol{l}$, then the brightness of any point of this surface is proportional to $\max\{0, f(u, v)\}$—this is the core of the surface lighting model studied by Johann Heinrich Lambert in 1760. Isolines of our shape function are curves of constant brightness on the surface; these characteristic lines are called **isophotes**.

If $\boldsymbol{p}$ is a regular parametrisation of class C^n, then the Gauss map, defined with the first-order derivatives of $\boldsymbol{p}$, is a mapping of class C^{n-1} and, therefore, also the function f is of class C^{n-1}. Hence, the magnification index for the shape function used to obtain the isophotes is equal to 1. On the left side of Figure 6.5 we can see isophotes with points of discontinuity of the tangent line, while the isophotes on the right side are smooth. Thus, isophotes are an appropriate tool to distinguish surfaces of classes G^1 and G^2.

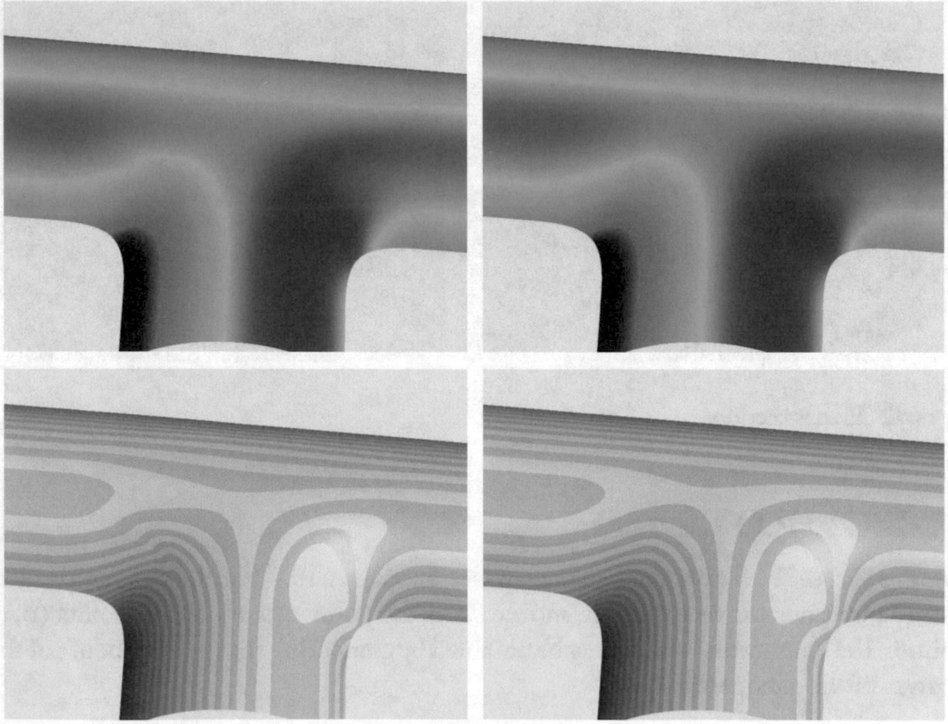

Figure 6.5: Isophotes

6.4 REFLECTION LINES

We can imagine a surface $\mathcal{M}$ as a curved perfect mirror. Let ℓ be a fixed line in the space. We can describe it in parametric form: $\ell = \{\boldsymbol{a} + t\boldsymbol{t} : t \in \mathbb{R}\}$. A viewer whose eye is positioned at a point $\boldsymbol{e}$ may see the reflection of the line ℓ in this mirror. The set of points of the surface $\mathcal{M}$, at which the viewer sees the reflection of points of ℓ is, hardly surprisingly, called the **reflection line** (Klass [1980]). Is there a shape function, whose isolines correspond to reflection lines? And how to construct it?

At first we find a function $\tilde{f}$ which determines just one reflection line. For a point (u, v) of the domain A, we define the vectors

$$\boldsymbol{v} = \boldsymbol{e} - \boldsymbol{p}, \quad \boldsymbol{r} = 2\langle \boldsymbol{n}, \boldsymbol{v}\rangle \boldsymbol{n} - \boldsymbol{v}, \quad \text{and} \quad \boldsymbol{s} = \boldsymbol{a} - \boldsymbol{p},$$

where $\boldsymbol{n} = \boldsymbol{n}(u, v)$ is a unit normal vector of the surface $\mathcal{M}$ at the point $\boldsymbol{p} = \boldsymbol{p}(u, v)$. The vector $\boldsymbol{r}$ is the image of $-\boldsymbol{v}$ reflected in the plane tangent to the surface at $\boldsymbol{p}$. Any light coming to the point $\boldsymbol{p}$ from the direction of $\boldsymbol{r}$ will be reflected by the mirror towards the point $\boldsymbol{e}$ (Figure 6.6).

If the point $\boldsymbol{p}(u, v)$ belongs to the reflection line, the vectors $\boldsymbol{r}$, $\boldsymbol{s}$ and $\boldsymbol{t}$ are linearly dependent, i.e., $\det[\boldsymbol{r}, \boldsymbol{s}, \boldsymbol{t}] = 0$. Therefore, the reflection line corresponds to the set of zeros of the function

$$\tilde{f}(u, v) = \det[\boldsymbol{r}, \boldsymbol{s}, \boldsymbol{t}].$$

It is a shape function; its other isolines do not determine reflection lines but some other curves, whose interpretation does not seem to be very clear.

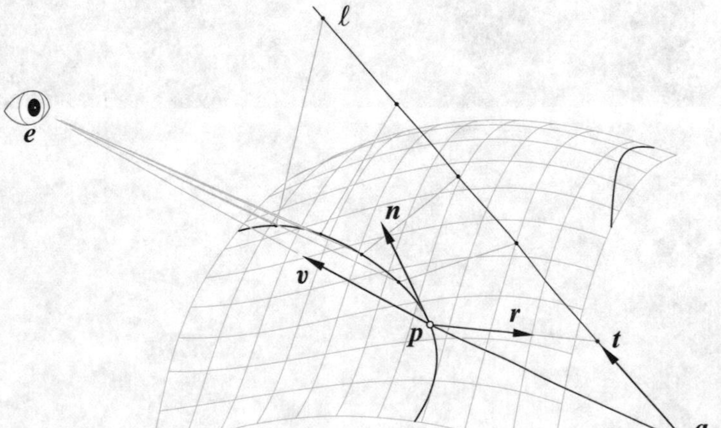

Figure 6.6: A reflection line

Before we proceed, let us note that the linear dependence of the vectors $\boldsymbol{r}$, $\boldsymbol{s}$ and $\boldsymbol{t}$ is equivalent to the linear dependence of $-\boldsymbol{r}$, $\boldsymbol{s}$ and $\boldsymbol{t}$. Our geometric condition may, therefore, lead to obtaining curves not being reflection lines in the physical sense; this "reflection" would cause the light to pass through the mirror and go in the opposite direction (a curve corresponding to such a "reflection" appears also on the surface shown in Figure 6.6).

Now our goal is to construct a shape function whose each isoline defines either a reflection line, or a "pseudoreflection line", as discussed above. These curves will be determined by an infinite family of parallel coplanar straight lines. Having the parametric representation of a plane, $\pi = \{\, \boldsymbol{a} + s\boldsymbol{u} + t\boldsymbol{t} \colon s, t \in \mathbb{R} \,\}$, where $\boldsymbol{a}$ is a point and $\boldsymbol{u}$ and $\boldsymbol{t}$ are linearly independent vectors, we can use the parameter s to specify the line, passing through the point $\boldsymbol{a} + s\boldsymbol{u}$ and having the direction of the vector $\boldsymbol{t}$. The equality

$$\boldsymbol{p} + r\boldsymbol{r} = \boldsymbol{a} + s\boldsymbol{u} + t\boldsymbol{t}$$

may be rewritten as a system of linear equations in the form $A\boldsymbol{x} = \boldsymbol{s}$, where

$$A = [\boldsymbol{r}, -\boldsymbol{u}, -\boldsymbol{t}], \quad \boldsymbol{x} = [r, s, t]^T \quad \text{and} \quad \boldsymbol{s} = \boldsymbol{a} - \boldsymbol{p}.$$

Figure 6.7: Reflection lines

Figure 6.8: Reflection lines outdoors

To evaluate the shape function f, we set up and solve this system[1] and we take $f(u, v) = s$.

The Gauss map, used to define the functions $\tilde{f}$ and f, is expressed with the first-order derivatives of $\boldsymbol{p}$. We can prove, by a straightforward calculation, that the magnification index in this case is 1. Therefore, reflection lines, as well as isophotes, are an appropriate tool to reveal the discontinuities of surface curvature. On a surface of class G^1, which is not of class G^2, these characteristic lines are continuous, but they have points of discontinuity of the tangent line that are visible on images.

Reflection lines were used to evaluate designs by car stylists even before the computer era. A wooden or clay prototype model painted with a glossy lacquer was placed in a room with fluorescent tubes collocated on the walls and the ceiling, and then examined from various points. How important a part of the design process it is, and how much it reveals about the surface shape, one can see in a simple outdoor experiment (Figure 6.8).

[1]The shape function f constructed here might be written explicitly using the well known Cramer formula. I am *not* using it, no way, because in general using determinants in numerical computations, for a number of reasons, is indeed a poor idea. In computer programs we need reliable numerical algorithms, like the Gaussian elimination, which is much better even for such small systems of linear equations.

6.5 HIGHLIGHT LINES

Characteristic lines similar to reflection lines, named **highlight lines** (Beier and Chen [1994]), are independent of the viewer's position. This property makes it possible to take a closer look at such a line or examine its shape from a number of points—changes of the reflection lines caused by moving the viewer might be undesirable, and fixing a viewer position to define reflection lines and looking at them from another point is unnatural.

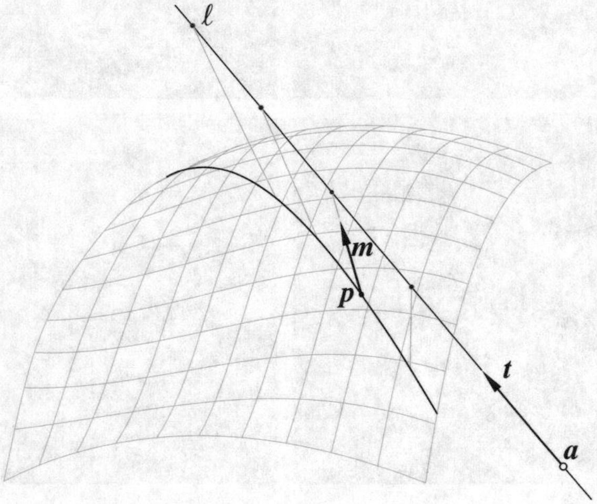

Figure 6.9: A highlight line

A highlight line on a surface $\mathcal{M}$ is defined with a line ℓ in the space; a point $\boldsymbol{p}(u, v)$ is a point of the highlight line if the normal line of the surface at this point intersects ℓ.[2] A shape function, whose set of zeros is the highlight line corresponding to the line ℓ given by a point $\boldsymbol{a}$ and a vector $\boldsymbol{t}$, may be written using the formula

$$\tilde{f}(u, v) = \det[\boldsymbol{m}, \boldsymbol{s}, \boldsymbol{t}],$$

where $\boldsymbol{m}$ is a (not necessarily unit) normal vector of the surface and $\boldsymbol{s} = \boldsymbol{a} - \boldsymbol{p}(u, v)$.

As in the case of reflection lines, isolines of the function $\tilde{f}$ other than the set of zeros do not determine highlight lines. Having a family of parallel coplanar lines, we can define a shape function f, whose every isoline determines a highlight line. The equality

$$\boldsymbol{p} + r\boldsymbol{m} = \boldsymbol{a} + s\boldsymbol{u} + t\boldsymbol{t}$$

is a system of linear equations $A\boldsymbol{x} = \boldsymbol{s}$, where

$$A = [\boldsymbol{m}, -\boldsymbol{u}, -\boldsymbol{t}],$$

[2] A highlight line is a special case of a reflection line, with the viewer position $\boldsymbol{e}$ located on the line ℓ.

Figure 6.10: Highlight lines

and the unknown vector $\boldsymbol{x} = [r, s, t]^T$. After solving this system, we take $f(u, v) = s$. Due to the fact that the normal vector $\boldsymbol{m}$ is expressed with the first-order partial derivatives of the parametrisation, the magnification index of this shape function is 1. The discontinuities of the tangent line of the highlight lines in Figure 6.10 indicate discontinuities of curvature of the surface.

6.6 SURFACE CURVATURES

The mean curvature and the Gaussian curvature (see Section A.10.2) are "ready to use" shape functions carrying plenty of information about the surface shape. They may be evaluated using Formulae (A.52) and (A.53). A consequence of the presence of the second-order derivatives of the parametrisation in the formulae for the coefficients of matrices of the second fundamental form, used to obtain the curvatures, is that the magnification index of the two shape functions is 2.

Images of the mean and Gaussian curvatures rendered with a continuous palette reveal discontinuities of the surface curvature very clearly. The characteristic lines—isolines of the mean or Gaussian curvature, visible on the pictures obtained with a discontinuous palette—are bro-

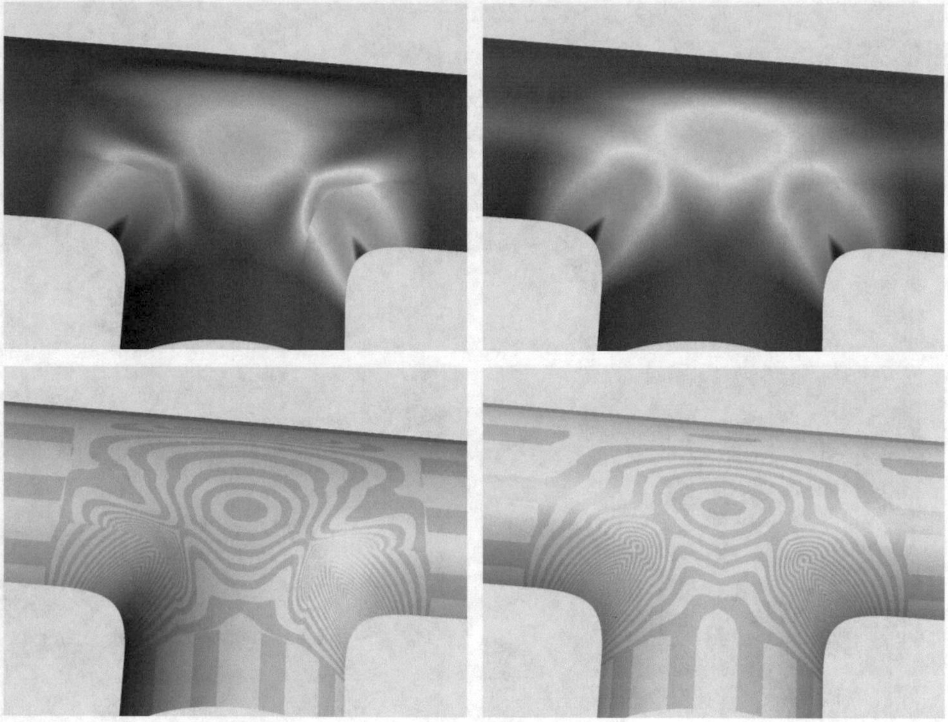

Figure 6.11: Mean curvature

ken at the points of discontinuity of the surface curvature. Any discontinuity of the Gaussian or mean curvature obviously indicates a discontinuity of the surface curvature. One can ask a reverse question: do these shape functions always reveal curvature discontinuities? In other words: is it possible that the mean or Gaussian curvature is continuous on a surface of class G^1, which is not of class G^2?

The answer is the following: if two smooth patches sharing a boundary curve and having a common tangent plane at each point of this curve (i.e., forming a surface of class G^1) do not form a surface of class G^2, then the mean curvature of the surface is discontinuous at this curve. The first proof of this fact was given by Barbara Putz [1992] (see also Putz [1996]). In [1996] Xiuzi Ye proved it independently, together with the following property: if the curvature of the surface is discontinuous, then the Gaussian curvature may still be continuous at the common curve of the patches, if the normal curvature of the two patches in the direction tangent to their common curve is zero. Otherwise, any discontinuity of the surface curvature translates to the discontinuity of the Gaussian curvature. These properties of the mean and Gaussian curvatures are known under the name given by Ye: the Linkage Curve Theorem.

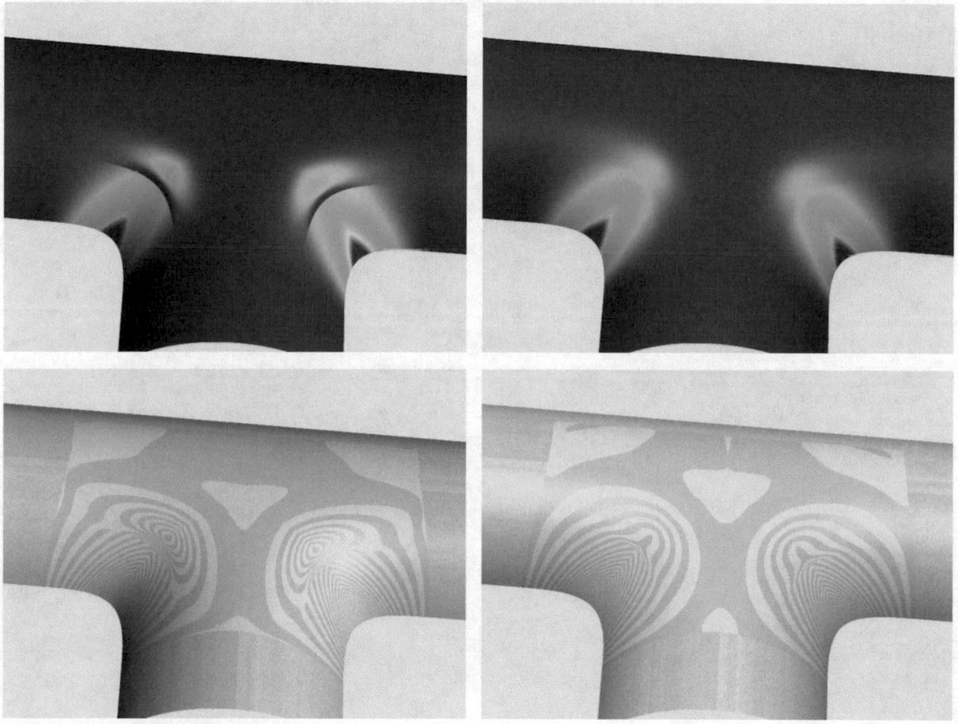

Figure 6.12: Gaussian curvature

APPENDIX A

Background

Basic information necessary for our study of geometric continuity, gathered here, includes properties of parametric curves and patches and their most popular representations. Other relevant topics are polynomial interpolation problems, approximation properties of splines, elements of differential geometry and Fàa di Bruno's formula.

A.1 LAGRANGE AND HERMITE INTERPOLATION

Let $u_0, \dots, u_n$ be given numbers. If each of them is different from the others, we can specify arbitrary numbers $f_0, \dots, f_n$ and look for a function h such that $h(u_i) = f_i$ for $i = 0, \dots, n$. This is called a Lagrange interpolation problem. The existence and uniqueness of its solution depends on additional conditions to be satisfied by the function h. In the linear space $\mathbb{R}[\cdot]_n$ the solution is unique; in other words, there exists a unique polynomial h of degree at most n, satisfying the interpolation conditions considered above.

We obtain a more general problem by allowing some (or all) interpolation knots $u_0, \dots, u_n$ to coincide. Obviously, we cannot specify two different function values for the same argument. But if a number u_i appears r times in the sequence $u_0, \dots, u_n$ (we say that the knot u_i has the multiplicity r), then we can demand that the function to be found at the knot u_i and its derivatives up to the order $r - 1$ take prescribed values. This interpolation problem, bearing the name of Charles Hermite, also has a unique solution in the space of real polynomials of degree at most n. Examples are shown in Figure A.1.

Functions of interpolation may be searched also in other function spaces, e.g. of splines or trigonometric polynomials; the existence of solutions depends on algebraic properties of those spaces. Having a Lagrange interpolation problem and a basis $\{g_0, \dots, g_n\}$ of a function space, we can write the following system of linear equations:

$$\begin{bmatrix} g_0(u_0) & \dots & g_n(u_0) \\ \vdots & & \vdots \\ g_0(u_n) & \dots & g_n(u_n) \end{bmatrix} \begin{bmatrix} a_0 \\ \vdots \\ a_n \end{bmatrix} = \begin{bmatrix} f_0 \\ \vdots \\ f_n \end{bmatrix}. \tag{A.1}$$

Its solution, if it exists, is a vector of coefficients of the function $h = \sum_{i=0}^{n} a_i g_i$, which satisfies the interpolation conditions. We can modify this approach to solve Hermite interpolation problems but, as a too general method, it does not work particularly well in many cases. Below we show a better algorithm of finding algebraic polynomials of interpolation.

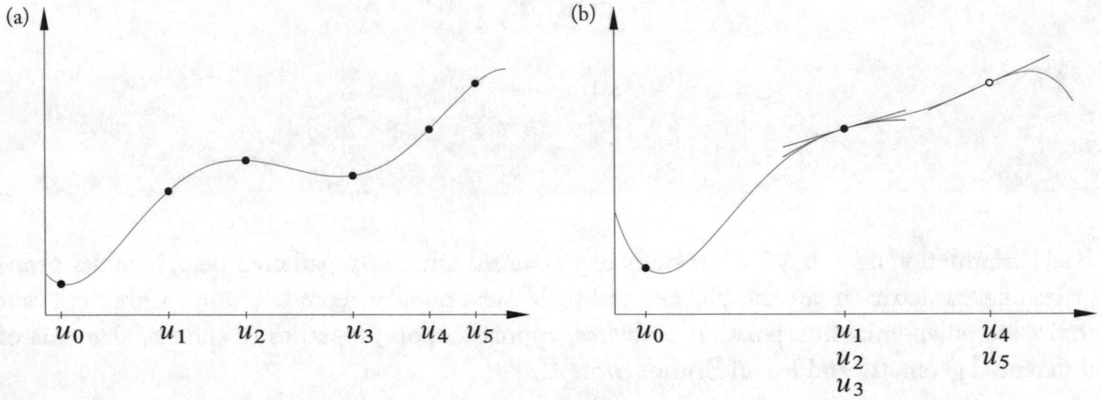

Figure A.1: Polynomial interpolation: (a) Lagrange, (b) Hermite. Interpolation conditions are shown as prescribed points of the graphs, tangent lines and an osculating parabola

A.1.1 THE DIVIDED DIFFERENCES ALGORITHM

The matrix of the system (A.1) written for the power basis $\{1, x, \ldots, x^n\}$ is full; it may be obtained with $O(n^2)$ operations and then the system may be solved with $O(n^3)$ operations, e.g. with the Gaussian elimination. This computational cost may be reduced by using a different basis. Let $p_0, \ldots, p_n$ be polynomials defined as follows:

$$\begin{aligned} p_0(x) &= 1, \\ p_1(x) &= x - u_0, \\ p_2(x) &= (x - u_0)(x - u_1), \\ &\vdots \\ p_n(x) &= (x - u_0)(x - u_1) \ldots (x - u_{n-1}). \end{aligned}$$

The degree of the polynomial p_k is k; the set $\{p_0, \ldots, p_n\}$ is a basis of the space $\mathbb{R}[\cdot]_n$, called **Newton's basis**. If we use it to set up the system (A.1), we obtain a lower triangular matrix and the system may be solved with $O(n^2)$ operations. It is then possible to find the coefficients in power basis with $O(n^2)$ operations but usually it is not worth doing. Having the knots $u_0, \ldots, u_{n-1}$, the coefficients in the Newton basis, say, $b_0, \ldots, b_n$, and a number x, we can evaluate the polynomial $h(x) = \sum_{i=0}^{n} b_i p_i(x)$ with $O(n)$ operations.

Below we describe another algorithm of computing the coefficients $b_0, \ldots, b_n$ of the polynomial of interpolation. If all knots are simple, i.e., of multiplicity one, then we can define **divided differences** of a function f by the following formulae:

$$f[u_i] \stackrel{\text{def}}{=} f(u_i), \tag{A.2}$$

$$f[u_i, \ldots, u_{i+k}] \stackrel{\text{def}}{=} \frac{f[u_i, \ldots, u_{i+k-1}] - f[u_{i+1}, \ldots, u_{i+k}]}{u_i - u_{i+k}}. \tag{A.3}$$

The divided differences are connected with derivatives in the way described by the following theorem, proved in many books about numerical methods:

Theorem A.1 *If a function f is of class C^n in an interval $[a,b]$, and $u_i,\dots,u_{i+n} \in [a,b]$, then there exists a number $\xi \in [a,b]$ such that*

$$f[u_i,\dots,u_{i+n}] = \frac{f^{(n)}(\xi)}{n!}. \tag{A.4}$$

A consequence of this theorem is the continuity of the k-th order divided difference with respect to the knots $u_i,\dots,u_{i+k}$, if the function f is of class C^k. It allows one to extend the definition of the divided differences to the case when *all* knots coincide:

$$f[\underbrace{u_i,\dots,u_i}_{k+1}] \stackrel{\text{def}}{=} \frac{f^{(k)}(u_i)}{k!}. \tag{A.5}$$

The divided differences are symmetric functions of their knots. This property makes it possible to define the divided difference for an arbitrary knot sequence, after sorting this sequence. In this way, if not all knots coincide, then $u_i \neq u_{i+n}$ and $f[u_i,\dots,u_{i+n}]$ may be obtained using Formula (A.3). Applying it recursively leads either to the divided differences of order 0, i.e., function values, which are given, or to the divided differences over all knots coincident, where Formula (A.5) may be used—provided that the function f has sufficiently many continuous derivatives at the multiple knot and the derivatives are given.

If the knots are fixed, the divided differences are linear functionals acting on f. Let us apply them to the elements of the corresponding Newton basis:

$$\begin{aligned}
p_k[x,u_0] &= \frac{(x-u_0)\dots(x-u_{k-1}) - (u_0-u_0)\dots(u_0-u_{k-1})}{x-u_0} \\
&= (x-u_1)\dots(x-u_{k-1}), \\
p_k[x,u_0,u_1] &= \frac{p_k[x,u_0]-p_k[u_0,u_1]}{x-u_1} = (x-u_2)\dots(x-u_{k-1}), \\
&\ \ \vdots \\
p_k[x,u_0,\dots,u_{k-1}] &= 1.
\end{aligned}$$

The last equality holds in particular for $x = u_k$. The divided differences of order higher than k of the polynomial p_k are equal to zero. But if we set $x = u_i$ for the divided difference of order $i < k$, then we also obtain $p_k[u_0,\dots,u_i] = 0$. Therefore, if we apply the divided difference to a linear combination of the elements of the Newton basis, $h(x) = \sum_{i=0}^{n} b_i p_i(x)$, then we obtain

$$h[u_0,\dots,u_k] = \sum_{i=0}^{n} b_i p_i[u_0,\dots,u_k] = b_k.$$

Hence, the divided differences of a function f are coefficients of its polynomial of interpolation with given knots, in the Newton basis defined with these knots. To solve a Lagrange interpolation problem one can use Algorithm A.1, which uses Formulae (A.2) and (A.3).

Algorithm A.1 Computing divided differences with all knots of multiplicity 1

Input : knots $u_0, \ldots, u_n$ in the array u, corresponding function values in the array y

```
for ( j = 1; j <= n; j++ )
  for ( i = n; i >= j; i-- )
    y[i] = (y[i]-y[i-1])/(u[i]-u[i-j]);
```

Output : divided differences $f[u_0], f[u_0, u_1], \ldots, f[u_0, \ldots, u_n]$ in the array y

Algorithm A.2, solving Hermite interpolation problems by computing the divided differences, will work with a monotone knot sequence. If the number i is the first position of a knot in the sequence, then the i-th element of the array y must be $f(u_i)$, y[i+1] $= f'(u_i)$, y[i+2] $= f''(u_i)$ and so on. In the auxiliary array m the algorithm stores the information about the order of derivatives given at the corresponding positions in the array y. During the computations, the numbers in the array m are decreased. Dividing the derivatives in the array y by the number j, which is the order of divided differences computed in the j-th execution of the loop, results in an effective division of the k-th order derivative by $k!$.

Algorithm A.2 The divided differences algorithm for multiple knots

Input : knots $u_0, \ldots, u_n$ in the array u, function values and derivatives in the array y

```
m[0] = 0;
for ( i = 1; i <= n; i++ )
  m[i] = u[i] == u[i-1] ? m[i-1]+1 : 0;
for ( j = 1; j <= n; j++ )
  for ( i = n; i >= j; i-- )
    if ( m[i] == 0 )
      y[i] = (y[i]-y[i-1])/(u[i]-u[i-j]);
    else {
      y[i] /= j;
      m[i]--;
    }
```

Output : divided differences $f[u_0], f[u_0, u_1], \ldots, f[u_0, \ldots, u_n]$ in the array y

Divided differences have many properties which make them a useful tool in numerous applications. To study B-spline functions, we need the following:

Theorem A.2 ***Leibniz formula:*** *if* $f(x) = g(x)h(x)$ *and the involved divided differences exist (i.e., the functions* g *and* h *are of class* C^k *in a neighbourhood of each knot of multiplicity* $k+1$*), then*

$$f[x_i, \ldots, x_{i+n}] = \sum_{j=0}^{n} g[x_i, \ldots, x_{i+j}]h[x_{i+j}, \ldots, x_{i+n}]. \tag{A.6}$$

A proof of this property may be found e.g. in de Boor [1978]. Note that the formula for the n-th order derivative of the product of functions is a special case of (A.6).

A.2 BÉZIER CURVES

A.2.1 DEFINITION

A Bézier curve of degree n is a parametric curve represented by a sequence of $n+1$ points, $\boldsymbol{p}_0, \ldots, \boldsymbol{p}_n$, called the **control points**. The polyline consisting of n line segments whose consecutive vertices are the points $\boldsymbol{p}_0, \ldots, \boldsymbol{p}_n$ is often called the **control polygon.** The parametrisation is given by the formula

$$\boldsymbol{p}(t) = \sum_{i=0}^{n} \boldsymbol{p}_i B_i^n(t), \tag{A.7}$$

where the symbols B_i^n denote the **Bernstein basis polynomials**[1] of degree n:

$$B_i^n(t) = \binom{n}{i} t^i (1-t)^{n-i}. \tag{A.8}$$

These polynomials are linearly independent, nonnegative for $t \in [0,1]$, and the sum $B_0^n(t) + \cdots + B_n^n(t)$ for any t is equal to 1. It is convenient to take $B_i^n(t) = 0$ if $i < 0$ or $i > n$. The Bernstein polynomials satisfy the recursive formula

$$B_0^0(t) = 1, \tag{A.9}$$
$$B_i^n(t) = tB_{i-1}^{n-1}(t) + (1-t)B_i^{n-1}(t), \tag{A.10}$$

making it possible to obtain these polynomials from Pascal's triangle, like the binomial coefficients.

Usually, we assume that the domain of a Bézier curve is the interval [0, 1]. Then, the entire curve is located within the convex hull of the control polygon. This fact is known as the **convex**

[1]Introduced by Bernstein in [1912].

hull property of the Bézier representation of curves. In addition, the end points of the arc are the first and the last control point: $\boldsymbol{p}(0) = \boldsymbol{p}_0$ and $\boldsymbol{p}(1) = \boldsymbol{p}_n$.

The point $\boldsymbol{p}(t)$, for a given t, may be found with the well known **de Casteljau's algorithm** (Algorithm A.3), which produces also other results: representations of two pieces of the arc and points, which may be used to compute derivatives (Fig. A.2).

Algorithm A.3 De Casteljau's algorithm

Input : control points $\boldsymbol{p}_0, \ldots, \boldsymbol{p}_n$, number t

```
/* Denote p_i^(0) = p_i for i = 0,...,n */
for ( j = 1; j <= n; j++ )
  for ( i = 0; i <= n-j; i++ )
    p_i^(j) = (1 - t) p_i^(j-1) + t p_{i+1}^(j-1) ;
```

Output : point $\boldsymbol{p}(t) = \boldsymbol{p}_0^{(n)}$,
control points $\boldsymbol{p}_0^{(0)}, \ldots, \boldsymbol{p}_0^{(n)}$ of the arc over the interval $[0, t]$,
control points $\boldsymbol{p}_0^{(n)}, \ldots, \boldsymbol{p}_n^{(0)}$ of the arc over the interval $[t, 1]$,
points $\boldsymbol{p}_0^{(n-k)}, \ldots, \boldsymbol{p}_k^{(n-k)}$ such that $\frac{\mathrm{d}^k}{\mathrm{d}t^k}\boldsymbol{p}(t) = \frac{n!}{(n-k)!}\Delta^k \boldsymbol{p}_0^{(n-k)}$ (see Section A.2.2)

A.2.2 DERIVATIVES

Using the formula for the derivative of the Bernstein polynomials,

$$\frac{\mathrm{d}}{\mathrm{d}t} B_i^n(t) = n\big(B_{i-1}^{n-1}(t) - B_i^{n-1}(t)\big), \tag{A.11}$$

where (by convention) $B_i^{n-1}(t) = 0$ if $i < 0$ or $i \geq n$, one can derive the following formula:

$$\frac{\mathrm{d}}{\mathrm{d}t}\boldsymbol{p}(t) = \sum_{i=0}^{n-1} n(\boldsymbol{p}_{i+1} - \boldsymbol{p}_i) B_i^{n-1}(t). \tag{A.12}$$

If we denote the vectors $\Delta \boldsymbol{p}_i = \boldsymbol{p}_{i+1} - \boldsymbol{p}_i$, then we express the derivative in Bézier form:

$$\frac{\mathrm{d}}{\mathrm{d}t}\boldsymbol{p}(t) = \sum_{i=0}^{n-1} n \Delta \boldsymbol{p}_i B_i^{n-1}(t). \tag{A.13}$$

By repeating this calculation, we can find the higher order derivatives:

$$\frac{\mathrm{d}^k}{\mathrm{d}t^k}\boldsymbol{p}(t) = \sum_{i=0}^{n-k} \frac{n!}{(n-k)!} \Delta^k \boldsymbol{p}_i B_i^{n-k}(t), \tag{A.14}$$

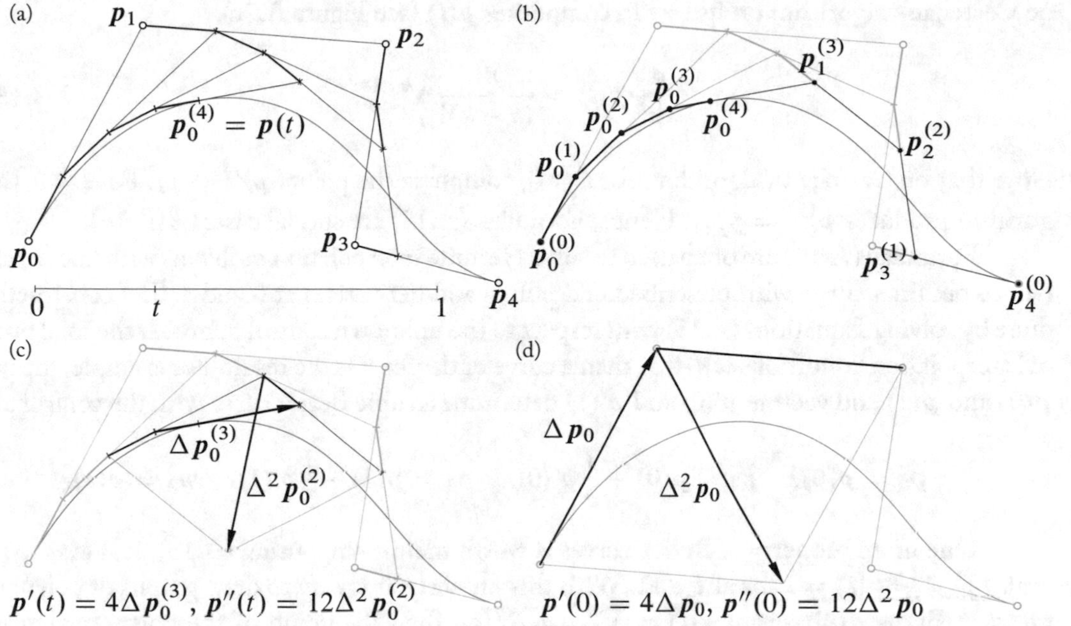

Figure A.2: De Casteljau's algorithm and its results for a quartic Bézier curve: (a) the point $\boldsymbol{p}(t)$, (b) the representation of arc pieces, (c) the first- and second-order derivatives at t, (d) the first- and second-order derivatives at 0

where Δ^k is the k-th order difference operator:

$$\Delta^k \boldsymbol{p}_i = \Delta^{k-1} \boldsymbol{p}_{i+1} - \Delta^{k-1} \boldsymbol{p}_i = \sum_{j=0}^{k} (-1)^{k-j} \binom{k}{j} \boldsymbol{p}_{i+j}.$$

The last derivative which may be non-zero is the n-th; our Bézier curve is a polynomial curve of degree n, and additionally we cannot define differences of a higher order, having only $n+1$ control points.

A direct consequence of Formula (A.14) is that the k-th order derivatives of the parametrisation $\boldsymbol{p}$ for $t = 0$ and $t = 1$ are determined respectively by the first $k+1$ and last $k+1$ control points (see Figure A.2d):

$$\frac{\mathrm{d}^k}{\mathrm{d}t^k} \boldsymbol{p}(0) = \frac{n!}{(n-k)!} \Delta^k \boldsymbol{p}_0 \quad \text{and} \quad \frac{\mathrm{d}^k}{\mathrm{d}t^k} \boldsymbol{p}(1) = \frac{n!}{(n-k)!} \Delta^k \boldsymbol{p}_{n-k}. \tag{A.15}$$

Taking $k = 1$, we conclude that if $\boldsymbol{p}_1 \neq \boldsymbol{p}_0$ and $\boldsymbol{p}_{n-1} \neq \boldsymbol{p}_n$, then the curve at its end points is tangent to the first and the last line segment of the control polygon. Furthermore, for any $t \in \mathbb{R}$, the k-th order derivative of the parametrisation $\boldsymbol{p}$ may be expressed with the points obtained by de Casteljau's algorithm on its way to computing $\boldsymbol{p}(t)$ (see Figure A.2c):

$$\frac{\mathrm{d}^k}{\mathrm{d}t^k}\boldsymbol{p}(t) = \frac{n!}{(n-k)!}\Delta^k \boldsymbol{p}_0^{(n-k)}. \tag{A.16}$$

Note that de Casteljau's algorithm, for $t = 0$, computes the points $\boldsymbol{p}_i^{(j)} = \boldsymbol{p}_i$. For $t = 1$ the algorithm produces $\boldsymbol{p}_i^{(j)} = \boldsymbol{p}_{j+i}$. Hence, Formulae (A.15) are special cases of (A.16).

Formulae (A.15) are often used to solve Hermite interpolation problems with knots 0 and 1, i.e., to obtain a curve with prescribed end points and derivatives at 0 and 1. The construction is done by solving Equations (A.15) with respect to the unknown control points. If the total number of interpolation conditions is $n + 1$, then a curve of degree n is the result. For example, the points $\boldsymbol{p}(0)$ and $\boldsymbol{p}(1)$ and vectors $\boldsymbol{p}'(0)$ and $\boldsymbol{p}'(1)$ determine a cubic Bézier curve with the control points

$$\boldsymbol{p}_0 = \boldsymbol{p}(0), \quad \boldsymbol{p}_1 = \boldsymbol{p}(0) + \frac{1}{3}\boldsymbol{p}'(0), \quad \boldsymbol{p}_2 = \boldsymbol{p}(1) - \frac{1}{3}\boldsymbol{p}'(1), \quad \boldsymbol{p}_3 = \boldsymbol{p}(1).$$

One more property of Bézier curves is worth mentioning: using (A.15), it is easy to prove that $\sum_{i=0}^{n} \frac{i}{n} B_i^n(t) = t$ for all $t \in \mathbb{R}$. With this property, it is easy to draw graphs of polynomials given in Bernstein bases: if $p(t) = \sum_{i=0}^{n} a_i B_i^n(t)$, then the graph of the polynomial p is the Bézier curve whose control points are $\boldsymbol{p}_i = [\frac{i}{n}, a_i]^T$ for $i = 0, \ldots, n$ (see e.g. Figure 2.6).

A.2.3 ALGEBRAIC OPERATIONS

Polynomials form a ring, an algebraic structure with two well known operations of addition and multiplication. Given vectors of coefficients of two polynomials in a fixed basis (e.g. power or Bernstein), we can add or subtract the vectors to perform the addition and subtraction of the polynomials. If $a(t) = A_n t^n + \cdots + A_1 t + A_0$ and $b(t) = B_m t^m + \cdots + B_1 t + B_0$, then the coefficients of the product $c(t) = a(t)b(t) = C_{n+m} t^{n+m} + \cdots + C_1 t + C_0$ of the two polynomials may be computed using Algorithm A.4.

Algorithm A.4 Multiplication of polynomials in power basis or in scaled Bernstein bases

```
for ( k = 0; k <= n+m; k++ ) C[k] = 0;
for ( i = 0; i <= n; i++ )
  for ( j = 0; j <= m; j++ ) C[i+j] += A[i]*B[j];
```

Let $b_i^n(t) \stackrel{\text{def}}{=} t^i(1-t)^{n-i}$. The polynomials $b_0^n, \ldots, b_n^n$ form a basis of the linear space of real polynomials of degree not greater than n, denoted by $\mathbb{R}[\cdot]_n$. We call this basis the **scaled Bernstein basis** of degree n. If

$$a(t) = \sum_{i=0}^{n} a_i B_i^n(t) = \sum_{i=0}^{n} A_i b_i^n(t),$$

then $A_i = \binom{n}{i} a_i$ for $i = 0, \dots, n$. If $(A_0, \dots, A_n)$ and $(B_0, \dots, B_m)$ are vectors representing the polynomials $a(t)$ and $b(t)$ in the scaled Bernstein bases of degrees n and m, then Algorithm A.4 operating on these vectors produces the vector of coefficients $(C_0, \dots, C_{n+m})$, representing the product $c(t) = a(t)b(t)$ in the scaled Bernstein basis of degree $n + m$.

Given the coefficients of the polynomials $a(t)$ and $b(t)$ in the (ordinary) Bernstein bases of degrees n and m, we can easily find the coefficients of their product in the Bernstein basis of degree $n + m$; it suffices to pass to the scaled bases, do the multiplication and then convert the result to the Bernstein basis. This is done by Algorithm A.5.

Algorithm A.5 Multiplication of polynomials in Bernstein bases

```
for ( i = 0; i <= n; i++ ) A[i] = (n i)*a[i];
for ( j = 0; j <= m; j++ ) B[j] = (m j)*b[j];
for ( k = 0; k <= n+m; k++ ) C[k] = 0;
for ( i = 0; i <= n; i++ )
  for ( j = 0; j <= m; j++ ) C[i+j] += A[i]*B[j];
for ( k = 0; k <= n+m; k++ ) c[k] = C[k]/(n+m k);
```

To add or subtract polynomials, we need their coefficients in the same basis; when the sum of two polynomials given in Bernstein bases of different degrees is needed, the procedure of **degree elevation** may be used. All coefficients of a constant polynomial in the Bernstein basis of (any) degree m are equal to that constant. Hence, degree elevation is just the multiplication of the given polynomial $a(t)$ by $b(t) = \sum_{i=0}^{m} B_j^m(t) = 1$. If the degrees of representations of two polynomials to be added are different, then Algorithm A.5 may be used to align the degrees before adding the vectors of coefficients of the polynomials.

The method of doing algebraic operations outlined above may be applied to polynomials as well as to polynomial vector functions—as long as the multiplication of the coefficients makes sense.

Example A.3 Consider the construction of three vector functions, say, $\boldsymbol{T}(t)$, $\boldsymbol{N}(t)$ and $\boldsymbol{B}(t)$, which describe the unnormalised elements of the Frenet frame (see Section A.10.1) of a given Bézier curve $\boldsymbol{p}(t)$ in $\mathbb{R}^3$. For a given t, the Frenet frame consists of the unit vectors

$$\boldsymbol{t}(t) = \frac{\boldsymbol{p}'(t)}{\|\boldsymbol{p}'(t)\|_2}, \quad \boldsymbol{n}(t) = \frac{\boldsymbol{p}''(t) - \langle \boldsymbol{p}''(t), \boldsymbol{t}(t)\rangle \boldsymbol{t}(t)}{\|\boldsymbol{p}''(t) - \langle \boldsymbol{p}''(t), \boldsymbol{t}(t)\rangle \boldsymbol{t}(t)\|_2}, \quad \boldsymbol{b}(t) = \frac{\boldsymbol{p}'(t) \wedge \boldsymbol{p}''(t)}{\|\boldsymbol{p}'(t) \wedge \boldsymbol{p}''(t)\|_2}.$$

Given a Bézier curve $\boldsymbol{p}(t)$, for $\boldsymbol{T}(t)$ we can take the derivative using Formula (A.13), or we can ignore the constant factor n and take $\boldsymbol{T}(t) = \sum_{i=0}^{n-1} \Delta \boldsymbol{p}_i B_i^{n-1}(t) = \frac{1}{n}\boldsymbol{p}'(t)$. The normal vector, $\boldsymbol{n}(t)$, is obtained by the orthonormalisation of the second-order derivative, $\boldsymbol{p}''(t)$, to $\boldsymbol{t}(t)$. To obtain a polynomial vector function, we drop the denominator and we get

$$\boldsymbol{N}(t) = \frac{1}{n(n-1)}\big(\langle \boldsymbol{T}(t), \boldsymbol{T}(t)\rangle \boldsymbol{p}''(t) - \langle \boldsymbol{p}''(t), \boldsymbol{T}(t)\rangle \boldsymbol{T}(t)\big).$$

The scalar products $\langle \boldsymbol{T}(t), \boldsymbol{T}(t)\rangle$ and $\langle \boldsymbol{p}''(t), \boldsymbol{T}(t)\rangle$ in this formula are polynomials of t (of degrees $2n-2$ and $2n-3$ respectively), which we denote $P(t)$ and $Q(t)$. Their coefficients may be obtained using scalar multiplications of the vector coefficients of the functions $\boldsymbol{T}(t)$ and $\frac{1}{n(n-1)}\boldsymbol{p}''(t) = \sum_{i=0}^{n-2} \Delta^2 \boldsymbol{p}_i B_i^{n-2}(t)$. To obtain the curve $\boldsymbol{N}(t)$, we have to compute the products $\frac{1}{n(n-1)}\boldsymbol{p}''(t)P(t)$ and $\boldsymbol{T}(t)Q(t)$. This is done by multiplying vector coefficients of the appropriate curves by scalar coefficients of $P(t)$ and $Q(t)$. In this way, we obtain two polynomial vector functions of the same degree $3n-4$; hence, their subtraction may be done without degree elevation. The entire computation is done by Algorithm A.6. An example of a result is shown in Figure A.3.

Algorithm A.6 Finding a Bézier representation of the unnormalised Frenet frame

Input : Control points $\boldsymbol{p}_0, \dots, \boldsymbol{p}_n$ of a Bézier curve in $\mathbb{R}^3$, $n \geq 2$

```
/* compute the first and second order differences */
/* and transform to the scaled bases */
for ( i = 0; i <= n-1; i++ ) { d[i] = p_{i+1} − p_i;   D[i] = (n−1 choose i) d[i]; }
for ( i = 0; i <= n-2; i++ ) S[i] = (n−2 choose i)(d[i+1] − d[i]);
/* do the multiplications and subtraction */
for ( k = 0; k <= 2*n-2; k++ ) P[k] = 0;
for ( i = 0; i <= n-1; i++ )
  for ( j = 0; j <= n-1; j++ ) P[i+j] += ⟨D[i], D[j]⟩;
for ( k = 0; k <= 2*n-3; k++ ) { Q[k] = 0;   B[k] = 0; }
for ( i = 0; i <= n-2; i++ )
  for ( j = 0; j <= n-1; j++ )
    { Q[i+j] += ⟨S[i], D[j]⟩;   B[i+j] += D[j]∧S[i]; }
for ( k = 0; k <= 3*n-4; k++ ) N[k] = 0;
for ( i = 0; i <= 2*n-2; i++ )
  for ( j = 0; j <= n-2; j++ ) N[i+j] += P[i]*S[j];
for ( i = 0; i <= 2*n-3; i++ )
  for ( j = 0; j <= n-1; j++ ) N[i+j] -= Q[i]*D[j];
/* transform from the scaled to the Bernstein bases */
for ( k = 0; k <= 2*n-3; k++ ) B[k] /= (2n−3 choose k);
for ( k = 0; k <= 3*n-4; k++ ) N[k] /= (3n−4 choose k);
```

Output : Control points $\boldsymbol{d}_0, \dots, \boldsymbol{d}_{n-1}$, $\boldsymbol{N}_0, \dots, \boldsymbol{N}_{3n-4}$ and $\boldsymbol{B}_0, \dots, \boldsymbol{B}_{2n-3}$ of the curves $\boldsymbol{T}(t)$, $\boldsymbol{N}(t)$ and $\boldsymbol{B}(t)$

Needless to say, it is usually more practical to find the vectors of the Frenet frame for any given t by computing the first and second-order derivatives of the parametrisation $\boldsymbol{p}$ at t and then computing the Frenet frame vectors at that point. The explicit Bézier representations of the

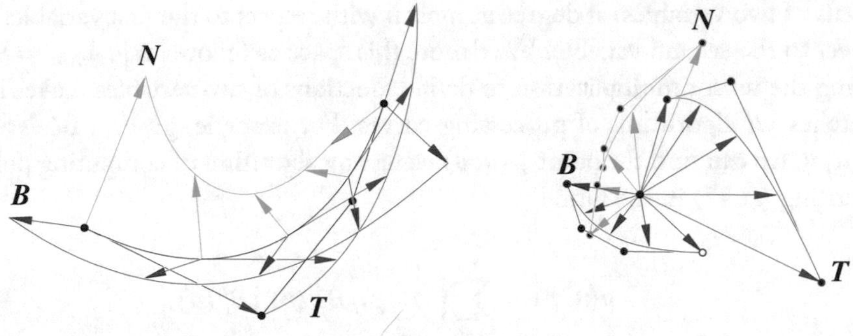

Figure A.3: Bézier representation of the Frenet frame of a cubic Bézier curve

curves $\boldsymbol{T}(t)$, $\boldsymbol{N}(t)$ and $\boldsymbol{B}(t)$ may be useful on the rare occasions when the convex hull property of Bézier curves is used to estimate the directions of the vectors forming the Frenet frame for the entire arc. However, this example is a good illustration of the variety of algebraic operations in Bernstein bases. Similar algebraic operations are essential in constructions of patches whose junctions are smooth, discussed in Chapters 3–5.

A.3 BÉZIER PATCHES

Tensor product Bézier patches are defined with the formula

$$\boldsymbol{p}(u,v) = \sum_{i=0}^{n}\sum_{j=0}^{m} \boldsymbol{p}_{ij} B_i^n(u) B_j^m(v), \tag{A.17}$$

with the **control points** $\boldsymbol{p}_{ij}$ being vertices of the **control net** and the Bernstein polynomials used to define the Bézier curves. Usually, we assume that u and v range between 0 and 1 and, thus, the domain of the patch is the unit square, $[0, 1] \times [0, 1]$.

The term "tensor product" refers to the way of multiplying functions: having two functions, f and g (which may have different domains), we define a function of two variables, say, h, with the formula

$$h(u,v) = f(u)g(v).$$

We often consider functions as *vectors*, i.e., elements of linear vector spaces, and then we do not want to be distracted by the notation, which requires naming function arguments. In that case we can use the more handy notation $h = f \otimes g$.

The domain of the function h is the Cartesian product of the domains of f and g. The sets of Bernstein polynomials $\{\, B_i^n : i = 0, \dots, n \,\}$ and $\{\, B_j^m : j = 0, \dots, m \,\}$ are bases of linear vector spaces $\mathbb{R}[\cdot]_n$ and $\mathbb{R}[\cdot]_m$ whose elements are polynomials of degrees up to n and m respectively. The set of tensor products of all pairs $B_i^n \otimes B_j^m$ is a basis of the linear space whose elements are polynomials of two variables, of degree at most n with respect to the first variable, and at most m with respect to the second variable. We denote this space as follows: $\mathbb{R}[\cdot,\cdot]_{n,m} = \mathbb{R}[\cdot]_n \otimes \mathbb{R}[\cdot]_m$.

Using the tensor multiplication to define functions of two variables makes it easy to apply to the patches *all* algorithms of processing curves. For example, given a Bézier patch and the numbers u, v, we can find the point $\boldsymbol{p}(u,v)$ using any algorithm of computing points of a curve. After rewriting (A.17) in the form

$$\boldsymbol{p}(u,v) = \sum_{i=0}^{n}\Bigl(\sum_{j=0}^{m} \boldsymbol{p}_{ij} B_j^m(v)\Bigr) B_i^n(u),$$

we compute the points of $n+1$ Bézier curves of degree m, $\boldsymbol{q}_i = \sum_{j=0}^{m} \boldsymbol{p}_{ij} B_j^m(v)$, and then the point $\boldsymbol{p}(u,v) = \sum_{i=0}^{n} \boldsymbol{q}_i B_i^n(u)$. Alternatively, we can write

$$\boldsymbol{p}(u,v) = \sum_{j=0}^{m}\Bigl(\sum_{i=0}^{n} \boldsymbol{p}_{ij} B_i^n(u)\Bigr) B_j^m(v)$$

and proceed in the obvious way.

The tensor product definition makes it possible to easily compute derivatives of the parametrisation (A.17). To write the formulae in a simple form, we need *two* difference operators, defined as follows:

$$\Delta_1 \boldsymbol{p}_{ij} = \boldsymbol{p}_{i+1,j} - \boldsymbol{p}_{ij}, \quad \Delta_2 \boldsymbol{p}_{ij} = \boldsymbol{p}_{i,j+1} - \boldsymbol{p}_{ij}.$$

To a control net with $n+1$ columns and $m+1$ rows, which represents a Bézier patch of degree (n,m), we can apply powers of the above two operators

$$\Delta_1^k \boldsymbol{p}_{ij} = \Delta_1^{k-1} \boldsymbol{p}_{i+1,j} - \Delta_1^{k-1} \boldsymbol{p}_{ij} = \sum_{p=0}^{k} (-1)^{k-p} \binom{k}{p} \boldsymbol{p}_{i+p,j},$$

$$\Delta_2^l \boldsymbol{p}_{ij} = \Delta_2^{l-1} \boldsymbol{p}_{i,j+1} - \Delta_2^{l-1} \boldsymbol{p}_{ij} = \sum_{q=0}^{l} (-1)^{l-q} \binom{l}{q} \boldsymbol{p}_{i,j+q},$$

where $k \le n$ and $l \le m$. It is easy to verify that these operators are commutative, i.e., the result of their composition is the same, no matter which one was applied first. So, we have

$$\Delta_1^k \Delta_2^l \boldsymbol{p}_{ij} = \sum_{p=0}^{k}\sum_{q=0}^{l} (-1)^{k+l-p-q} \binom{k}{p}\binom{l}{q} \boldsymbol{p}_{i+p,j+q}.$$

The first-order partial derivatives of the Bézier patch may be expressed by the formulae

$$\frac{\partial}{\partial u}\boldsymbol{p}(u,v) = \sum_{i=0}^{n-1}\sum_{j=0}^{m} n\Delta_1\boldsymbol{p}_{ij}\, B_i^{n-1}(u)B_j^m(v), \tag{A.18}$$

$$\frac{\partial}{\partial v}\boldsymbol{p}(u,v) = \sum_{i=0}^{n}\sum_{j=0}^{m-1} m\Delta_2\boldsymbol{p}_{ij}\, B_i^{n}(u)B_j^{m-1}(v), \tag{A.19}$$

which are special cases of the following:

$$\frac{\partial^{k+l}}{\partial u^k \partial v^l}\boldsymbol{p}(u,v) = \sum_{i=0}^{n-k}\sum_{j=0}^{m-l} \frac{n!}{(n-k)!}\frac{m!}{(m-l)!}\Delta_1^k\Delta_2^l\boldsymbol{p}_{ij}\, B_i^{n-k}(u)B_j^{m-l}(v). \tag{A.20}$$

De Casteljau's algorithm, used to compute the point $\boldsymbol{p}(u,v)$, may be generalised as follows: let $\boldsymbol{p}_{ij}^{(0,0)} = \boldsymbol{p}_{ij}$ for $i = 0,\ldots,n$ and $j = 0,\ldots,m$. Suppose that we have a net of points $\boldsymbol{p}_{ij}^{(k,l)}$, for $i = 0,\ldots,n-k$ and $j = 0,\ldots,m-l$. If $l < n$, then we can compute

$$\boldsymbol{p}_{ij}^{(k+1,l)} = (1-u)\boldsymbol{p}_{ij}^{(k,l)} + u\,\boldsymbol{p}_{i+1,j}^{(k,l)}.$$

If $l < m$, then we can compute

$$\boldsymbol{p}_{ij}^{(k,l+1)} = (1-v)\boldsymbol{p}_{ij}^{(k,l)} + v\,\boldsymbol{p}_{i,j+1}^{(k,l)}.$$

Again, it is easy to check that to obtain the points $\boldsymbol{p}_{ij}^{(k+1,l+1)}$, we have to make the above two steps the necessary number of times, but the ordering of these steps is irrelevant. In particular, to obtain the point $\boldsymbol{p}_{00}^{(n,m)} = \boldsymbol{p}(u,v)$, we have to make the "$u$" step n times and the "v" step m times. The points obtained on the way may be used to compute the partial derivatives of the parametrisation $\boldsymbol{p}$:

$$\frac{\partial^{k+l}}{\partial u^k \partial v^l}\boldsymbol{p}(u,v) = \frac{n!}{(n-k)!}\frac{m!}{(m-l)!}\Delta_1^k\Delta_2^l\boldsymbol{p}_{00}^{(n-k,m-l)}. \tag{A.21}$$

An example is shown in Figure A.4. By analogy to Bézier curves, if we take $u = 0$, $u = 1$, $v = 0$ or $v = 1$, then we can obtain derivatives of Bézier patches at the boundary. Also, it is possible to construct Bézier patches with a prescribed boundary and partial derivatives, which is discussed in Section A.9.

A.4 B-SPLINE CURVES

There are a number of equivalent definitions of B-spline functions and curves: using the truncated power function, the Mansfield–de Boor–Cox formula, knot insertion, blossoming and simplexes, to name the most popular. Also, there are a variety of notations and conventions used in publications about B-splines, which may be confusing.[2] Geometric spline curves, described in Chapter 2, are obtained by generalising B-spline curves.

[2]Having made my choice, I apologise to the readers whose favourite notation is different.

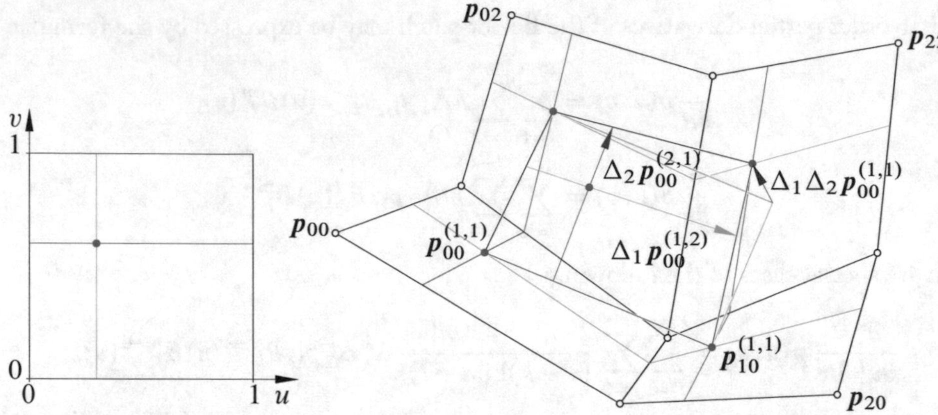

Figure A.4: Computing the first-order partial derivatives and the mixed partial derivative of a Bézier patch of degree (2, 2): $\boldsymbol{p}_u = 2\Delta_1 \boldsymbol{p}_{00}^{(1,2)}$, $\boldsymbol{p}_v = 2\Delta_2 \boldsymbol{p}_{00}^{(2,1)}$, $\boldsymbol{p}_{uv} = 4\Delta_1\Delta_2 \boldsymbol{p}_{00}^{(1,1)}$

A.4.1 NORMALISED B-SPLINE FUNCTIONS

Definition A.4 *The function defined by the formula*

$$(t-u)_+^n \stackrel{\text{def}}{=} \begin{cases} (t-u)^n & \text{if } t \geq u, \\ 0 & \text{else} \end{cases} \tag{A.22}$$

*is called the **truncated power function** of degree n.*

We may fix one of the two variables, t or u, as a parameter and, thus, obtain a function of one variable, for example $f_t^n(u) = (t-u)_+^n$, or $g_u^n(t) = (t-u)_+^n$. If $n > 0$, then the truncated power function is continuous together with derivatives up to the order $n-1$. The function $(t-u)_+^0$ is undefined at $t = u$. By convention, we assume that in this case $(t-u)_+^0 = 1$; however, the simple formulae

$$\frac{\mathrm{d}^k}{\mathrm{d}u^k} f_t^n(u) = (-1)^k \frac{n!}{(n-k)!}(t-u)_+^{n-k}, \qquad \frac{\mathrm{d}^k}{\mathrm{d}t^k} g_u^n(t) = \frac{n!}{(n-k)!}(t-u)_+^{n-k} \tag{A.23}$$

must be used with caution. If $t = u$, then derivatives of order $k \geq n$ of the functions f_t and g_u do not exist, and the formulae above are incorrect.

Definition A.5 *Let $u_i, \ldots, u_{i+n+1}$ be a non-decreasing sequence of numbers called **knots**. A **normalised B-spline function** of degree n with these knots is given by the formula*

$$N_i^n(t) \stackrel{\text{def}}{=} (-1)^{n+1}(u_{i+n+1} - u_i) f_t^n[u_i, \ldots, u_{i+n+1}], \tag{A.24}$$

where $f_t^n[u_i, \ldots, u_{i+n+1}]$ is the divided difference of order $n+1$ of the function $f_t^n(u)$.

The function $f_t^n(u)$ is equal to zero for all $t < u_i$, and then $N_i^n(t) = 0$. On the other hand, if $t > u_{i+n+1}$, then the function $f_t^n(u)$ is a polynomial of the variable u. The divided difference of order $n+1$ of any polynomial of degree n or lower is equal to zero. Thus, we obtained the following property:

Property A.6 *The function N_i^n defined with Formula (A.24) may take non-zero values only between its smallest and greatest knot, i.e., in the interval $[u_i, u_{i+n+1}]$.*

As we can see, if $u_i = \cdots = u_{i+n+1}$, then the factor $(u_{i+n+1} - u_n)$ in (A.24) is zero. The function N_i^n with these knots is zero except at the knot u_i, where it is undefined; it is convenient to assume that its value at u_i is also zero, which means that a B-spline function whose all knots coincide is the zero function.

A non-decreasing sequence of knots $u_0, \ldots, u_N$ makes it possible to define B-spline functions $N_0^n, \ldots, N_{N-n-1}^n$. To obtain at least one function, we need $N > n$. We usually impose the condition of **progressivity** on knot sequences: there must be $u_i < u_{i+n+1}$ for $i = 0, \ldots, N-n-1$. All B-spline functions of degree n defined with such a progressive knot sequence are non-zero.

In Figure A.5 we can see examples of B-spline functions of degree 1, 2 and 3, defined with the same knot sequence. The knot sequence here is strictly increasing, and all functions are continuous. Moreover, the functions of degrees greater than 1 have the first-order derivative continuous. Looking at the graphs, we can make more observations, which will now be proved as properties of B-spline functions.

Property A.7 *Functions N_i^n satisfy the following* ***Mansfield–de Boor–Cox formula***[3]*:*

$$N_i^0(t) = \begin{cases} 1 & \text{for } t \in [u_i, u_{i+1}), \\ 0 & \text{else,} \end{cases} \tag{A.25}$$

$$N_i^n(t) = \frac{t - u_i}{u_{i+n} - u_i} N_i^{n-1}(t) + \frac{u_{i+n+1} - t}{u_{i+n+1} - u_{i+1}} N_{i+1}^{n-1}(t) \quad \text{for } n > 0. \tag{A.26}$$

Proof. Function N_i^0 is defined with $f_t^0(u) = (t-u)_+^0$, which gives us

$$N_i^0(t) = -(u_{i+1} - u_i)\frac{(t-u_i)_+^0 - (t-u_{i+1})_+^0}{u_i - u_{i+1}},$$

and it suffices to check three cases: $t < u_i$, $u_i \le t < u_{i+1}$ and $u_{i+1} \le t$.

If $n > 0$, then the function $f_t^n(u)$ is the product of the polynomial $g(u) = t - u$ and the function $h(u) = (t-u)_+^{n-1}$. The relation among divided differences of two functions and their product is described by the Leibniz formula (A.6). All divided differences of order higher than 1 of the polynomial $g(u)$ are equal to zero; hence, this formula in our case has only two terms:

$$f_t^n[u_i, \ldots, u_{i+n+1}] = g[u_i]h[u_i, \ldots, u_{i+n+1}] + g[u_i, u_{i+1}]h[u_{i+1}, \ldots, u_{i+n+1}],$$

[3]See de Boor [1972] and Cox [1972].

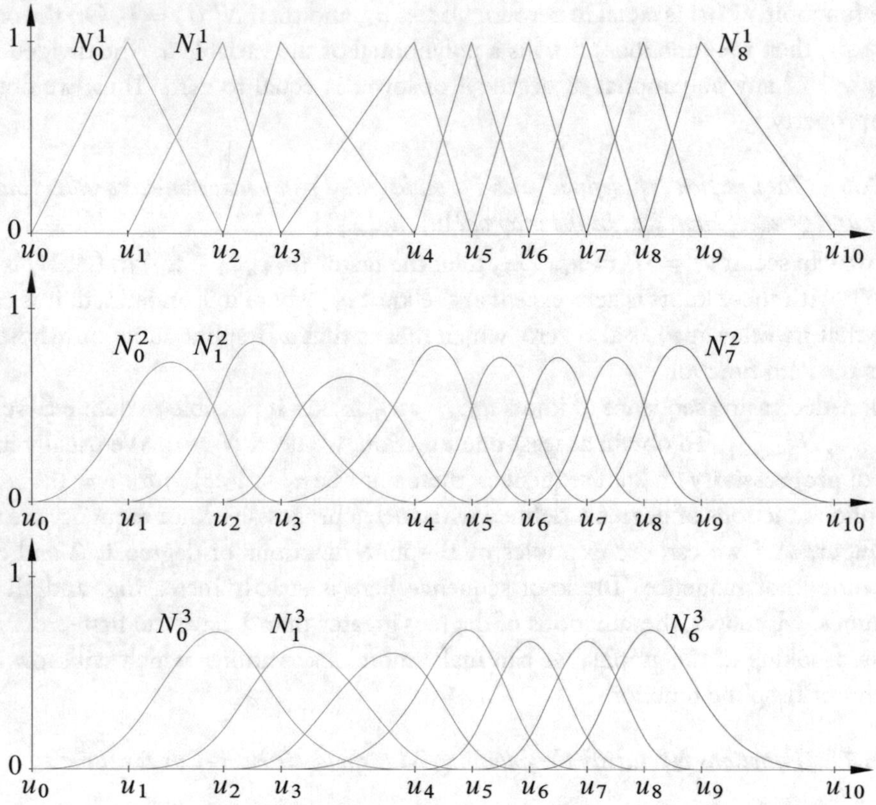

Figure A.5: Graphs of B-spline functions

and there is $g[u_i, u_{i+1}] = -1$. After taking an inductive assumption that

$$h[u_i, \dots, u_{i+n}] = \frac{(-1)^n N_i^{n-1}(t)}{u_{i+n} - u_i}, \quad h[u_{i+1}, \dots, u_{i+n+1}] = \frac{(-1)^n N_{i+1}^{n-1}(t)}{u_{i+n+1} - u_{i+1}},$$

we can calculate

$$N_i^n(t) = (-1)^{n+1}(u_{i+n+1} - u_i) \times \\ \times \left((t - u_i) \frac{\frac{(-1)^n N_i^{n-1}(t)}{u_{i+n} - u_i} - \frac{(-1)^n N_{i+1}^{n-1}(t)}{u_{i+n+1} - u_{i+1}}}{u_i - u_{i+n+1}} - \frac{(-1)^n N_{i+1}^{n-1}(t)}{u_{i+n+1} - u_{i+1}} \right).$$

The proof is completed by reordering the expression on the right hand side. □

Figure A.6 shows the Mansfield–de Boor–Cox formula in action; two B-spline functions of degree $n-1$ are multiplied by polynomials of degree 1, taking respectively the values 0 and 1, and

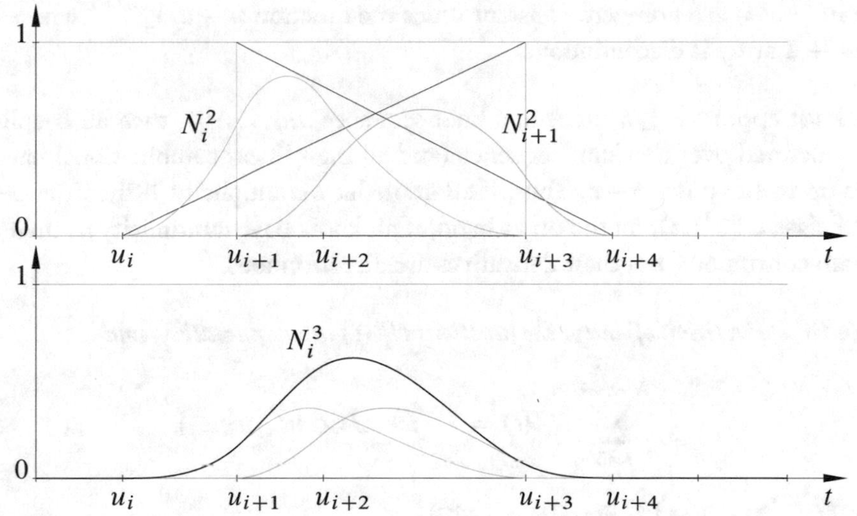

Figure A.6: Graphs of functions in the Mansfield–de Boor–Cox formula

1 and 0 at the end points of the intervals, in which these functions are non-zero. In general, if $u_i = \cdots = u_{i+n} < u_{i+n+1}$ or $u_i < u_{i+1} = \cdots = u_{i+n+1}$, then N_i^{n-1} or N_{i+1}^{n-1} is the zero function and the term with it vanishes.

Property A.8 *In each interval $(u_j, u_{j+1}) \subset [u_i, u_{i+n+1}]$ the function $N_i^n(t)$ is a polynomial of degree n.*

Proof. The fact that the function $N_i^n(t)$ in (u_j, u_{j+1}) is a polynomial of degree at most n is an immediate consequence of the Mansfield–de Boor–Cox formula. To prove that the degree is equal to n, we can check that the n-th order derivative of $N_i^n(t)$ is non-zero, which may be done using the formula for the derivative of B-spline functions discussed later (see Properties A.12 and A.14). □

Property A.9 *If the number of appearances of the number u_j in the sequence $u_i, \ldots, u_{i+n+1}$ is r, then the function $N_i^n(t)$ in a neighbourhood of u_j has continuous derivatives up to the order $n - r$ (and no more).*

Proof. The divided difference of a function is a linear combination of the values of this function and its derivatives at knots; if the multiplicity of a knot u_j is r, then the derivatives up to the order $r - 1$ are involved. If $r \leq n$, then the derivatives of order less than r of the function $f_t^n(u)$ are (up to constant factors—see Formulae (A.23)) truncated power functions of degree greater than $n - r$, and they are continuously differentiable $n - r$ times. One of the terms of the divided difference in (A.24) is a non-zero constant times the function $(t-u_j)_+^{n-r+1}$ whose derivative of order $n-r+1$ at u_j is discontinuous. □

If a knot appears $r \le n$ times in a knot sequence $u_0, \dots, u_N$, then all B-spline functions of degree n defined over this knot sequence (and all their linear combinations) have continuous derivatives up to the order $n-r$. Thus, if all knots have multiplicity 1, then the B-spline functions are of class $C^{n-1}(\mathbb{R})$; on the other hand, if all knots have multiplicity n, then the B-spline functions are continuous, but their derivatives are discontinuous.

Property A.10 ***Partition of unity:*** *the functions $N_i^n(t)$ are nonnegative, and*

$$\sum_{i=0}^{N-n-1} N_i^n(t) = 1 \quad \textit{for all } t \in [u_n, u_{N-n}).$$

Moreover, if $t \in [u_k, u_{k+1}) \subset [u_n, u_{N-n})$, then

$$\sum_{i=k-n}^{k} N_i^n(t) = 1.$$

Proof. A simple inductive proof using the Mansfield–de Boor–Cox formula is left as an exercise for the reader. □

Property A.11 ***Convex hull property:*** *if $t \in [u_k, u_{k+1}) \subset [u_n, u_{N-n})$, then the value of the spline function*

$$s(t) = \sum_{i=0}^{N-n-1} d_i N_i^n(t) = \sum_{i=k-n}^{k} d_i N_i^n(t)$$

is a convex combination of the coefficients $d_{k-n}, \dots, d_k$.

Proof. This property is an immediate consequence of Property A.10. □

The last two properties are the reason for introducing the factor $(-1)^{n+1}(u_{i+n+1}-u_n)$ in the definition of B-spline functions, which are, therefore, called normalised.[4] Knot sequences $u_0, \dots, u_N$ used in most applications are such that $N > 2n$; this makes them long enough to obtain at least one interval in which the sum of B-spline functions equals 1. In theoretical considerations it may also be convenient to take an infinite sequence $\dots, u_{-2}, u_{-1}, u_0, u_1, u_2 \dots$. Then, if $\lim_{i\to-\infty} u_i = -\infty$ and $\lim_{i\to\infty} u_i = \infty$, the normalised B-spline functions make a partition

[4]Hence the symbol N_i^n.

of unity in the entire set of real numbers. Conclusions found using this trick (e.g. in Section A.4.3) are useful because of the "locality" of B-spline functions; between any two consecutive knots precisely $n+1$ B-spline functions of degree n are non-zero.

Property A.12 *The derivative of a B-spline function of degree $n>0$, at the points t other than knots of multiplicity n, is described by the formula*

$$\frac{\mathrm{d}}{\mathrm{d}t}N_i^n(t) = \frac{n}{u_{i+n}-u_i}N_i^{n-1}(t) - \frac{n}{u_{i+n+1}-u_{i+1}}N_{i+1}^{n-1}(t). \tag{A.27}$$

Proof. To obtain the derivative of a B-spline function N_i^n, we can replace the truncated power function $f_t^n(u)$ in Formula (A.24) by its derivative with respect to the parameter t: $n(t-u)_+^{n-1} = nf_t^{n-1}(u)$. Then we obtain the equality

$$\begin{aligned} N_i^{n\prime}(t) &= (-1)^{n+1}(u_{i+n+1}-u_i)nf_t^{n-1}[u_i,\dots,u_{i+n+1}] \\ &= (-1)^{n+1}(u_{i+n+1}-u_i)\frac{nf_t^{n-1}[u_i,\dots,u_{i+n}] - nf_t^{n-1}[u_{i+1},\dots,u_{i+n+1}]}{u_i-u_{i+n+1}}, \end{aligned}$$

which, after reordering, gives us (A.27). □

Exercise. Derive the formulae for the derivative of a function $s(t)=\sum_{i=0}^{N-n-1} d_i N_i^n(t)$:

$$s'(t) = \sum_{i=0}^{N-n-2} \frac{n(d_{i+1}-d_i)}{u_{i+n+1}-u_{i+1}}N_{i+1}^{n-1}(t) \quad \text{for } t\in(u_n,u_{N-n}), \tag{A.28}$$

$$s'(t) = \sum_{i=k-n}^{k-1} \frac{n(d_{i+1}-d_i)}{u_{i+n+1}-u_{i+1}}N_{i+1}^{n-1}(t) \quad \text{for } t\in(u_k,u_{k+1})\subset(u_n,u_{N-n}), \tag{A.29}$$

using Properties A.10 and A.12.[5]

Property A.13 ***Local linear independence:*** *if $u_k<u_{k+1}$, then the B-spline functions $N_{k-n}^n,\dots,N_k^n$ restricted to the interval (u_k,u_{k+1}) are linearly independent.*

Proof. It suffices to prove that if the value of a function $s(t)=\sum_{i=k-n}^{k} d_i N_i^n(t)$ at all points of the interval (u_k,u_{k+1}) is zero, then $d_{k-n}=\dots=d_k=0$. For $n=0$ this is obvious. For $n>0$ we make an assumption that the functions $N_{k-n+1}^{n-1},\dots,N_k^{n-1}$ restricted to the interval (u_k,u_{k+1}) are linearly independent. If the function s vanishes in (u_k,u_{k+1}), then its derivative is also zero in this interval. By Formula (A.29) and the inductive assumption, $d_{i+1}-d_i=0$ for $i=k-n,\dots,k-1$, hence $d_{k-n}=\dots=d_k$. But, by the convex hull property, $\sum_{i=k-n}^{k} d_k N_i^n(t)=d_k$ for all $t\in(u_k,u_{k+1})$. Hence, $d_k=0$ and, thus, the induction step has been made. □

[5]Note that Formula (A.28) is incorrect if t is a knot of multiplicity n.

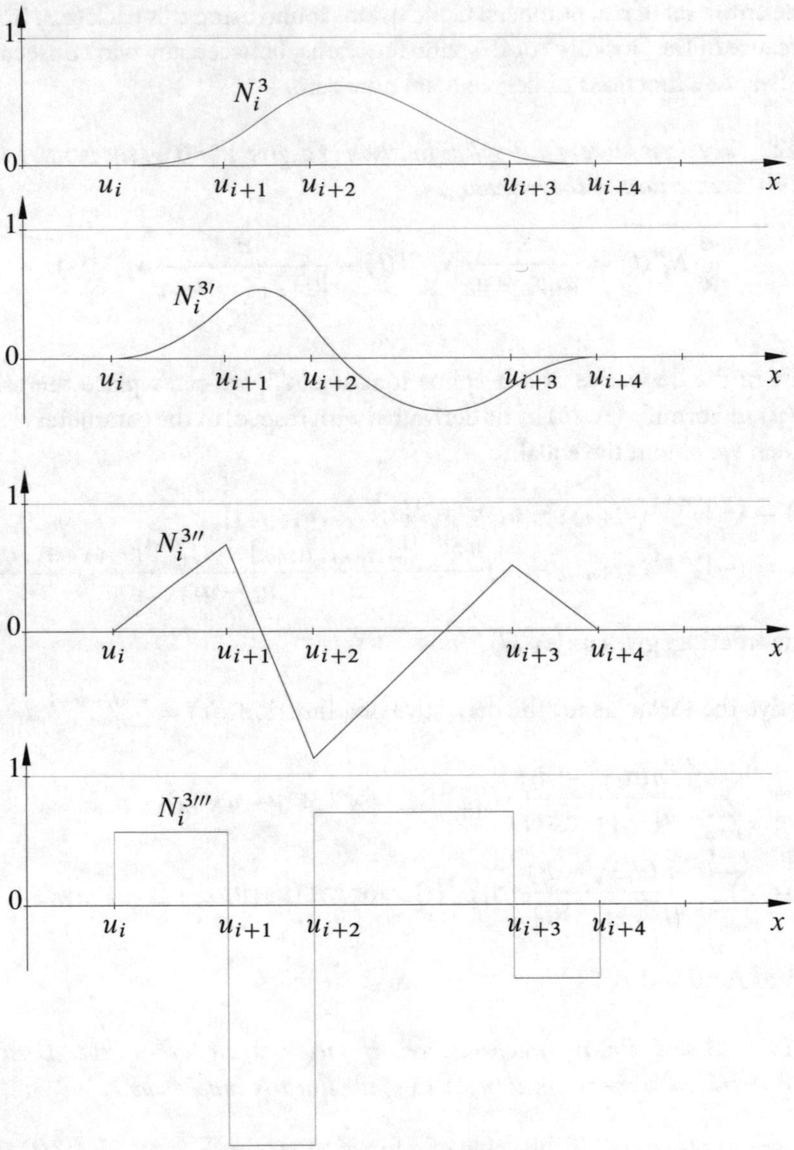

Figure A.7: A cubic B-spline function and its derivatives

Property A.14 ***Minimal support:*** *there is no non-zero spline function of degree* n *with knots* $u_i \leq \cdots \leq u_{i+n+1}$ *having continuous derivatives of order* $n - r$ *at each knot of multiplicity* r*, whose support (closure of the set of points, at which the function is non-zero) is a subset of the interval* $[u_i, u_{i+n+1}]$*, being the support of* N_i^n*.*

Proof. We prove it only for the case of all knots of multiplicity 1. The cases $n = 0$ and $n = 1$ are obvious. Let $n > 1$. Suppose that a function f contradicting the theorem, exists; f is a spline function of degree n of class C^{n-1} whose all knots are chosen from the set $\{u_i, \dots, u_{i+n+1}\}$. There is $f(u_i) = f'(u_i) = f(u_{i+n+1}) = f'(u_{i+n+1}) = 0$ and without loss of generality we may assume that $f(t_1^{(0)}) > 0$ for some $t_1^{(0)} \in (u_i, u_{i+n+1})$. The function f increases and then decreases; hence, due to the continuity of f' there exist points $t_1^{(1)}, t_2^{(1)}$ such that $u_i < t_1^{(1)} < t_1^{(0)} < t_2^{(1)} < u_{i+n+1}$ and $f'(t_1^{(1)}) > 0$, $f'(t_2^{(1)}) < 0$.

Using induction we apply the same reasoning to the continuous derivatives of f, i.e., up to the order $n-1$. The conclusion is that there exist points $t_1^{(n-1)} < \dots < t_n^{(n-1)}$ such that the values of the $n-1$-st order derivative of f at these points have alternating signs (i.e., $f^{(n-1)}(t_1^{(n-1)}) > 0$, $f^{(n-1)}(t_2^{(n-1)}) < 0$ etc.). But the function $f^{(n-1)}$ is a continuous spline of degree 1, whose knots must be chosen from the set $\{u_i, \dots, u_{i+n+1}\}$. Its derivative, $f^{(n)}$, is a piecewise constant function, whose values must have alternating signs in at least $n+1$ intervals bounded by the knots. That many intervals may be obtained only by taking all $n+2$ knots of the B-spline function N_i^n (see Figure A.7) and, thus, the support of f cannot be smaller than the interval $[u_i, u_{i+n+1}]$.
□

Property A.15 *If* $u_i < u_{i+1} = \dots = u_{i+n} < u_{i+n+1}$, *then* $N_i^n(u_{i+1}) = 1$.

Proof. By Property A.10, $\sum_{j=i-n}^{i} N_j^n(u_{i+1}) = 1$. By Property A.6, if $j \neq i$, then $N_j^n(u_{i+1}) = 0$ and $N_i^n(u_{i+1})$ is the only non-zero term of the sum. □

Property A.16 *If* $u_{k-n+1} = \dots = u_k < u_{k+1} = \dots = u_{k+n}$ *and* $t \in (u_k, u_{k+1})$, *then* $N_{k-n+i}^n(t) = B_i^n(s)$, *where* $s = (t - u_k)/(u_{k+1} - u_k)$.

Proof. For knots and the arguments t and s as above, Formulae (A.25) and (A.26) become respectively (A.9) and (A.10). □

The Mansfield–de Boor–Cox formulae (A.25), (A.26) may be used to evaluate B-spline functions for a given argument t; this is done by Algorithm A.7. This formula is also at the foundation of Algorithm A.8 for evaluating a function $s(t) = \sum_i d_i N_i^n(t)$ for a given t.

A.4.2 NONUNIFORM B-SPLINE CURVES

A B-spline curve of degree n is defined by specifying a non-decreasing knot sequece $u_0, \dots, u_N$, where $N > 2n$, and the **control points** $\boldsymbol{d}_0, \dots, \boldsymbol{d}_{N-n-1}$. The knot sequence ought to be progressive, i.e., satisfy the condition $u_{i+n+1} > u_i$ for $i = 0, \dots, N-n-1$. The curve has the

Algorithm A.7 De Boor's algorithm of evaluating B-spline functions

Input : Degree n, knots $u_{k-n}, \dots, u_{k+n+1}$, argument $t \in [u_k, u_{k+1})$

```
b[k] = 1;
for ( j = 1; j <= n; j++ ) {
  β = (u_{k+1} − t)/(u_{k+1} − u_{k−j+1});
  b[k-j] = β*b[k-j+1];
  for ( i = k-j+1; i < k; i++ ) {
    α = 1 − β;
    β = (u_{i+j+1} − t)/(u_{i+j+1} − u_{i+1});
    b[i] = α*b[i] + β*b[i+1];
  }
  b[k] *= (1 − β);
}
```

Output : Values of the functions N_i^n at t in the array b, for $i = k - n, \dots, k$

Algorithm A.8 De Boor's algorithm of evaluating a function represented in B-spline basis

Input : Degree n, knots $u_{k-n}, \dots, u_{k+n+1}$, coefficients $d_{k-n}, \dots, d_k$, argument $t \in [u_k, u_{k+1})$

```
/* denote d_i^(0) = d_i for i = k − n,...,k */
for ( j = 1; j <= n; j++ )
  for ( i = k-n+j; i <= k; i++ ) {
    α = (t − u_i)/(u_{i+n+1−j} − u_i);
    d_i^(j) = (1 − α)*d_{i−1}^(j−1) + α*d_i^(j−1);
  }
```

Output : $d_k^{(n)} = s(t)$

parametrisation

$$s(t) = \sum_{i=0}^{N-n-1} \boldsymbol{d}_i N_i^n(t), \quad t \in [u_n, u_{N-n}). \tag{A.30}$$

The domain of this parametrisation is the interval in which the sum of the B-spline functions N_i^n defined over the knot sequence $u_0, \dots, u_N$ is equal to 1.[6] For t in this interval, one can use Algorithm A.8 to compute the point $\boldsymbol{s}(t)$, replacing the coefficients d_i by the control points $\boldsymbol{d}_i$.

The properties of B-spline functions given in the previous section translate to the well-known properties of B-spline curves. Among the most important are the affine invariance of the representation (the image of the curve under an affine transformation f is represented by

[6]In practice we take the closed interval $[u_n, u_{N-n}]$, and assume that $\boldsymbol{s}(u_{N-n}) = \lim_{t \nearrow u_{N-n}} \boldsymbol{s}(t)$.

the control points $f(\boldsymbol{d}_0), \ldots, f(\boldsymbol{d}_{N-n-1})$), the convex hull property (if $t \in [u_k, u_{k+1})$, then the point $\boldsymbol{s}(t)$ is a convex combination of the points $\boldsymbol{d}_{k-n}, \ldots, \boldsymbol{d}_k$) and the derivative continuity rule (at a knot u_k of multiplicity $r < n$ the parametrisation $\boldsymbol{s}$ has $n - r$ continuous derivatives). From (A.28) we obtain the formula

$$\boldsymbol{s}'(t) = \sum_{i=0}^{N-n-2} \frac{n(\boldsymbol{d}_{i+1} - \boldsymbol{d}_i)}{u_{i+n+1} - u_{i+1}} N_{i+1}^{n-1}(t) \quad \text{for } t \in [u_n, u_{N-n}), \tag{A.31}$$

which makes it possible to estimate the directions of tangent lines using the convex hull property and (used recursively) to evaluate higher order derivatives or to impose Hermite interpolation conditions on spline curves.

Inserting knots

Given a B-spline curve represented with a knot sequence and control points, one can find representations of this curve with additional knots. There are a number of algebraically equivalent methods of doing it. The most flexible is the method invented by Boehm [1980] (Algorithm A.9). It constructs a representation of the curve with one additional knot. If the sequence $\hat{u}_0, \ldots, \hat{u}_{N+1}$ is obtained by inserting the number $t \in [u_n, u_{N-n})$ to the sequence $u_0, \ldots, u_N$ in such a place that the monotonicity is preserved, and the B-spline functions determined by this new sequence are denoted by $\hat{N}_0^n, \ldots, \hat{N}_{N-n}^n$, then

$$\boldsymbol{s}(t) = \sum_{i=0}^{N-n-1} \boldsymbol{d}_i N_i^n(t) = \sum_{i=0}^{N-n} \hat{\boldsymbol{d}}_i \hat{N}_i^n(t),$$

where $\hat{\boldsymbol{d}}_i$ are the control points computed by the algorithm.

Algorithm A.9 Boehm's knot insertion

Input: Degree n, knots $u_0, \ldots, u_N$, control points $\boldsymbol{d}_0, \ldots, \boldsymbol{d}_{N-n-1}$, number $t \in [u_n, u_{N-n})$

```
... /* Find k such that t ∈ [u_k, u_{k+1}) */
... /* If t = u_k , then find r such that u_{k-r} < u_{k-r+1} = t , else take r = 0 */
for ( i = N-n-1; i >= k-r; i-- ) d̂_{i+1} = d_i ;
for ( i = k-r; i >= k-n+1; i-- ) d̂_i = ((u_{i+n} − t)d_{i−1} + (t − u_i)d_i)/(u_{i+n} − u_i);
for ( i = k-n; i >= 0; i-- ) d̂_i = d_i ;
for ( i = N; i > k; i-- ) û_{i+1} = u_i ;
û_{k+1} = t ;
for ( i = k; i >= 0; i-- ) û_i = u_i ;
```

Output: Knots $\hat{u}_0, \ldots, \hat{u}_{N+1}$, control points $\hat{\boldsymbol{d}}_0, \ldots, \hat{\boldsymbol{d}}_{N-n}$

One of the most important properties of this algorithm is that if a number of new knots are inserted, the final effect does not depend on the ordering of these knots. There are a number of applications of knot insertion. It turns out that the points computed by Algorithm A.8 for a given j are identical to the points obtained when the number t is inserted for the j-th time by Algorithm A.9. Another application is finding the Bézier representations of the polynomial arcs of the curve. By Property A.16, inserting knots so as to obtain a representation with all knots of multiplicity $n+1$ results in obtaining the control polygon of the spline curve, made of Bézier control polygons of the arcs.[7] Such a representation of a B-spline curve is often used to draw the curve using a procedure of rendering Bézier curves.

Removing knots and algebra of splines

If a spline curve of degree n is represented in B-spline basis defined for a knot sequence $\hat{u}_0, \ldots, \hat{u}_{N+1}$, with a knot $\hat{u}_{k+1}$ of multiplicity $r+1$, such that $\hat{u}_k \le \hat{u}_{k+1} < \hat{u}_{k+2}$ and $n < k < N-n$, and the parametrisation has the derivative of order $n-r$ continuous at the knot $\hat{u}_{k+1}$, then this knot is removable, i.e., there exists a representation of this curve with the knot sequence $(u_0, \ldots, u_N) = (\hat{u}_0, \ldots, \hat{u}_k, \hat{u}_{k+2}, \ldots, \hat{u}_{N+1})$.

Algorithm A.9 is based on a linear dependency between the points $\boldsymbol{d}_0, \ldots, \boldsymbol{d}_{N-n-1}$ and the points $\hat{\boldsymbol{d}}_0, \ldots, \hat{\boldsymbol{d}}_{N-n}$ obtained by knot insertion:

$$\begin{aligned} \boldsymbol{d}_i &= \hat{\boldsymbol{d}}_i && \text{for } i \le k-n, \\ \frac{u_{i+n}-\hat{u}_{k+1}}{u_{i+n}-u_i}\boldsymbol{d}_{i-1} + \frac{\hat{u}_{k+1}-u_i}{u_{i+n}-u_i}\boldsymbol{d}_i &= \hat{\boldsymbol{d}}_i && \text{for } i = k-n+1, \ldots, k-r, \\ \boldsymbol{d}_{i-1} &= \hat{\boldsymbol{d}}_i && \text{for } i > k-r. \end{aligned} \tag{A.32}$$

Given the control points $\hat{\boldsymbol{d}}_i$ representing an arbitrary B-spline curve we can write these equations with the intention of finding unknown control points $\boldsymbol{d}_i$. The number of equations is the number of unknown control points plus 1. This system of equations is consistent if the knot $\hat{u}_k$ is removable and the curve has a shorter representation. To obtain a good accuracy, it is best to compute the points $\boldsymbol{d}_{k-n}, \ldots, \boldsymbol{d}_{k-r}$ by solving a linear least-squares problem.

Given a B-spline curve, one can remove all "superfluous" knots and obtain the "minimal" representation of the curve. It may be useful as a step of the **degree elevation** algorithm, which is the following: suppose that the degree of the initial representation is n and a representation of degree $n+k$ is needed. In the first step (Fig. A.8a) knots are inserted so as to obtain the representation with all knots of degree $n+1$; the resulting control polygon consists of Bézier control polygons of the polynomial arcs of the curve. Then, degree elevation (to $n+k$) is applied to these arcs as described in Section A.2.3. The B-spline representation of the curve with the control polygon made of the Bézier representations of degree $n+k$ of the arcs has the knot

[7]In principle it suffices to insert each knot up to the multiplicity n; the last vertex of the control polygon of each Bézier arc is then the first control point of the next arc. However, a spline curve may have knots of multiplicity greater than n and then the curve does not have to be continuous. Then, we have to obtain formally disconnected Bézier control polygons, and the multiplicity of all knots after inserting must be $n+1$. In practice, it happens with B-spline representations of derivatives of B-spline curves (see Formula (A.31)).

sequence with all knots of multiplicity $n + k + 1$ (Fig. A.8b). By Property A.9, it is possible to remove all knots inserted in the first step (Fig. A.8c).

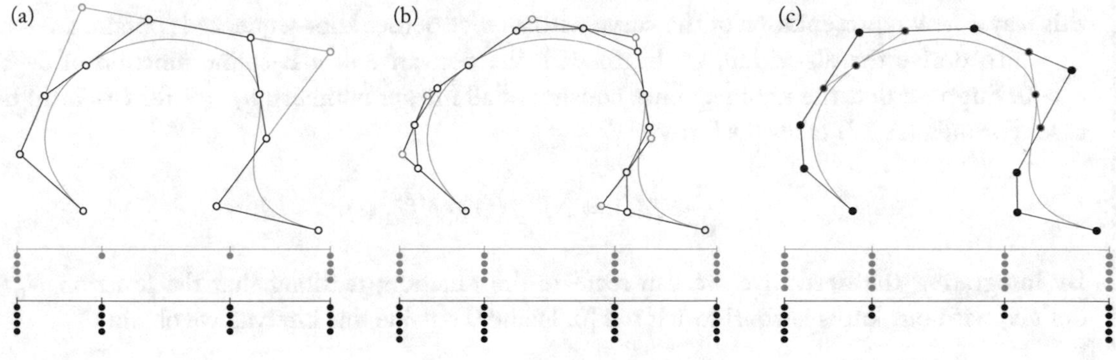

Figure A.8: Degree elevation for a B-spline curve from 3 to 4

Another application is the implementation of algebraic operations (multiplication and addition) for spline functions. The multiplication may be done using the piecewise Bézier representation of the arguments. Note that the product of two functions having at a point continuous derivatives up to the order k and l is guaranteed to have the continuous derivative of order $\min\{k, l\}$ at that point. This fact makes it possible to determine the removable knots of the piecewise Bézier representation of the product, using Property A.9.

Convergence of repeated knot insertion

If an infinite sequence of numbers $v_1, v_2, \ldots$ is dense in the interval $[u_n, u_{N-n})$, then the process of inserting these numbers as new knots produces an infinite sequence of representations of a given curve $\boldsymbol{s}$, with the control polylines made of increasing numbers of shorter and shorter line segments. These polylines converge to the curve. It may be proved (see Cohen and Schumaker [1985]) that the distance between the polyline and the curve is estimated by the expression Ch^2, where the constant C depends on the curve and h is the maximal distance between consecutive knots of its representation. This is a fast convergence; often it suffices to insert relatively few knots in order to obtain a polyline being a very good approximation of the spline curve. This fact is exploited by the methods of generating surfaces via mesh refinement discussed in Section A.6.

A.4.3 B-SPLINE CURVES WITH UNIFORM KNOTS

Plenty of properties of B-spline curves become simpler if we impose the restriction that for all i, $u_{i+1} - u_i = h$, where h is a fixed positive constant. An immediate consequence is that all B-spline functions are identical up to translations of the argument:

$$N_i^n(t) = N_{i+k}^n(t + kh) \quad \text{for all } k \in \mathbb{Z}, t \in \mathbb{R}.$$

The restriction imposed on the knot sequence makes it possible to use special algorithms. One of them is the algorithm given by Lane and Riesenfeld [1980], which is a special method of knot insertion; a new knot is inserted in the middle of each interval ended by two old knots. In this way a new representation of the curve with a twice denser knot sequence is obtained.

To derive the algorithm, we begin with the derivative of a B-spline function of degree $n > 0$. Suppose that the knot sequence consists of all integer numbers: $u_i = i$ for $i \in \mathbb{Z}$. In this case, Formula (A.27) takes the form

$$\frac{\mathrm{d}}{\mathrm{d}t} N_i^n(t) = N_i^{n-1}(t) - N_{i+1}^{n-1}(t).$$

By integrating the derivative, we can recreate the function; recalling that the function $N_0^0(t)$ defined with our knots is equal to 1 if $t \in [0, 1)$ and 0 outside this interval, we obtain

$$\begin{aligned} N_i^n(t) &= \int_{-\infty}^{t} N_i^{n-1}(u) - N_{i+1}^{n-1}(u)\,\mathrm{d}u = \int_{t-1}^{t} N_i^{n-1}(u)\,\mathrm{d}u \\ &= \int_{\mathbb{R}} N_i^{n-1}(t-u) N_0^0(u)\,\mathrm{d}u. \end{aligned}$$

The final expression describes the operation called convolution of two functions, namely, N_i^{n-1} and N_0^0. In the short form it is written as follows: $N_i^n = N_i^{n-1} * N_0^0$.

The functions $M_i^n(t) \stackrel{\text{def}}{=} N_0^n(2t - i)$ are B-spline functions defined with a sequence of uniform knots being integer multiples of $\frac{1}{2}$. The support of the function M_i^n is the interval $[\frac{i}{2}, \frac{i+n+1}{2}]$. Let $\boldsymbol{c}_i$ be points defined for all integer i. For all $j = 0, 1, 2, \ldots$ there exists a B-spline curve of degree j with these control points:

$$\boldsymbol{s}^{(j)}(t) = \sum_{i \in \mathbb{Z}} \boldsymbol{c}_i N_i^j(t).$$

Having the points $\boldsymbol{c}_i$, we are going to find the points representing the curve $\boldsymbol{s}^{(n)}$ with the B-spline functions M_i^n, i.e., with the twice denser knot sequence; note that each curve $\boldsymbol{s}^{(j)}$ has its own control points $\boldsymbol{d}_i^{(j)}$ corresponding to the denser knots.

We begin with the curve of degree 0, which is just the sequence of the control points. There is $N_i^0(t) = M_{2i}^0(t) + M_{2i+1}^0(t)$ and

$$\boldsymbol{s}^{(0)}(t) = \sum_{i \in \mathbb{Z}} \boldsymbol{c}_i N_i^0(t) = \sum_{i \in \mathbb{Z}} \boldsymbol{c}_i \big(M_{2i}^0(t) + M_{2i+1}^0(t)\big) = \sum_{i \in \mathbb{Z}} \boldsymbol{d}_i^{(0)} M_i^0(t);$$

hence, $\boldsymbol{d}_{2i}^{(0)} = \boldsymbol{d}_{2i+1}^{(0)} = \boldsymbol{c}_i$ for all $i \in \mathbb{Z}$. The new representation of the curve $\boldsymbol{c}^{(0)}$ may be obtained by **doubling** of the original control points, $\boldsymbol{c}_i$.

Let $j > 0$. The function M_i^j may also be obtained as the convolution of B-spline functions of lower degree—there is $M_i^j = 2M_i^{j-1} * M_0^0$ for all i. Note that $N_0^0 = M_0^0 + M_1^0$. We use these facts to calculate

$$\begin{aligned}
s^{(j)}(t) &= \sum_{i\in\mathbb{Z}} c_i N_i^j(t) = \int_{\mathbb{R}} \sum_{i\in\mathbb{Z}} c_i N_i^{j-1}(t-u) N_0^0(u)\,\mathrm{d}u \\
&= \int_{\mathbb{R}} \sum_{i\in\mathbb{Z}} \boldsymbol{d}_i^{(j-1)} M_i^{j-1}(t-u)\big(M_0^0(u) + M_1^0(u)\big)\,\mathrm{d}u \\
&= \frac{1}{2}\sum_{i\in\mathbb{Z}} \boldsymbol{d}_i^{(j-1)}\big(M_i^j(t) + M_{i+1}^j(t)\big) \\
&= \sum_{i\in\mathbb{Z}} \frac{1}{2}\big(\boldsymbol{d}_{i-1}^{(j-1)} + \boldsymbol{d}_i^{(j-1)}\big) M_i^j(t).
\end{aligned}$$

The result of this calculation is the following: $\boldsymbol{d}_i^{(j)} = \frac{1}{2}(\boldsymbol{d}_{i-1}^{(j-1)} + \boldsymbol{d}_i^{(j-1)})$. The computation of midpoints between $\boldsymbol{d}_{i-1}^{(j-1)}$ and $\boldsymbol{d}_i^{(j-1)}$ is called **averaging**.

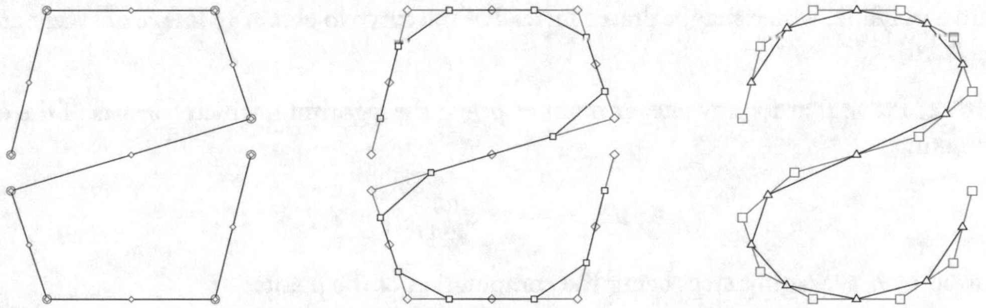

Figure A.9: Steps of the Lane–Riesenfeld algorithm

The Lane–Riesenfeld algorithm producing the control points of a spline curve of degree n corresponding to a twice denser uniform knot sequence consists of the doubling operation followed by n averaging steps. The result of the first averaging is the initial polygon with each line segment halved by a new vertex (Fig. A.9). In practice, we have a finite sequence of control points, say, $\boldsymbol{c}_0, \dots, \boldsymbol{c}_{N-n-1}$, and the parametrisation whose domain is the interval $[n, N-n]$. The Lane–Riesenfeld algorithm produces the control points $\boldsymbol{d}_n^{(n)}, \dots, \boldsymbol{d}_{2N-2n-1}^{(n)}$.[8]

The control polygon is an approximation of the B-spline curve it represents. Repeating the knot insertion produces a sequence of control polygons of the same curve, and the distance between any of those polygons and the curve (assuming the same parametrisation) is approximately proportional to the square of the distance between the knots. In particular, the control polygon produced by the Lane–Riesenfeld algorithm, corresponding to a twice denser knot sequence, is

[8]By renumbering the points, i.e., taking $\boldsymbol{d}_i = \boldsymbol{d}_{i+n}^{(n)}$, we can obtain a sequence of control points $\boldsymbol{d}_0, \dots, \boldsymbol{d}_{2N-3n-1}$, which may be convenient in a computer implementation. Note that the actual distance h between the knots is irrelevant and we can define a new parametrisation of the curve with the integer knots $0, \dots, 2N-2n$ and with the renumbered control points. The domain of the new parametrisation is the interval $[n, 2N-3n)$.

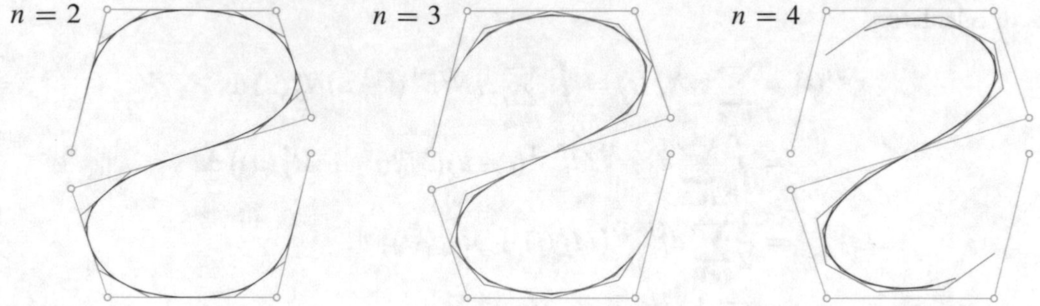

Figure A.10: Convergence of control polygons to a quadratic, cubic and quartic spline curve

about four times closer to the curve than the original polygon. By repeating the algorithm, we obtain a sequence of polylines which converges to the curve. Usually several iterations suffice to obtain a polyline, which may be drawn instead of the curve to obtain an image of excellent quality (Fig. A.10).

Exercise. Prove that for any integer number $p \geq 2$ the algorithm, which consists of the **copying** step, taking

$$\boldsymbol{d}_{pi}^{0} = \cdots = \boldsymbol{d}_{pi+p-1}^{(0)} = \boldsymbol{c}_i,$$

followed by n **averaging** steps being the computation of the points

$$\boldsymbol{d}_i^{(j)} = \frac{1}{p}(\boldsymbol{d}_{i-p+1}^{(j-1)} + \cdots + \boldsymbol{d}_i^{(j-1)}) \quad \text{for } j = 1, \ldots, n,$$

produces a new representation of a B-spline curve of degree n with uniform knots having the control points $\boldsymbol{c}_i$. The new representation corresponds to a uniform knot sequence p times denser than the original one.

A.5 TENSOR PRODUCT B-SPLINE PATCHES

The relation between **tensor product B-spline patches** and B-spline curves is analogous to that between Bézier patches and curves. In the formula

$$\boldsymbol{s}(u, v) = \sum_{i=0}^{N-n-1} \sum_{j=0}^{M-m-1} \boldsymbol{d}_{ij} N_i^n(u) N_j^m(v) \tag{A.33}$$

there are **control points** $\boldsymbol{d}_{ij}$ and two families of normalised B-spline functions, N_i^n and N_j^m, defined with two (in general different, even if $n = m$) non-decreasing knot sequences $u_0, \ldots, u_N$

and $v_0, \ldots, v_M$, where $N > 2n$ and $M > 2m$. The domain of the parametrisation $\boldsymbol{s}$ is the rectangle $[u_n, u_{N-n}) \times [v_m, v_{M-m})$, i.e., the Cartesian product of the intervals in which the B-spline functions $N_0^n, \ldots, N_{N-n-1}^n$ and $N_0^m, \ldots, N_{M-m-1}^m$ make partitions of unity.[9]

Algorithms of processing B-spline curves may be used to B-spline patches; in particular, it is possible to compute a point $\boldsymbol{s}(u, v)$ (which may be done by computing a number of points of B-spline curves) or perform knot insertion, degree elevation and computation of partial derivatives. Formula (A.31) allows us to write the formulae

$$\frac{\partial}{\partial u}\boldsymbol{s}(u, v) = \sum_{i=0}^{N-n-2}\sum_{j=0}^{M-m-1} \frac{n(\boldsymbol{d}_{i+1,j} - \boldsymbol{d}_{ij})}{u_{i+n+1} - u_{i+1}} N_{i+1}^{n-1}(u) N_j^m(v), \tag{A.34}$$

$$\frac{\partial}{\partial v}\boldsymbol{s}(u, v) = \sum_{i=0}^{N-n-1}\sum_{j=0}^{M-m-2} \frac{m(\boldsymbol{d}_{i,j+1} - \boldsymbol{d}_{ij})}{v_{j+m+1} - v_{j+1}} N_i^n(u) N_{j+1}^{m-1}(v), \tag{A.35}$$

which may be used to compute the partial derivatives of the parametrisation. These formulae may also be used to construct B-spline patches satisfying interpolation conditions at the boundary, which is a step of the constructions described in Section 3.6. The control points of a patch with a prescribed boundary curve and cross-boundary derivative may be obtained as in the example below. Suppose that the degree of a patch with respect to u is 3. Let the boundary curve of this patch, corresponding to $u = u_3$, be given in the form of a B-spline curve $\boldsymbol{c}$. Moreover, let a B-spline curve $\boldsymbol{d}$ specify the partial derivative with respect to u at $u = u_3$. Both curves have the same domain $[a, b]$. The first step is to find compatible representations of both curves, i.e., representations of the same degree m and with the same knot sequence $v_0, \ldots, v_M$. If the original representations are of different degrees, then degree elevation of one curve is necessary; also, knot insertion may have to be applied (see Section A.4.2). After these preparations, we have the control points $\boldsymbol{c}_j$ and $\boldsymbol{d}_j$ such that

$$\boldsymbol{c}(v) = \sum_{j=0}^{M-m-1} \boldsymbol{c}_j N_j^m(v), \qquad \boldsymbol{d}(v) = \sum_{j=0}^{M-m-1} \boldsymbol{d}_j N_j^m(v).$$

Due to Property A.15, it is convenient to take $u_1 = u_2 = u_3$. As the multiplicity of the knot u_3, which is the lower boundary of the range of the parameter u, is equal to the degree of the patch with respect to this parameter, fixing $u = u_3$ in (A.33) and (A.34) gives us the following formulae:

$$\boldsymbol{d}_{0j} = \boldsymbol{c}_j, \tag{A.36}$$

$$\boldsymbol{d}_{1j} = \boldsymbol{d}_{0j} + \frac{u_4 - u_1}{3}\boldsymbol{d}_j. \tag{A.37}$$

[9]We usually take the closed rectangle $[u_n, u_{N-n}] \times [v_m, v_{M-m}]$; see the footnote on page 180.

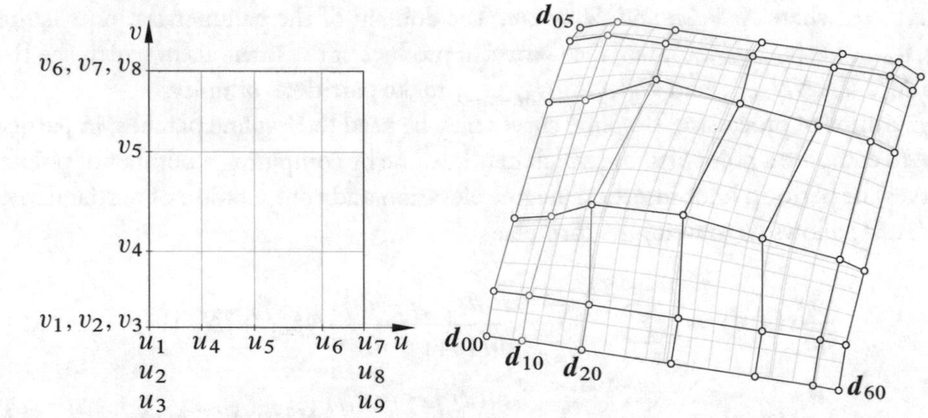

Figure A.11: A bicubic B-spline patch and its knot sequences. The first three columns of the control net determine the boundary curve and the cross-boundary derivatives of order 1 and 2 for $u = u_3$

Using these formulae, we construct the first two columns of the control net. The other control points may be chosen arbitrarily. If the second-order cross-boundary derivative of the patch $\boldsymbol{s}$ is specified in addition to the interpolation conditions considered above, then the control points in the third column have to be

$$\boldsymbol{d}_{2j} = \boldsymbol{d}_{1j} + (u_5 - u_2)\Big(\frac{1}{3}\boldsymbol{d}_j + \frac{u_4 - u_2}{6}\boldsymbol{e}_j\Big), \tag{A.38}$$

where $\boldsymbol{e}_j$ are the control points of the curve specifying the second-order cross-boundary derivative $\boldsymbol{e}(v) = \sum_{j=0}^{M-m-1} \boldsymbol{e}_j N_j^m(v)$. This formula is derived from (A.34) used recursively. An example of a B-spline patch in Figure A.11 is drawn using colours to indicate the columns of the control net involved in the interpolation problem considered here. In a similar way, we can use formulae for control points in the last two or three columns and in the first or last two or three rows of the control net of a B-spline patch of interpolation to given boundary curves and cross-boundary derivatives. If higher order derivatives are to be interpolated, patches of a higher degree must be used to ensure the continuity of these derivatives.

A.6 MESHES AND GENERALISED B-SPLINE SURFACES

Tensor product B-spline patches may be defined with uniform knot sequences. It is possible to generalise the definition and obtain surfaces with arbitrary topology, represented by meshes. The constructions described in Chapter 5 may be used to fill polygonal holes in such surfaces, and in this section we focus on the information necessary to develop these constructions.

A.6.1 B-SPLINE PATCHES WITH UNIFORM KNOTS

A B-spline patch defined with the knot sequences $u_0, \ldots, u_N$ and $v_0, \ldots, v_M$ such that $u_{i+1} - u_i = \text{const}$, $v_{j+1} - v_j = \text{const}$ may be processed using all algorithms applicable in the general case. In addition, algorithms assuming uniform knot sequences may be used. The Lane–Riesenfeld algorithm described in Section A.4.3 may be used to double the density of each knot sequence—either rows or columns of the control net of the patch may be processed just like control polygons of B-spline curves of degree n or m.

Note that if doubling or averaging, applied to rows of the control net, is followed by doubling or averaging done in columns, then the two operations may be done in the reversed order; the final result is the same. This property is the basis of the two-dimensional variant of the Lane–Riesenfeld algorithm for tensor product B-spline patches with uniform knots and such that $n = m$. Let the control points of the patch be denoted by $\boldsymbol{c}_{ik}$. In the first step of the algorithm, these points are copied:

$$\boldsymbol{d}^{(0)}_{2i,2k} = \boldsymbol{d}^{(0)}_{2i+1,2k} = \boldsymbol{d}^{(0)}_{2i,2k+1} = \boldsymbol{d}^{(0)}_{2i+1,2k+1} = \boldsymbol{c}_{ik}.$$

Each row and column of the control net produces two rows or columns; this **doubling** operation is followed by n **averaging steps**:

$$\boldsymbol{d}^{(j)}_{ik} = \frac{1}{4}\big(\boldsymbol{d}^{(j-1)}_{i-1,k-1} + \boldsymbol{d}^{(j-1)}_{i,k-1} + \boldsymbol{d}^{(j-1)}_{i-1,k} + \boldsymbol{d}^{(j-1)}_{i,k}\big),$$

for $j = 1, \ldots, n$. The points $\boldsymbol{d}^{(n)}_{ik}$ form a representation of the B-spline patch of degree (n, n) whose original control points are $\boldsymbol{c}_{ik}$; the new representation has both knot sequences uniform, but twice denser. Iterating this algorithm produces a sequence of control nets converging to the spline patch represented by all these nets.

A.6.2 MESH REPRESENTATION OF SURFACES

To obtain a surface with arbitrary topology, we replace the control net of a B-spline patch whose vertices may easily be stored in a rectangular array by a **mesh**. A mesh is an object consisting of **vertices, edges** and **facets** satisfying the conditions described below. A vertex is a control point of the surface with a given location in space. An edge is a line segment ended at both sides by vertices. A facet is a closed polyline made of at least three edges; one can see a facet as a surface (a polygon or a soap membrane filling the frame made of the edges). The following conditions must be satisfied:

- An edge may belong to either one or two facets; in the former case it will be called a **boundary edge**, while in the latter case it is called **internal**.
- A facet made of k edges must have k different vertices; note that different vertices may have the same location in space. Different vertices (and edges and facets) have different numbers

(or other identifiers) which distinguish them even if their locations coincide. Two facets may have at most one common edge.

- A vertex may have zero or two incident boundary edges and accordingly it is called an **internal** or a **boundary vertex**. A common vertex of a number of facets (which have edges incident with this vertex) must be surrounded by these facets in a simple way, i.e., crossing subsequent edges incident with the vertex, one must be able to go around it and visit all the facets having this vertex. An internal vertex ought to be incident with at least three edges.

Other restrictions may be imposed, e.g. the orientability of the mesh (which precludes the presence of a Möbius band made of facets in the mesh). A mesh, as described above, may be seen as a surface made of polygons; facets whose vertices are not coplanar may be divided into triangles and then rendered on a picture or used to manufacture an object, whose surface is made of the triangles, using a 3D printer. However, for the purposes of this book we are more interested in curved smooth surfaces represented by meshes.

A.6.3 MESH REFINEMENT

A smooth surface may be obtained by repeating iteratively an operation of **mesh refinement**. A variety of such operations are described in detail in the book by Prautzsch, Boehm and Paluszny [2002]. Here we consider just one mesh refinement operation, a generalisation of the Lane–Riesenfeld algorithm for tensor product patches (see Section A.6.1).

The operation consists of a step called doubling, followed by n averaging steps; the number n is the degree of polynomial patches which make the surface, e.g. with $n = 3$ we obtain a piecewise bicubic surface.

The mesh obtained in the **doubling** step has facets corresponding to all vertices, edges and facets of the original mesh. A facet corresponding to an edge is a quadrilateral; its two opposite edges have the vertices located at the locations of the end points of the original edge and, consequently, the other two edges are contracted to points. In Figure A.12a these facets are drawn as narrow trapeziums in order to better show the idea. If the original edge is a boundary edge, its corresponding facet is attached to the mesh and one of its edges becomes a boundary edge of the resulting mesh. If the original edge is internal, then the facet is inserted between the copies of the two facets sharing the edge.

The facets of the original mesh are copied to the final mesh; note, however, that as a facet is defined as a closed polyline made of edges, the edges and the vertices of the copy will have new identifiers, even if the new vertices have the same locations in space. Finally, the facets corresponding to the vertices of the original mesh are contracted to points, i.e., all vertices of a facet corresponding to a vertex have the same location. An internal vertex incident with k edges produces a facet with k edges. A boundary vertex incident with k edges may also produce a k-gonal facet, though it may be a $k+1$-gon, as in Figure A.12; in this way, for a boundary vertex inci-

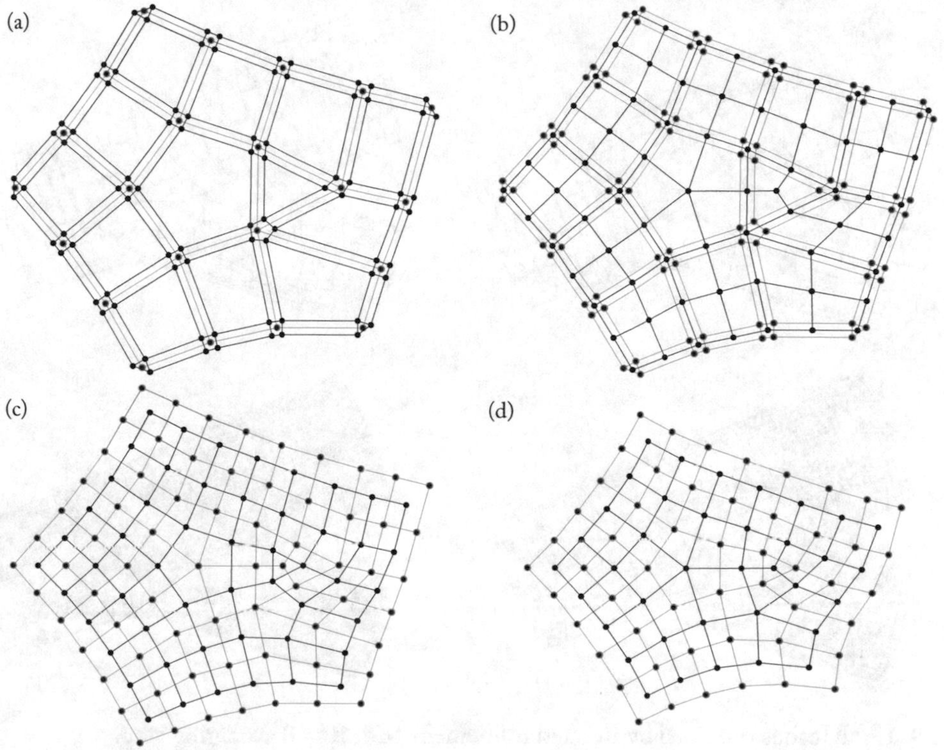

Figure A.12: Doubling (a) and three subsequent steps of averaging (b), (c), (d). The initial meshes are blue, the new ones are black

dent with two edges we obtain a triangular facet, and if $k > 2$, then the number of edges of a facet corresponding to a boundary vertex is irrelevant.

The operation of **averaging** produces a mesh whose vertices correspond to the facets of the original mesh, whose facets correspond to the internal vertices of the original mesh and whose edges correspond to the original internal edges. Had the original mesh been a planar graph, this operation would produce the dual graph. Each vertex of the new mesh is located at the gravity centre of the vertices of the corresponding original facet (assuming that all vertices have the same weight).

We can see the effect of iterating the refinement operation in Figure A.13. Just as the control polygons obtained by iterating the Lane–Riesenfeld algorithm converge to the B-spline curve represented by these polygons, the sequence of meshes converges to a limiting surface. If $n = 2$ or $n = 3$, then the refinement operation is a variant of the algorithm described by Doo and Sabin [1978], or the algorithm of Catmull and Clark [1978] respectively. The refinement operation may be modified in order to enhance the shape of the limiting surface. It is done by

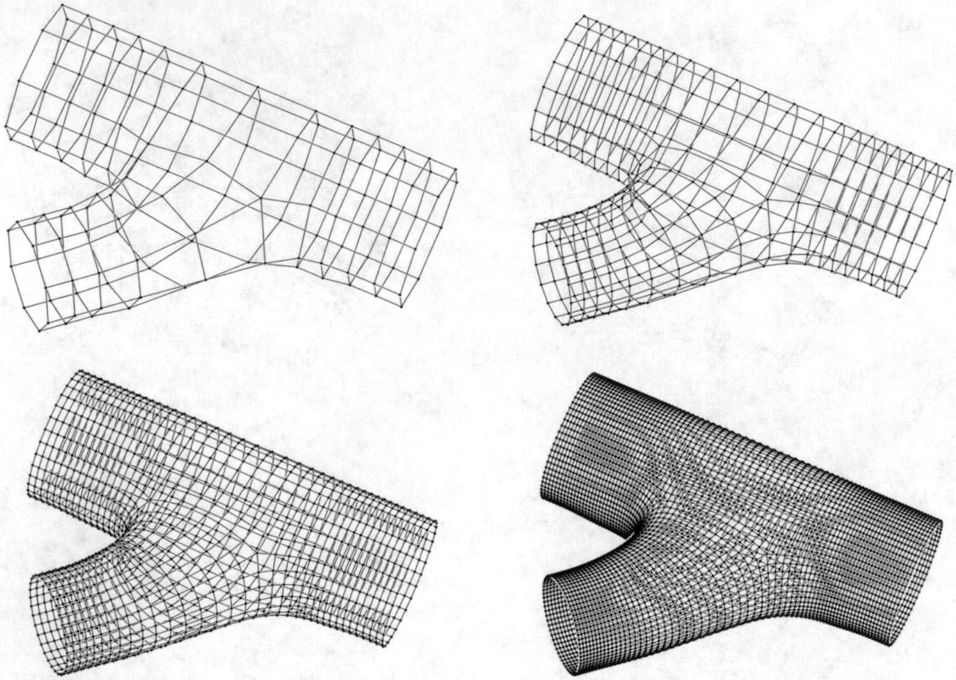

Figure A.13: Meshes obtained by iterated refinement with $n = 3$ averaging steps

changing the method of computing positions of vertices in a vicinity of the extraordinary elements of the resulting mesh; here this subject is not developed.

A.6.4 EXTRAORDINARY ELEMENTS OF MESHES

As we can see in Figure A.12a, the number of edges incident with any internal vertex of a mesh obtained by doubling is equal to 4. The facets being copies of the original facets preserve their numbers of edges, and a facet corresponding to the original internal vertex incident with k edges is a k-gon. On the other hand, the averaging operation produces a mesh whose facets preserve the number of edges incident with the corresponding original (internal) vertices. The number of edges incident with a vertex produced by averaging is equal to the number of internal edges of the corresponding original facet.

Traditionally, the name **extraordinary element** is used to denote an internal vertex incident with $k \neq 4$ edges, or a non-quadrilateral facet. The number of extraordinary elements of a mesh obtained by doubling or averaging cannot exceed the number of extraordinary elements of the mesh being the argument of any of the two operations. All extraordinary elements of a mesh obtained by doubling are facets, because any of its internal vertices is incident with four edges. An extraordinary facet is either a copy of a non-quadrilateral facet or it corresponds to an extraordinary vertex.[10] On the other hand, all extraordinary elements of the mesh produced by averaging correspond to extraordinary elements of the mesh subject to this operation: the internal vertices with k incident edges are transformed to facets with k edges, and the k-gonal facets with all edges internal are transformed to vertices incident with k edges. The number of extraordinary elements may even decrease: if a facet has at least one boundary edge, then it will be transformed to a boundary vertex, which is not an extraordinary element. Thus, if the refinement operation is the doubling followed by n averaging steps, then the number of extraordinary elements of the refined mesh cannot be greater than the number of extraordinary elements of the original mesh. If n is odd, then all extraordinary elements of the refined mesh are vertices, and if n is even, then all extraordinary elements are facets (see Figure A.12).

A.6.5 THE LIMITING SURFACE

If the vertices of an initial mesh form a rectangular array with at least $n + 1$ rows and columns (see Figure A.14), and if all its edges join neighbouring vertices in each row and column, so that all the facets are quadrilateral, then the mesh is a control net of a tensor product B-spline patch. Suppose that this patch is defined with uniform (i.e., equidistant) knot sequences. As we know from Sections A.4.3 and A.6.1, the refinement operation with n averaging steps will produce a new control net of the same patch, with twice denser knot sequences. Formally, each polynomial patch of degree (n, n) being part of the B-spline patch will be divided into quarters. Iterating the refinement operation will produce a sequence of meshes converging to this B-spline patch.

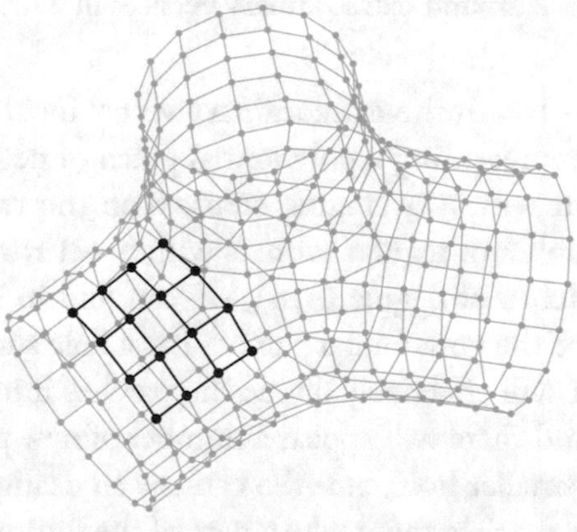

Figure A.14: A control net of a bicubic patch in a mesh

The refinement operation applied iteratively to an arbitrary mesh (*with* extraordinary elements) also produces a sequence of meshes which converges to a limiting surface. If the initial mesh contains a "rectangular" submesh representing a B-spline patch, the subsequent meshes will

[10] We ignore non-quadrilateral facets corresponding to boundary vertices of the original mesh subjected to doubling, because the subsequent averaging step will transform them back to boundary vertices.

contain control nets of this B-spline patch (representing it with denser uniform knot sequences), being a part of the limiting surface. It remains to be seen what the extraordinary elements turn into.

The refinement operation produces a mesh whose extraordinary elements—internal vertices incident with three or more than four edges (if n is odd) or non-quadrilateral facets (if n is even)—are surrounded by quadrangular facets. A part of the mesh made of the extraordinary element and r layers of quadrangular facets around it is called here the **Sabin net** of radius r (see Loop and DeRose [1990]). Examples are shown in Figure A.15.

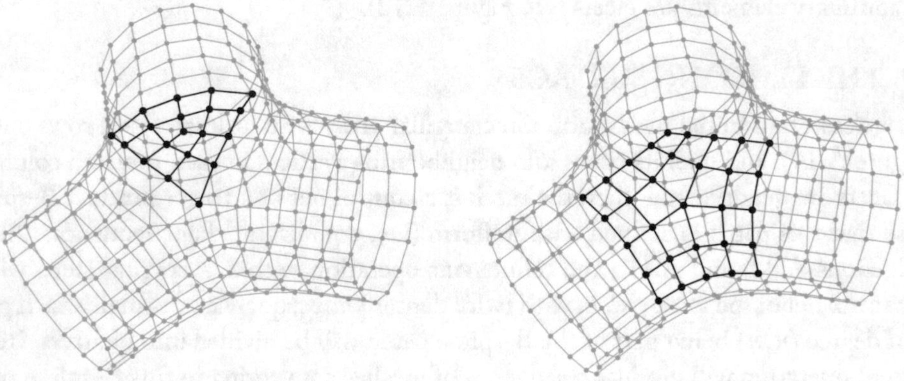

Figure A.15: Sabin nets of radius 2 around extraordinary vertices in a mesh

A Sabin net of radius $n - 1$ with the extraordinary vertex incident with k edges does not contain any regular submeshes representing a polynomial patch of degree (n, n), but if it is subject to the refinement operation with n averaging steps, then the result is a Sabin net of radius $\frac{3}{2}(n - 1)$. Apart from a number of regular subnets with $n + 1$ rows and columns of vertices which represent polynomial patches of degree (n, n), we can find in it a (smaller) Sabin net of radius $n - 1$. The surface made of the polynomial patches has a hole surrounded by these patches. Another iteration of refinement will effectively divide the patches into quarters (and apart from that, it will not change them), and there will appear a number of new patches filling a part of the hole and surrounding another, smaller hole, etc. We can see an example in Figure A.16, which shows the boundaries of bicubic patches that make a part of the limiting surface of the sequence of meshes, whose first four are in Figure A.13. Iterating the refinement operation produces a sequence of meshes convergent to the limiting surface, which, therefore, consists of a countable set of polynomial patches getting smaller and smaller as we approach the limit of the sequence of the corresponding extraordinary elements of the meshes.

Exercise. Find the formula for the radius of the Sabin net obtained by refinement with n averaging steps of a Sabin net of radius $n - 1$ with an extraordinary facet, for even n.

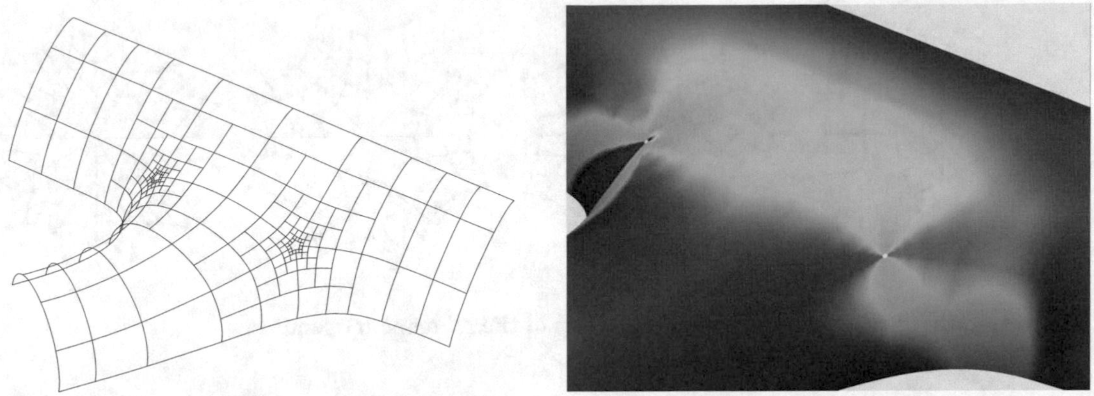

Figure A.16: Bicubic patches surrounding extraordinary points in the limiting surface and mean curvature image

A natural question is what is the order of geometric continuity of the limiting surface? Each pair of polynomial patches having a common curve have the same cross-boundary derivatives up to the order $n-1$, but at the extraordinary point there is a singularity. Consider the following operation whose argument is a Sabin net of radius n: the mesh is refined and its part—a Sabin net built in just the same way—is extracted (Fig. A.17). Having a fixed scheme of vertex numbering, the same in both nets, we can write the computation of coordinates of the new vertices using a square matrix A_{nk}, whose dimensions are equal to the number of vertices of both Sabin nets. There is $\boldsymbol{r} = A_{nk}\boldsymbol{a}$, where $\boldsymbol{a}$ and $\boldsymbol{r}$ are column matrices made of positions of the vertices of the argument and result.

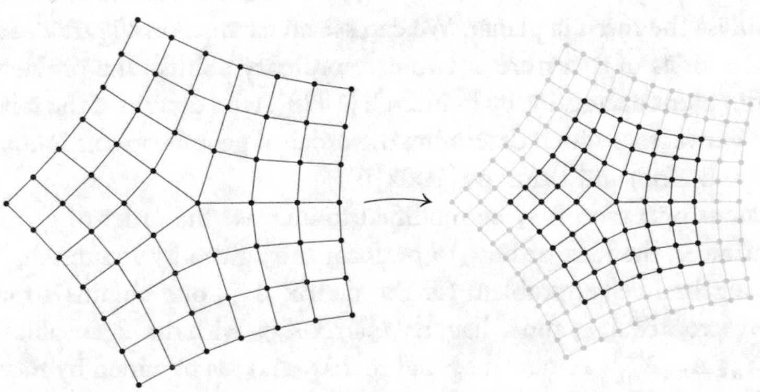

Figure A.17: A planar Sabin net being an eigenvector of the refinement operation

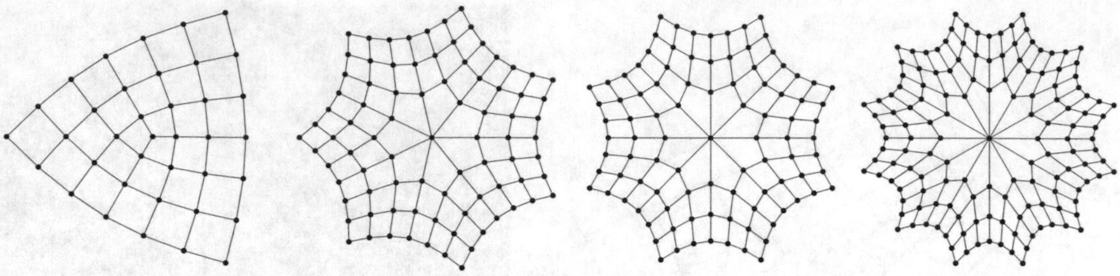

Figure A.18: Further examples of eigenvectors of the refinement operation

The coefficients of the matrix A_{nk} are nonnegative and their sum in each row is 1, which is a consequence of the fact that each vertex of the result is a convex combination of the vertices of the argument. The greatest eigenvalue of the matrix A_{nk}, λ_1, is equal to 1. The second greatest eigenvalue of A_{nk}, λ_2, has the multiplicity 2. Suppose that the central vertex of the nets in Figure A.17 is located at the origin of the coordinate system in the plane. Two vectors, made respectively of the x and y coordinates of the vertices of the argument, are the eigenvectors of the matrix $A_{3,5}$, related to the operation shown on the picture. Therefore, the result is the Sabin net obtained by a homotetia of the initial net; the scaling factor is the double eigenvalue, $\lambda_2 = \lambda_3$. In general, the order of geometric continuity of the limiting surface around the extraordinary point depends on the eigenvalues and eigenvectors of the matrix which describes the refinement operation, as in the example above. Doo and Sabin were aware of that in [1978]. Nevertheless, the question "what is the actual class of geometric continuity of the limiting surface around the extraordinary point" was an open problem until Reif [1995] made some progress. It turns out that if $n = 3$ and the polynomial patches are regular, then the surface is of class G^1, but it is not of class G^2 unless the mesh is planar. We can see an example in Figure A.16, with an image indicating discontinuities of curvature at two extraordinary points. The problem was solved for a wide class of refinement operations by Prautzsch [1998], who described the relations among the eigenvalues and eigenvectors which determine the order of geometric continuity in explicit form (see also Prautzsch, Boehm and Paluszny [2002]).

The refinement operation may be modified to increase the order of geometric continuity of the limiting surface. The idea of how to perform this, given by Prautzsch, is the following: after solving the algebraic eigenproblem for the matrix A_{nk}, one obtains a matrix X_{nk} whose columns are eigenvectors of A_{nk}, and a diagonal matrix Λ_{nk}, with the eigenvalues on the diagonal. There is $A_{nk} = X_{nk}\Lambda_{nk}X_{nk}^{-1}$. A new diagonal matrix, $\hat{\Lambda}_{nk}$, is obtained by modifying Λ_{nk}; the modification is done by decreasing offending eigenvalues (the eigenvalues $\lambda_1 = 1$ and $\lambda_2 = \lambda_3$ are left unchanged). The matrix $\hat{A}_{nk} = X_{nk}\hat{\Lambda}_{nk}X_{nk}^{-1}$, which has the eigenvectors of A_{nk} and modified eigenvalues, is used to compute the positions of the vertices of the Sabin net in the refined mesh.

Piecewise bicubic limiting surfaces obtained by iterating the refinement operation made of doubling and three averaging steps, modified as above, are of class G^2, but they exhibit (rather small, but visible on images of curvature) undulations around the extraordinary points. We do not discuss these issues further, as we are more interested in filling polygonal holes in surfaces with a finite number of regular patches. However, the Sabin nets being eigenvectors of the refinement operation (like these shown in Figures A.17, A.18 and 5.4b) are often a good choice for domain nets used in the constructions described in Chapter 5.

A.7 RATIONAL CURVES AND PATCHES

Curves and surfaces, whose parametrisations are described (piecewise) by polynomials, are insufficient in some applications. One of the most important restrictions is the lack of the possibility to represent circular arcs; the only conic curves having polynomial parametrisations are parabolas. One might represent a circle with a parametrisation made of trigonometric functions, but it is also possible to use (piecewise) rational parametrisations to represent circles and arcs of any conic curve. The possibility of representing circles allows us also to represent surfaces of revolution by (piecewise) rational parametrisations.

An in-depth study of rational curves and patches is beyond the scope of this book, and interested readers may find it in Farin [1999]. Here we focus on the properties directly related with geometric continuity. Two or three rational functions, which make a parametrisation of a curve or a patch, are given by numerators and denominators being polynomials (the denominators must be non-zero). It is always possible to represent these rational functions with a common denominator. Hence, a planar rational curve may be represented by three polynomials, while a rational curve or surface in the three-dimensional space may be represented by four polynomials, of one or two variables respectively.

A **homogeneous curve** or a **homogeneous patch** is a curve or a patch in the three- or four-dimensional space, described by these three or four polynomials. Homogeneous curves or patches are often used as internal representations of rational curves or patches in computer programs. By convention, the denominator describes the last coordinate of the homogeneous curve or patch. The relation between a planar rational curve $\boldsymbol{p}$ and its homogeneous representation $\boldsymbol{P}$ is described by the formulae

$$\boldsymbol{p}(t) = \begin{bmatrix} x(t) \\ y(t) \end{bmatrix} = \begin{bmatrix} X(t)/W(t) \\ Y(t)/W(t) \end{bmatrix}, \qquad \boldsymbol{P}(t) = \begin{bmatrix} X(t) \\ Y(t) \\ W(t) \end{bmatrix}.$$

This relation is shown in Figure A.19a; the rational curve is located in the plane $W = 1$ and it is the image of the homogeneous curve in the central projection on that plane.

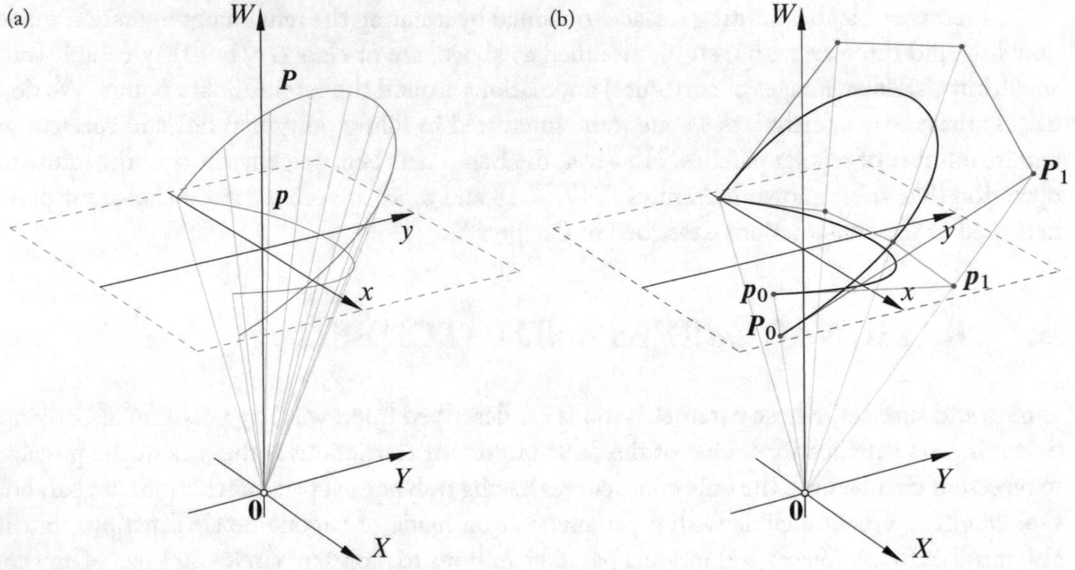

Figure A.19: A rational curve and its homogeneous representation (a), a rational Bézier curve and its homogeneous representation (b)

A homogeneous curve represented in the Bernstein polynomial basis of degree n is a representation of a **rational Bézier curve** of degree n:

$$\boldsymbol{P}(t) = \sum_{i=0}^{n} \boldsymbol{P}_i B_i^n(t), \qquad \boldsymbol{p}(t) = \frac{\sum_{i=0}^{n} w_i \boldsymbol{p}_i B_i^n(t)}{\sum_{i=0}^{n} w_i B_i^n(t)}.$$

The **control points** $\boldsymbol{p}_0, \ldots, \boldsymbol{p}_n$ of the rational curve are accompanied by the coefficients $w_0, \ldots, w_n$, called **weights**; they are the coordinates $W_0, \ldots, W_n$ of the control points $\boldsymbol{P}_0, \ldots, \boldsymbol{P}_n$ of the homogeneous Bézier curve. If the rational curve is planar, then the control points of the homogeneous Bézier curve are

$$\boldsymbol{P}_i = \begin{bmatrix} X_i \\ Y_i \\ W_i \end{bmatrix} = \begin{bmatrix} w_i x_i \\ w_i y_i \\ w_i \end{bmatrix}, \quad \text{where} \quad \boldsymbol{p}_i = \begin{bmatrix} x_i \\ y_i \end{bmatrix} = \begin{bmatrix} X_i / W_i \\ Y_i / W_i \end{bmatrix}.$$

Note that if all weights are the same, then the denominator is a constant and the curve $\boldsymbol{p}$ is a polynomial Bézier curve with the control points $\boldsymbol{p}_0, \ldots, \boldsymbol{p}_n$.

The homogeneous representations of rational Bézier patches and rational spline curves and patches are defined in a similar way. The computation of a point of a rational curve or patch may be done by computing the point of the homogeneous curve or patch; its first two or three coordinates are then divided by the last one to obtain the Cartesian coordinates of the result.

Also, the first step of evaluating derivatives of rational curves or patches may be the computation of the derivatives of the homogeneous representation. Let $\boldsymbol{r}(t) = w(t)\boldsymbol{p}(t)$, where $w(t)$ is the denominator; the function $\boldsymbol{r}$ is obtained by rejecting the last component of the homogeneous curve, i.e., the denominator. The formulae for subsequent derivatives of the parametrisation $\boldsymbol{p}$ may be obtained as follows:

$$
\begin{aligned}
\boldsymbol{r}' &= w'\boldsymbol{p} + w\boldsymbol{p}' &&\Rightarrow\quad \boldsymbol{p}' = \frac{1}{w}(\boldsymbol{r}' - w'\boldsymbol{p}),\\
\boldsymbol{r}'' &= w''\boldsymbol{p} + 2w'\boldsymbol{p}' + w\boldsymbol{p}'' &&\Rightarrow\quad \boldsymbol{p}'' = \frac{1}{w}(\boldsymbol{r}'' - 2w'\boldsymbol{p}' - w''\boldsymbol{p}),\\
\boldsymbol{r}''' &= w'''\boldsymbol{p} + 3w''\boldsymbol{p}' + 3w'\boldsymbol{p}'' + w\boldsymbol{p}''' &&\Rightarrow\quad \boldsymbol{p}''' = \frac{1}{w}(\boldsymbol{r}''' - 3w'\boldsymbol{p}'' - 3w''\boldsymbol{p}' - w'''\boldsymbol{p})
\end{aligned}
$$

etc.

Formulae for partial derivatives of rational patches may be obtained in a similar way, which is left as an exercise. Relations between rational curves or patches whose junctions are smooth are written in Chapters 2 and 3 in the form of equations for derivatives of the homogeneous representations.

A.8 SPLINE CURVES OF INTERPOLATION

In this section, we consider splines made of polynomials obtained by solving the Hermite interpolation problem with two knots of multiplicity r. In the space of polynomials of degree less than $2r$ such a problem has a unique solution; note that $2r - 1$ is an odd number.

A.8.1 BLENDING POLYNOMIALS

At first, we take a look at the case of $r = 2$. Using cubic Bernstein polynomials, we define four cubic polynomials

$$
\begin{aligned}
H_{00}(t) &= B_0^3(t) + B_1^3(t), & H_{10}(t) &= B_2^3(t) + B_3^3(t),\\
H_{10}(t) &= \tfrac{1}{3}B_1^3(t), & H_{11}(t) &= -\tfrac{1}{3}B_2^3(t).
\end{aligned}
\tag{A.39}
$$

Using Formulae (A.8) and (A.11), we can prove that the matrix

$$
\begin{bmatrix}
H_{00}(0) & H'_{00}(0) & H_{00}(1) & H'_{00}(1)\\
H_{01}(0) & H'_{01}(0) & H_{01}(1) & H'_{01}(1)\\
H_{10}(0) & H'_{10}(0) & H_{10}(1) & H'_{10}(1)\\
H_{11}(0) & H'_{11}(0) & H_{11}(1) & H'_{11}(1)
\end{bmatrix}
$$

is the 4×4 identity matrix. Therefore, the linear combination

$$
p(t) = a_0 H_{00}(t) + b_0 H_{01}(t) + a_1 H_{10}(t) + b_1 H_{11}(t)
$$

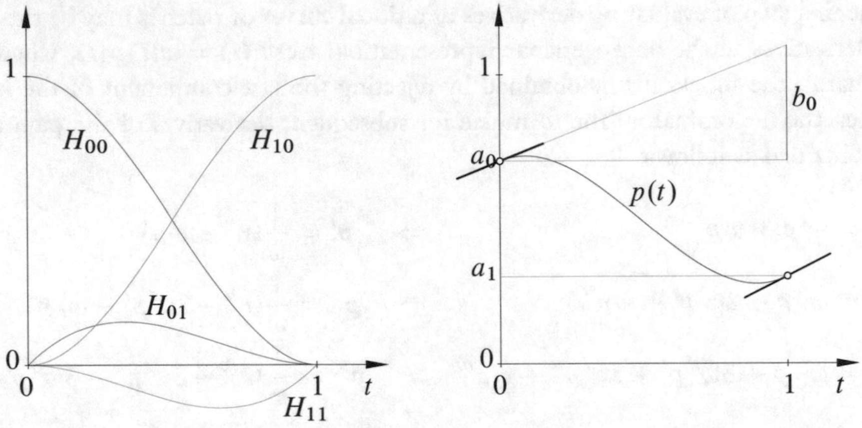

Figure A.20: Cubic polynomials for solving Hermite interpolation problems

is a cubic polynomial satisfying the conditions (see Figure A.20)

$$p(0) = a_0, \quad p'(0) = b_0, \quad p(1) = a_1, \quad p'(1) = b_1.$$

In a similar way, for any positive integer r we can define the set of $2r$ polynomials of degree $2r-1$: $\{H_{ij} : i = 0, 1,\ j = 0, \ldots, r-1\}$; the j-th order derivative[11] of the function H_{ij} at the point i is equal to 1. All the other derivatives of order less than r at the point i are zero, and all derivatives of order $0, \ldots, r-1$ at the point $1-i$ are also zero. For $r = 3$, using the Bernstein basis polynomials of degree 5, we can write

$$\begin{aligned}
H_{00}(t) &= B_0^5(t) + B_1^5(t) + B_2^5(t), & H_{10}(t) &= B_3^5(t) + B_4^5(t) + B_5^5(t), \\
H_{01}(t) &= \tfrac{1}{5}B_1^5(t) + \tfrac{2}{5}B_2^5(t), & H_{11}(t) &= -\tfrac{2}{5}B_3^5(t) - \tfrac{1}{5}B_4^5(t), \\
H_{02}(t) &= \tfrac{1}{20}B_2^5(t), & H_{12}(t) &= \tfrac{1}{20}B_3^5(t).
\end{aligned} \tag{A.40}$$

Their linear combination

$$p(t) = a_0 H_{00}(t) + b_0 H_{01}(t) + c_0 H_{02}(t) + a_1 H_{10}(t) + b_1 H_{11}(t) + c_1 H_{12}(t)$$

is the unique polynomial of degree ≤ 5 satisfying the conditions

$$p(0) = a_0,\ p'(0) = b_0,\ p''(0) = c_0,\ p(1) = a_1,\ p'(1) = b_1,\ p''(1) = c_1.$$

For any positive integer r, the basis of the space $\mathbb{R}[\cdot]_{2r-1}$ made of those polynomials is a perfect tool for immediately solving Hermite interpolation problems with two knots, 0 and 1, of multiplicity r.

[11]The function value may be considered as the derivative of order 0. It is convenient, so we take it up.

A.8.2 CUBIC SPLINES OF INTERPOLATION—HERMITE FORM

Let $u_0, \ldots, u_N$ be an increasing sequence of numbers which we take for knots of interpolation *and* for knots of a spline function. We are going to find a function s of class $C^2[u_0, u_N]$ such that it is a polynomial of degree at most 3 between any two consecutive knots, and for $i = 0, \ldots, N$ it takes a prescribed value a_i at u_i.

With each interval $[u_i, u_{i+1}]$, we associate four cubic polynomials:

$$\begin{aligned} H_{i,00}(x) &= H_{00}(t), \qquad & H_{i,10}(x) &= H_{10}(t), \\ H_{i,01}(x) &= h_i H_{01}(t), \qquad & H_{i,11}(x) &= h_i H_{11}(t), \end{aligned} \tag{A.41}$$

where $h_i = u_{i+1} - u_i, t = (x - u_i)/h_i$ and $H_{00}, H_{01}, H_{10}, H_{11}$ are the cubic polynomials given by Formulae (A.39). We may think of these four polynomials as the "local" basis for solving Hermite interpolation problems with double knots u_i, u_{i+1}. For arbitrary numbers $a_0, \ldots, a_N$ and $b_0, \ldots, b_N$, the function s, such that for $x \in [u_i, u_{i+1}]$

$$s(x) = a_i H_{i,00}(x) + b_i H_{i,01}(x) + a_{i+1} H_{i,10}(x) + b_{i+1} H_{i,11}(x),$$

is of class $C^1[u_0, u_N]$ and $s(u_i) = a_i$, $s'(u_i) = b_i$ for $i = 0, \ldots, N$ (see Figure A.21).

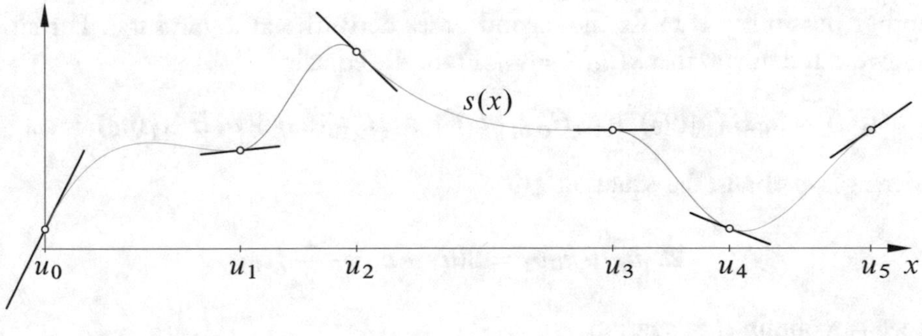

Figure A.21: Cubic spline function of interpolation of class C^1

The continuity of the second-order derivative of the function s, for given numbers $a_0, \ldots, a_N$, i.e., function values at the knots, is achieved by choosing the numbers $b_0, \ldots, b_N$ which are the first-order derivatives of s at the knots. The continuity of s'' at the knot u_i, where $i \in \{1, \ldots, N-1\}$, takes place when the two polynomials, say, p_{i-1} and p_i, which describe the function s in the intervals $[u_{i-1}, u_i]$ and $[u_i, u_{i+1}]$ respectively, have the same second-order derivative at u_i. The equation $p''_{i-1}(u_i) = p''_i(u_i)$, written with use of the "local" bases, is

$$\begin{aligned} a_{i-1} H''_{i-1,00}(u_i) + b_{i-1} H''_{i-1,01}(u_i) + a_i H''_{i-1,10}(u_i) + b_i H''_{i-1,11}(u_i) = \\ a_i H''_{i,00}(u_i) + b_i H''_{i,01}(u_i) + a_{i+1} H''_{i,10}(u_i) + b_{i+1} H''_{i,11}(u_i). \end{aligned}$$

After substituting the second-order derivatives

$$\begin{aligned} H''_{i,00}(u_i) &= -\frac{6}{h_i^2}, & H''_{i-1,00}(u_i) &= \frac{6}{h_{i-1}^2}, \\ H''_{i,01}(u_i) &= -\frac{4}{h_i}, & H''_{i-1,01}(u_i) &= \frac{2}{h_{i-1}}, \\ H''_{i,10}(u_i) &= \frac{6}{h_i^2}, & H''_{i-1,10}(u_i) &= -\frac{6}{h_{i-1}^2}, \\ H''_{i,11}(u_i) &= -\frac{2}{h_i}, & H''_{i-1,11}(u_i) &= \frac{4}{h_{i-1}} \end{aligned}$$

and reordering, we obtain the following equation:

$$h_i b_{i-1} + 2(h_{i-1} + h_i)b_i + h_{i-1}b_{i+1} = 3\left(h_i \frac{a_i - a_{i-1}}{h_{i-1}} + h_{i-1}\frac{a_{i+1} - a_i}{h_i}\right). \quad \text{(A.42)}$$

If the numbers $b_0, \ldots, b_N$ satisfy this equation for all $i \in \{1, \ldots, N-1\}$, then our spline of interpolation has the second-order derivative continuous in $[u_0, u_N]$. But the number of unknown variables in this system is greater than the number of equations. To obtain an interpolation problem with a unique solution, one has to take two additional equations. Usually these equations describe **end conditions**. In the simplest case one can fix arbitrarily b_0 and b_N, i.e., the first-order derivative of the function s at u_0 and u_N.

Another possibility is to fix the second-order derivative at u_0 and u_N. For an arbitrary number c_0, we can demand that $s(u_0)'' = c_0$. From the equality

$$s''(u_0) = a_0 H''_{0,00}(u_0) + b_0 H''_{0,01}(u_0) + a_1 H''_{i,10}(u_0) + b_1 H''_{0,11}(u_0) = c_0,$$

after reordering, we obtain the equation

$$2h_0 b_0 + h_0 b_1 = 3(a_1 - a_0) - \frac{1}{2}h_0^2 c_0. \quad \text{(A.43)}$$

Similarly, we can obtain the equation

$$h_{N-1}b_{N-1} + 2h_{N-1}b_N = 3(a_N - a_{N-1}) + \frac{1}{2}h_{N-1}^2 c_N,$$

which is equivalent to $s''(u_N) = c_N$. Taking $c_0 = c_N = 0$, i.e., $s''(u_0) = s''(u_N) = 0$ produces a function called **natural cubic spline.**

Other end conditions are considered and used in practice; more information may be found in textbooks, see e.g. Farin [2002]. At each end of the interval $[u_0, u_N]$ one can choose an end condition of a different kind. End conditions must be chosen so as to obtain a system of equations with a non-singular matrix.

One more possibility is to construct a **periodic spline of interpolation**: suppose that $a_N = a_0$. The assumption $b_N = b_0$ reduces (by one) the number of unknown variables. System of equations (A.42) is now one equation short. One can add to it the following:

$$h_0 b_{N-1} + 2(h_{N-1} + h_0) b_0 + h_{N-1} b_1 = 3\left(h_0 \frac{a_0 - a_{N-1}}{h_{N-1}} + h_{N-1} \frac{a_1 - a_0}{h_0} \right).$$

This equation is equivalent to $s''(u_0) = s''(u_N)$, and the spline, whose coefficients $a_0, \dots, a_{N-1}, a_N = a_0$ and $b_0, \dots, b_{N-1}, b_N = b_0$ satisfy it and Equations (A.42), may be periodically extended (with the period $T = u_N - u_0$) to obtain a function of class $C^2(\mathbb{R})$. Periodic spline functions are used in constructions of closed curves.

A.8.3 CUBIC B-SPLINES OF INTERPOLATION

A cubic spline of interpolation may be represented (and constructed directly) in B-spline form. Given the *interpolation knots* $u_0, \dots, u_N$, which form an increasing sequence, we choose a sequence of *knots of the spline*: $v_0, \dots, v_{N+6}$, which must be non-decreasing. We assume that $v_{i+3} = u_i$ for $i = 0, \dots, N$. It is convenient to take $v_0 = \cdots = v_2 = u_0$ and $v_{N+4} = \cdots = v_{N+6} = u_N$. The function has the form

$$s(t) = \sum_{i=0}^{N+2} d_i N_i^3(t),$$

where $N_0^3, \dots, N_{N+2}^3$ are cubic B-spline functions defined over the knot sequence $v_0, \dots, v_{N+6}$ and their values at any point may be computed using Algorithm A.7 (see page 180). Due to Property A.6 (p. 173), at each knot of interpolation, $u_i = v_{i+3}$, only three B-spline functions are non-zero, and the equation expressing the interpolation condition $s(u_i) = a_i$ binds only three coefficients of the function s:

$$N_i^3(u_i) d_i + N_{i+1}^3(u_i) d_{i+1} + N_{i+2}^3(u_i) d_{i+2} = a_i. \tag{A.44}$$

Now we have $N + 3$ unknown coefficients, $d_0, \dots, d_{N+2}$, and only $N + 1$ interpolation conditions described by Equations (A.44) for $i = 0, \dots, N$, so again we are two equations short. Any end condition useful in the construction of a spline in Hermite form may be expressed in terms of B-spline functions. For example, the condition $s'(u_0) = b_0$ is

$$N_0^{3\prime}(u_0) d_0 + N_1^{3\prime}(u_0) d_1 + N_2^{3\prime}(u_0) d_2 = b_0.$$

The coefficients $N_0^{3\prime}(u_0), N_1^{3\prime}(u_0), N_2^{3\prime}(u_0)$ of this equation may be obtained from Formula (A.28). With the knots $v_0 = \cdots = v_3 = u_0$, the equation above, after a rearrangement (left as an exercise), takes the form

$$-d_0 + d_1 = \frac{u_1 - u_0}{3} b_0.$$

With the same knots, the *alternative* end condition $s''(u_0) = c_0$ may be used in a similar exercise to derive the equation

$$\frac{1}{h_0}d_0 - \Big(\frac{1}{h_0} + \frac{1}{h_1}\Big)d_1 + \frac{1}{h_1}d_2 = \frac{h_0 + h_1}{6}c_0,$$

where $h_0 = u_1 - u_0$, $h_1 = u_2 - u_1$.

If a periodic spline function is to be constructed (with $a_0 = a_N$), it is better to take $v_i = u_{i-3+N} - T$ for $i = 0, 1, 2$ and $v_i = u_{i-3-N} + T$ for $i = N+4, N+5, N+6$, where $T = u_N - u_0$. With such knots $N^3_{N+i}(t+T) = N^3_i(t)$ for $i = 0, 1, 2$ and for all $t \in \mathbb{R}$. To obtain a spline function s such that $s(u_0) = s(u_N)$, $s'(u_0) = s'(u_N)$ and $s''(u_0) = s''(u_N)$ we simply take $d_N = d_0$, $d_{N+1} = d_1$ and $d_{N+2} = d_2$ (and we reject the superfluous equation (A.44) for $i = N$).

A.8.4 APPROXIMATION PROPERTIES

The facts proved in this section apply not only to scalar functions, but also to vector functions, i.e., spline parametrisations of curves. For the curves, the absolute value ($|\cdot|$) in the considerations below has to be replaced by some norm, e.g. Euclidean ($\|\cdot\|_2$).

Lemma A.17 *Let s be a cubic spline of class $C^2[a,b]$, with the knots $u_0 = a < u_1 < \cdots < u_N = b$. Let $c_i = s''(u_i)$ for $i = 0, \ldots, N$. The numbers $c_0, \ldots, c_N$ satisfy the equations*

$$\frac{h_{i-1}}{h_{i-1} + h_i}c_{i-1} + 2c_i + \frac{h_i}{h_{i-1} + h_i}c_{i+1} = 6s[u_{i-1}, u_i, u_{i+1}] \quad \textit{for } i = 1, \ldots, N-1, \qquad \text{(A.45)}$$

where $h_i = u_{i+1} - u_i$ for $i = 0, \ldots, N-1$.

Proof. Let p_{i-1} and p_i be cubic polynomials, which describe the function s in the intervals $[u_{i-1}, u_i]$ and $[u_i, u_{i+1}]$ respectively. Their second-order derivatives are polynomials of degree 1 such that $p''_{i-1}(u_{i-1}) = c_{i-1}$, $p''_{i-1}(u_i) = p''_i(u_i) = c_i$ and $p''_i(u_{i+1}) = c_{i+1}$. These interpolation conditions allow us to write

$$p''_{i-1}(t) = \frac{c_i - c_{i-1}}{h_{i-1}}u + c_i, \qquad p''_i(t) = \frac{c_{i+1} - c_i}{h_i}u + c_i,$$

where $u = t - u_i$. By integrating the expressions above twice, with the constants of integration $b_i = s'(u_i)$ and $a_i = s(u_i)$, we obtain

$$p_{i-1}(t) = \frac{c_i - c_{i-1}}{6h_{i-1}}u^3 + \frac{c_i}{2}u^2 + b_i u + a_i, \quad p_i(t) = \frac{c_{i+1} - c_i}{6h_i}u^3 + \frac{c_i}{2}u^2 + b_i u + a_i.$$

In the above we substitute $u = -h_{i-1}$ and $u = h_i$, which gives us the equalities

$$a_{i-1} = p_{i-1}(u_{i-1}) = -\frac{c_i - c_{i-1}}{6}h^2_{i-1} + \frac{c_i}{2}h^2_{i-1} - b_i h_{i-1} + a_i,$$
$$a_{i+1} = p_i(u_{i+1}) = \frac{c_{i+1} - c_i}{6}h^2_i + \frac{c_i}{2}h^2_i - b_i h_i + a_i.$$

Both of them may be solved with respect to b_i, which gives us

$$\frac{a_i - a_{i-1}}{h_{i-1}} + \frac{c_i h_{i-1}}{3} + \frac{c_{i-1}h_{i-1}}{6} = b_i = \frac{a_{i+1} - a_i}{h_i} - \frac{c_i h_i}{3} - \frac{c_{i+1}h_i}{6}.$$

After noticing divided differences of the function s and a simple reordering, we get

$$2h_{i-1}c_i + h_{i-1}c_{i-1} + 2h_i c_i + h_i c_{i+1} = 6\big(s[u_i, u_{i+1}] - s[u_{i-1}, u_i]\big).$$

Equation (A.45) is obtained by dividing both sides by $h_{i-1} + h_i = u_{i+1} - u_{i-1}$. □

Lemma A.18 *Let the numbers $c_0, \dots, c_N$ satisfy the system of linear equations*

$$w_i c_{i-1} + 2c_i + (1 - w_i)c_{i+1} = v_i \quad \text{for } i = 1, \dots, N-1,$$

where $0 \le w_i \le 1$. If there exists a constant K such that $|c_0| \le K$, $|c_N| \le K$ and $|v_i| \le K$ for $i = 1, \dots, N-1$, then $|c_i| \le K$ for $i = 0, \dots, N$.

Proof. Let $j \in \{0, \dots, N\}$ be chosen so that $|c_j| \ge |c_i|$ for all $i = 0, \dots, N$. If $j = 0$ or $j = N$, then the claim follows immediately. Otherwise, we can estimate

$$|c_j| \le 2|c_j| - w_j|c_{j-1}| - (1 - w_j)|c_{j+1}| \le |v_j|.$$

□

Exercise. Prove Lemma A.18 with the assumptions $|c_0| \le K$, $|c_N| \le K$ replaced by $c_0 = c_N$ and $w_0 c_{N-1} + 2c_0 + (1 - w_0)c_1 = v_0$, where $0 \le w_0 \le 1$.

Theorem A.19 *Let f be a function of class $C^2[a,b]$ and let s be a cubic spline with knots $u_0 = a < u_1 < \dots < n_N = b$, interpolating the function f at these knots. Let M_2 be a constant such that $|f''(t)| \le M_2$ for all $t \in [a,b]$ and $|s''(a)| \le 3M_2$ and $|s''(b)| \le 3M_2$. Then*

$$|f(t) - s(t)| \le \frac{1}{2}M_2 h^2, \qquad |f'(t) - s'(t)| \le 2M_2 h,$$

for all $t \in [a,b]$, where $h = \max\{h_0, \dots, h_{N-1}\}$ and $h_i = u_{i+1} - u_i$.

Proof. The function s is a spline of interpolation to f; therefore, $s[u_{i-1}, u_i, u_{i+1}] = f[u_{i-1}, u_i, u_{i+1}]$. By Theorem A.1, for each $i \in \{1, \dots, N-1\}$, there exists $\xi_i \in [u_{i-1}, u_{i+1}]$ such that $f[u_{i-1}, u_i, u_{i+1}] = f''(\xi_i)/2$; hence, $|6f[u_{i-1}, u_i, u_{i+1}]| \le 3M_2$. By Lemmas A.17 and A.18, $|s''(u_i)| \le 3M_2$ for $i = 0, \dots, N$. Therefore, the function $e(t) \overset{\text{def}}{=} f(t) - s(t)$, which has a continuous second-order derivative in $[a, b]$, satisfies the condition $|e''(t)| \le 4M_2$ for all $t \in [a, b]$. Let $t \in [u_i, u_{i+1}]$. Due to $e(u_i) = e(u_{i+1}) = 0$, there is

$$e[u_i, t, u_{i+1}] = \frac{e(t)}{(t - u_i)(t - u_{i+1})}, \quad \text{hence}$$
$$e(t) = e[u_i, t, u_{i+1}](t - u_i)(t - u_{i+1}) = \frac{e''(\eta_i)}{2}(t - u_i)(t - u_{i+1})$$

for some $\eta_i \in [u_i, u_{i+1}]$. There is also $|(t - u_i)(t - u_{i+1})| \le \frac{1}{4}h_i^2$, which implies the inequality $|e(t)| \le \frac{1}{2}M_2 h_i^2$.

The second inequality claimed by the theorem is an estimate of the derivative of the function e. If $a \le t_0 < t_1 \le b$, then

$$|e'(t_1) - e'(t_0)| = \left| \int_{t_0}^{t_1} e''(t)\,\mathrm{d}t \right| \le 4M_2(t_1 - t_0).$$

Suppose that $e'(u) > 2M_2h_i$ for some $u \in [u_i, u_{i+1}]$. This supposition and the inequality above imply that if $t < u$, then $e'(t) > 2M_2h_i - 4M_2(u - t)$ and if $t > u$, then $e'(t) > 2M_2h_i - 4M_2(t - u)$. Let $g_0 = u - u_i$ and $g_1 = u_{i+1} - u$. The numbers g_0 and g_1 are nonnegative and $g_0 + g_1 = h_i$. It follows that

$$0 = e(u_{i+1}) - e(u_i) = \int_{u_i}^{u_{i+1}} e'(t)\,\mathrm{d}t >$$
$$\int_{u_i}^{u} \bigl(2M_2h_i - 4M_2(u - t)\bigr)\,\mathrm{d}t + \int_{u}^{u_{i+1}} \bigl(2M_2h_i - 4M_2(t - u)\bigr)\,\mathrm{d}t =$$
$$2M_2\bigl((g_0 + g_1)^2 - g_0^2 - g_1^2\bigr) = 4M_2g_0g_1 \ge 0,$$

i.e., $0 > 0$, and this inconsistency precludes the supposition. A similar inconsistency follows from the supposition that $e'(t) < -2M_2h_i$, and the proof is complete. □

Exercise. Prove Theorem A.19 with a weaker assumption: assume that the function f is only of class $C^1[a, b]$, but its first-order derivative satisfies the Lipschitz condition, i.e., $|f'(t_1) - f'(t_0)| \le L_1|t_1 - t_0|$ for all $t_0, t_1 \in [a, b]$, with some constant L_1. To do it, take $M_2 = 2L_1$ and show that e' satisfies the Lipschitz condition with the constant $4M_2$.

Theorem A.20 *Let f be a function of class $C^2[a, b]$ whose second-order derivative satisfies the Lipschitz condition,*

$$|f''(t_1) - f''(t_0)| \le L_2|t_1 - t_0| \quad \text{for all } t_0, t_1 \in [a, b].$$

Let $h = \max\{h_0, \dots, h_{N-1}\}$, where $h_i = u_{i+1} - u_i$. Let s be a cubic spline function with the knots $u_0 = a < u_1 < \dots < u_N = b$, interpolating the function f at these knots and such that $|s''(a) -$

$f''(a)| \leq 3L_2h$, $|s''(b) - f''(b)| \leq 3L_2h$. *Then, for all* $t \in [a,b]$,

$$|f(t) - s(t)| \leq \frac{7}{16}L_2h^3, \qquad |f'(t) - s'(t)| \leq \frac{7}{4}L_2h^2, \qquad |f''(t) - s''(t)| \leq \frac{7}{2}L_2h.$$

Proof. Let $d_i = s''(u_i) - f''(u_i)$ for $i = 0, \ldots, N$. By replacing $c_i = s''(u_i)$ with $d_i + f''(u_i)$ and substituting $f[u_{i-1}, u_i, u_{i+1}]$ for $s[u_{i-1}, u_i, u_{i+1}]$ in Equation (A.45), we obtain

$$\frac{h_{i-1}}{h_{i-1}+h_i}d_{i-1} + 2d_i + \frac{h_i}{h_{i-1}+h_i}d_{i+1} = \\ f[u_{i-1}, u_i, u_{i+1}] - \frac{h_{i-1}}{h_{i-1}+h_i}f''(u_{i-1}) - 2f''(u_i) - \frac{h_i}{h_{i-1}+h_i}f''(u_{i+1}).$$

Let q_i denote the expression on the right-hand side. There exists $\xi_i \in [u_{i-1}, u_{i+1}]$ such that $6f[u_{i-1}, u_i, u_{i+1}] = 3f''(\xi_i)$, and we can write

$$q_i = \frac{h_{i-1}}{h_{i-1}+h_i}\big(f''(\xi_i) - f''(u_{i-1})\big) + 2\big(f''(\xi_i) - f''(u_i)\big) + \frac{h_i}{h_{i-1}+h_i}\big(f''(\xi_i) - f''(u_{i+1})\big).$$

From the Lipschitz condition it follows that

$$|f''(\xi_i) - f''(u_{i-1})| \leq L_2|\xi_i - u_{i-1}|, \quad |f''(\xi_i) - f''(u_i)| \leq L_2 \max\{h_{i-1}, h_i\}, \\ |f''(\xi_i) - f''(u_{i+1})| \leq L_2|\xi_i - u_{i+1}|.$$

Using these inequalities, we can estimate

$$\left|\frac{h_{i-1}}{h_{i-1}+h_i}\big(f''(\xi_i) - f''(u_{i-1})\big) + \frac{h_i}{h_{i-1}+h_i}\big(f''(\xi_i) - f''(u_{i+1})\big)\right| \leq \\ \frac{\max\{h_{i-1}, h_i\}}{h_{i-1}+h_i}L_2\big(|\xi_i - u_{i-1}| + |\xi_i - u_{i+1}|\big) = L_2 \max\{h_{i-1}, h_i\},$$

and conclude that $|q_i| \leq 3L_2h$. Then, by Lemma A.18, there is $|d_i| \leq 3L_2h$ for $i = 0, \ldots, N$.

The function $f''(t)$ is the sum of two terms, $f_1''(t)$ and $f_2''(t)$; the first term is defined as the spline of degree 1, interpolating f'' at the knots $u_0, \ldots, u_N$, which are zeros of the second term. We can estimate

$$|f''(t) - s''(t)| \leq |f_2''(t)| + |f_1''(t) - s''(t)|.$$

For $t \in [u_i, u_{i+1}]$, using the Lipschitz condition, we can estimate $|f_2''(t)| \leq \frac{L_2}{2}h_i \leq \frac{L_2}{2}h$ (Fig. A.22). The function $f_1''(t) - s''(t)$, being a spline of degree 1, takes its extremal values at the knots, which allows us to write $|f_1''(t) - s(t)| \leq \max\{|d_i|, |d_{i+1}|\} \leq 3L_2h$. By adding the two estimates, we prove that $|f''(t) - s''(t)| = |e''(t)| \leq \frac{7}{2}L_2h$.

The other two inequalities may be proved using the above, with calculations similar to those used in the proof of Theorem A.19. This is left to the reader as an exercise. □

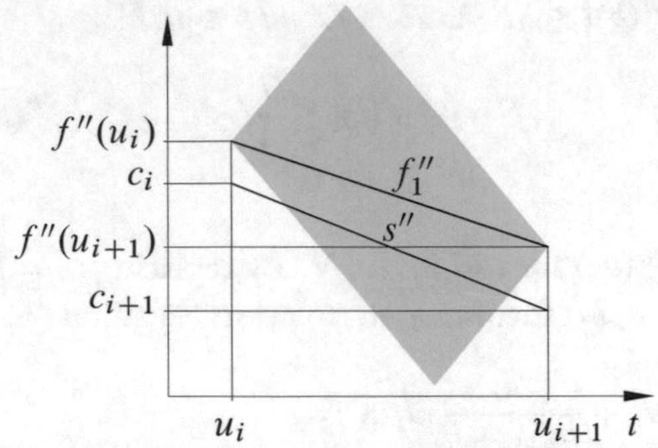

Figure A.22: Estimating $|f''_2(t)| = |f''(t) - f''_1(t)|$. The graph of f'' over the interval $[u_i, u_{i+1}]$ is contained in the parallelogram whose edges are inclined at the angles $\pm\alpha$, where $\tan\alpha = L_2$

Exercise. Prove modifications of Theorems A.19 and A.20 for periodic functions f (with the period $b - a$) and cubic splines of interpolation s such that $s(b) = s(a)$, $s'(b) = s'(a)$, $s''(b) = s''(a)$.

We often need to approximate a function f satisfying the assumptions of both theorems proved above. Theorem A.19 applies to a wider class of splines, being less restrictive to the end conditions. Consider a sequence of cubic splines of interpolation having knot sequences with decreasing maximal distances h between consecutive knots. If $f''(a) \neq 0$ or $f''(b) \neq 0$, then the natural cubic splines, whose second-order derivatives at a and b are zero, allow us to obtain a good approximation of f and f', but, as h tends to 0, the error of approximation of f'' by s'' is never less than $\max\{|f''(a)|, |f''(b)|\}$. However, if the splines are constructed with end conditions taken more carefully (one can even take $s''(a) = f''(a)$ and $s''(b) = f''(b)$), the approximation error of f'' by s'' may be reduced below an arbitrary level. This is essential in constructions of spline surfaces whose boundaries are attached to the boundary of an arbitrary surface in such a way that the junction looks like one of class G^2 (see Section 3.6.2). Also, as we can see, the rates of convergence of approximation errors of f by s and f' by s' to 0 are higher.

A.9 COONS PATCHES

Given interpolation conditions in the form of prescribed values and derivatives up to the order n at two knots, one can solve the Hermite interpolation problem and construct a polynomial of degree $2n + 1$ as in Section A.1 or A.8.1. With given points and derivative vectors, one obtains curves of interpolation. A two-dimensional analogy are Coons patches (see Farin [2002]), which interpolate given boundary curves and cross-boundary derivatives; the interpolation knots u_0, u_1, v_0, v_1 determine the rectangular domain $[u_0, u_1] \times [v_0, v_1]$ of the patch. We assume that $u_0 = v_0 = 0$ and $u_1 = v_1 = 1$.

A.9.1 BILINEARLY BLENDED COONS PATCHES

Let $\boldsymbol{c}_{00}$ and $\boldsymbol{c}_{10}$ be two given parametric curves whose domain (parameter range) is the interval $[0, 1]$. The parametric patch given by the formula

$$\boldsymbol{p}_1(u, v) = \boldsymbol{c}_{00}(u)(1 - v) + \boldsymbol{c}_{10}(u)v, \quad u, v \in [0, 1]$$

interpolates these curves, being parts of its boundary; they are the constant parameter curves of the patch $\boldsymbol{p}_1$, corresponding to $v = 0$ and $v = 1$. The polynomials $H_{00}(v) = 1 - v$ and $H_{10}(v) = v$ used to construct the patch $\boldsymbol{p}_1$ are called **blending functions**.

In a similar way, one can use the polynomials H_{00} and H_{10} to define a patch

$$\boldsymbol{p}_2(u, v) = H_{00}(u)\boldsymbol{d}_{00}(v) + H_{10}(u)\boldsymbol{d}_{10}(v),$$

whose boundaries are given curves $\boldsymbol{d}_{00}$ and $\boldsymbol{d}_{10}$; note that these are curves of constant parameter, $u = 0$ and $u = 1$ respectively, of the patch $\boldsymbol{p}_2$.

If the four curves considered above satisfy the equations $\boldsymbol{c}_{00}(0) = \boldsymbol{d}_{00}(0)$, $\boldsymbol{c}_{00}(1) = \boldsymbol{d}_{10}(0)$, $\boldsymbol{c}_{10}(0) = \boldsymbol{d}_{00}(1)$ and $\boldsymbol{c}_{10}(1) = \boldsymbol{d}_{10}(1)$, then they form a closed curve in space, and it is possible to define a surface patch whose boundary is this closed curve. The simplest way of constructing the patch, whose boundary consists of the curves $\boldsymbol{c}_{00}$, $\boldsymbol{c}_{10}$, $\boldsymbol{d}_{00}$ and $\boldsymbol{d}_{10}$, is by using the formula

$$\boldsymbol{p}(u, v) = \boldsymbol{p}_1(u, v) + \boldsymbol{p}_2(u, v) - \boldsymbol{p}_3(u, v), \tag{A.46}$$

where

$$\begin{aligned}\boldsymbol{p}_3(u, v) = {}& H_{00}(u)H_{00}(v)\boldsymbol{c}_{00}(0) + H_{00}(u)H_{10}(v)\boldsymbol{c}_{10}(0) + \\ & H_{10}(u)H_{00}(v)\boldsymbol{c}_{00}(1) + H_{10}(u)H_{10}(v)\boldsymbol{c}_{10}(1), \quad u, v \in [0, 1].\end{aligned}$$

Because the blending functions are polynomials of degree 1, the patch $\boldsymbol{p}$ defined above is called a **bilinearly blended Coons patch** (Fig. A.23). Note that in general it is not a bilinear patch, as the curves used to define it need not be lines.

If two families of smooth curves form a net with quadrilateral holes, then one might fill each hole with a bilinearly blended Coons patch, thus obtaining a continuous surface of interpolation to the curves. Except for very special (and not very interesting) cases, this surface is not smooth; the given curves will form ridges on it. To obtain a smooth surface, one needs to take into account at least the first-order derivatives of the curves and the patches to be constructed.

A.9.2 BICUBICALLY AND BIQUINTICALLY BLENDED COONS PATCHES

To ensure the smoothness of junctions of patches filling holes in a quadrangular net of curves, one can specify cross-boundary derivatives along each curve. The parametrisation $\boldsymbol{c}_{00}$ of the boundary curve corresponding to $v = 0$ is accompanied by a vector function $\boldsymbol{c}_{01}$; the goal is to obtain a patch $\boldsymbol{p}$ such that $\frac{\partial}{\partial v}\boldsymbol{p}(u, 0) = \boldsymbol{c}_{01}(u)$ for all $u \in [0, 1]$. Similarly, a curve $\boldsymbol{c}_{11}$ determines the partial derivative $\frac{\partial}{\partial v}\boldsymbol{p}$ for $v = 1$. To define the patch $\boldsymbol{p}_1$, having the specified boundary curves

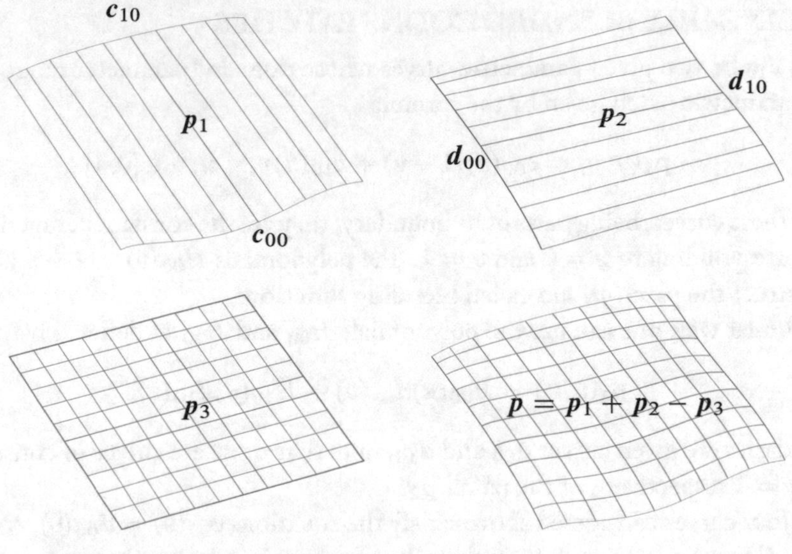

Figure A.23: A bilinearly blended Coons patch and its components

and cross-boundary derivatives corresponding to $v = 0$ and $v = 1$, we can use the cubic polynomials H_{00}, H_{01}, H_{10} and H_{11} defined by Formula (A.39) in order to construct cubic splines of interpolation. The patch $\boldsymbol{p}_1$ is given by the formula

$$\boldsymbol{p}_1(u,v) = \boldsymbol{c}_{00}(u)H_{00}(v) + \boldsymbol{c}_{01}(u)H_{01}(v) + \boldsymbol{c}_{10}(u)H_{10}(v) + \boldsymbol{c}_{11}(u)H_{11}(v), \quad u, v \in [0,1],$$

and similarly obtained is the patch $\boldsymbol{p}_2$, interpolating the boundary curves $\boldsymbol{d}_{00}$, $\boldsymbol{d}_{10}$ and cross-boundary derivatives $\boldsymbol{d}_{01}$, $\boldsymbol{d}_{11}$:

$$\boldsymbol{p}_2(u,v) = H_{00}(u)\boldsymbol{d}_{00}(v) + H_{01}(u)\boldsymbol{d}_{01}(v) + H_{10}(u)\boldsymbol{d}_{10}(v) + H_{11}(u)\boldsymbol{d}_{11}(v), \quad u, v \in [0,1].$$

The Coons patch is defined by Formula (A.46) with the patches $\boldsymbol{p}_1$ and $\boldsymbol{p}_2$ defined as above and the patch $\boldsymbol{p}_3$ considered below. Its construction is possible if the boundary curves and cross-boundary derivatives satisfy **compatibility conditions**. The four boundary curves have to form a closed curve, just like in the case of bilinearly blended Coons patches. At the point $(0,0)$ of the domain of the patch, this positional compatibility condition is the equality $\boldsymbol{c}_{00}(0) = \boldsymbol{d}_{00}(0)$. In addition, it is necessary to ensure that $\boldsymbol{d}_{01}(0) = \boldsymbol{c}'_{00}(0)$ and $\boldsymbol{c}_{01}(0) = \boldsymbol{d}'_{00}(0)$ and also $\boldsymbol{c}'_{01}(0) = \boldsymbol{d}'_{01}$. It is convenient to write the complete set of compatibility conditions at all four corners of a bicubically blended Coons patch in matrix form:

$$\begin{bmatrix} \boldsymbol{c}_{00}(0) & \boldsymbol{c}_{10}(0) & \boldsymbol{c}_{01}(0) & \boldsymbol{c}_{11}(0) \\ \boldsymbol{c}_{00}(1) & \boldsymbol{c}_{10}(1) & \boldsymbol{c}_{01}(1) & \boldsymbol{c}_{11}(1) \\ \boldsymbol{c}'_{00}(0) & \boldsymbol{c}'_{10}(0) & \boldsymbol{c}'_{01}(0) & \boldsymbol{c}'_{11}(0) \\ \boldsymbol{c}'_{00}(1) & \boldsymbol{c}'_{10}(1) & \boldsymbol{c}'_{01}(1) & \boldsymbol{c}'_{11}(1) \end{bmatrix} = \begin{bmatrix} \boldsymbol{d}_{00}(0) & \boldsymbol{d}_{00}(1) & \boldsymbol{d}'_{00}(0) & \boldsymbol{d}'_{00}(1) \\ \boldsymbol{d}_{10}(0) & \boldsymbol{d}_{10}(1) & \boldsymbol{d}'_{10}(0) & \boldsymbol{d}'_{10}(1) \\ \boldsymbol{d}_{01}(0) & \boldsymbol{d}_{01}(1) & \boldsymbol{d}'_{01}(0) & \boldsymbol{d}'_{01}(1) \\ \boldsymbol{d}_{01}(0) & \boldsymbol{d}_{11}(1) & \boldsymbol{d}'_{11}(0) & \boldsymbol{d}'_{11}(1) \end{bmatrix}.$$

Both sides of the equation above describe a matrix P, which consists of the corners and derivatives of the patches $\boldsymbol{p}_1$, $\boldsymbol{p}_2$, and also of the patch $\boldsymbol{p}_3$ defined with the formula

$$\boldsymbol{p}_3(u,v) = [H_{00}(u), H_{10}(u), H_{01}(u), H_{11}(u)]\, P \begin{bmatrix} H_{00}(v) \\ H_{10}(v) \\ H_{01}(v) \\ H_{11}(v) \end{bmatrix}.$$

Note that if all curves used to define the patches $\boldsymbol{p}_1$ and $\boldsymbol{p}_2$ are cubic, then $\boldsymbol{p}_1 = \boldsymbol{p}_2 = \boldsymbol{p}_3$ and the bicubically blended Coons patch is bicubic. However, these curves may be arbitrary; their parametrisations just have to satisfy the compatibility conditions. For practical reasons, we demand that the parametrisations be regular and have continuous derivatives.

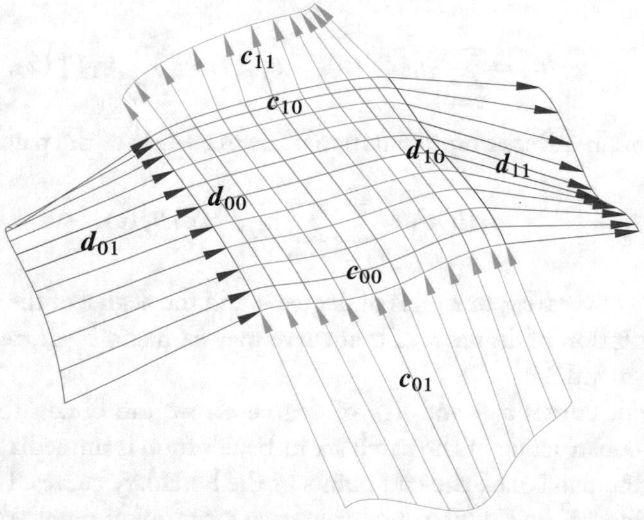

Figure A.24: A bicubically blended Coons patch and its boundary curves and cross-boundary derivatives

If the continuity of the first-order derivatives of Coons patches across their common boundaries is insufficient (as it gives us only the means to guarantee the tangent plane continuity of the surface made of these patches), we can use **biquintically blended Coons patches**. Each of its boundary curves is accompanied with two vector functions which describe the cross-boundary derivatives of order 1 and 2. The second-order cross-boundary derivatives below are denoted by the symbols $\boldsymbol{c}_{02}$ etc. The patch $\boldsymbol{p}_1$, defined as a part of the construction, is given by the formula

$$\begin{aligned}\boldsymbol{p}_1(u,v) = {} & \boldsymbol{c}_{00}(u)H_{00}(v) + \boldsymbol{c}_{01}(u)H_{01}(v) + \boldsymbol{c}_{02}(u)H_{02}(v) + \\ & \boldsymbol{c}_{10}(u)H_{10}(v) + \boldsymbol{c}_{11}(u)H_{11}(v) + \boldsymbol{c}_{12}(u)H_{12}(v), \quad u, v \in [0,1],\end{aligned}$$

where $H_{00}, \ldots, H_{12}$ are the polynomials of degree 5 defined with Formulae (A.40).

Exercise. Describe further steps of the construction of biquintically blended Coons patches, analogous to that of bicubically blended patches. Particular attention has to be paid to the compatibility conditions, which involve derivatives up to the order 2 of all 12 vector functions, which determine the patch, for u and v equal to 0 and 1.

Biquintically blended Coons patches are used to obtain surfaces of class G^2; an example is the construction in Chapter 5.

A.9.3 COONS PATCHES IN BÉZIER FORM

The polynomials used to define a bilinearly blended Coons patch are the Bernstein polynomials: $H_{00}(t) = B_0^1(t)$ and $H_{10}(t) = B_1^1(t)$. If the boundary curves $\boldsymbol{c}_{00}$ and $\boldsymbol{c}_{10}$ are Bézier curves of some degree n,

$$\boldsymbol{c}_{00}(u) = \sum_{i=0}^{n} \boldsymbol{p}_{i0} B_i^n(u), \quad \boldsymbol{c}_{10}(u) = \sum_{i=0}^{n} \boldsymbol{p}_{i1} B_i^n(u),$$

then it is easy to obtain a Bézier representation of degree $(n, 1)$ of the patch $\boldsymbol{p}_1$:

$$\boldsymbol{p}_1(u,v) = \sum_{i=0}^{n}\sum_{j=0}^{1} \boldsymbol{p}_{ij} B_i^n(u) B_j^1(v).$$

If one of the Bézier curves, $\boldsymbol{c}_{00}$ or $\boldsymbol{c}_{10}$, is of degree n and the degree of the other curve is smaller, the Bézier representation of degree n of that curve may be found by degree elevation in the way described in Section A.2.3.

Having Bézier curves $\boldsymbol{d}_{00}$ and $\boldsymbol{d}_{10}$ of degree m, we can obtain the Bézier patch $\boldsymbol{p}_2$ in a similar way. The construction of the patch $\boldsymbol{p}_3$ in Bézier form is immediate, as its control points are the corners of the patch, i.e., the end points of the boundary curves. However, to obtain the Bézier representation of the Coons patch, it is necessary to find representations of the patches $\boldsymbol{p}_1$, $\boldsymbol{p}_2$ and $\boldsymbol{p}_3$ which may be added and subtracted.

The problem is solved by degree elevation of the blending functions. The polynomials H_{00} and H_{10} of degree 1 are

$$H_{00}(v) = \sum_{j=0}^{m}\Big(1 - \frac{j}{m}\Big) B_j^m(v), \quad H_{10}(v) = \sum_{j=0}^{m} \frac{j}{m} B_j^m(v);$$

hence, the Bézier representation of degree (n, m) of $\boldsymbol{p}_1$ is the following:

$$\boldsymbol{p}_1(u, v) = \sum_{i=0}^{n}\sum_{j=0}^{m}\left(\left(1 - \frac{j}{m}\right)\boldsymbol{p}_{i0} + \frac{j}{m}\boldsymbol{p}_{i1}\right)B_i^n(u)B_j^m(v).$$

In the construction of $\boldsymbol{p}_2$, we need to represent the polynomials H_{00} and H_{10} in the Bernstein basis of degree n. And to construct $\boldsymbol{p}_3$, we need their representations of degree n and m.

The construction of Bézier representations of bicubically and biquintically blended Coons patches is based on the same principle, though the final formulae are not that simple. Given the curves $\boldsymbol{c}_{00}, \boldsymbol{c}_{10}$ and cross-boundary derivatives $\boldsymbol{c}_{01}, \boldsymbol{c}_{11}$ (and $\boldsymbol{c}_{02}, \boldsymbol{c}_{12}$), we may use degree elevation to obtain representations of all these curves in the same Bernstein basis. The blending functions $H_{00}, H_{10}, H_{01}, H_{11}$ (and H_{02}, H_{12}), given by Formulae (A.39) or (A.40), have to be represented in the Bernstein basis of degree m, where m is the maximal degree of the curves $\boldsymbol{d}_{00}$, $\boldsymbol{d}_{10}$ $\boldsymbol{d}_{01}$, $\boldsymbol{d}_{11}$ (and $\boldsymbol{d}_{02}$, $\boldsymbol{d}_{12}$). The tensor multiplication of the curves (whose parameter is u) and blending polynomials (whose parameter is v) gives us the formulae for the control points of $\boldsymbol{p}_1$. Describing the rest of the construction is an exercise.

A.10 CURVATURES OF CURVES AND SURFACES

Geometric continuity of curves and surfaces is inseparable from their curvatures, studied in any course of differential geometry. The reader is encouraged to find the background for the definitions, properties and derivations of formulae given in this section in other textbooks, e.g. do Carmo [1976], Klingenberg [1978].

A.10.1 CURVATURES OF CURVES

The length of a curve given in parametric form, whose parametrisation $\boldsymbol{c}(t)$ is of class $C^1[a, b]$, may be obtained by integrating the length (i.e., the Euclidean norm) of the derivative; the length of the arc between the points $\boldsymbol{c}(t_0)$ and $\boldsymbol{c}(t_1)$, where $a \le t_0 \le t_1 \le b$, is the integral

$$l(t_0, t_1) = \int_{t_0}^{t_1} \|\boldsymbol{c}'(t)\|_2 \, \mathrm{d}t.$$

We can extend this formula for $t_1 < t_0$ by taking $l(t_0, t_1) = -l(t_1, t_0)$. Let t_0 be fixed and let $s(t) = l(t_0, t)$; if the parametrisation $\boldsymbol{c}$ is regular, the function s is monotonically increasing and it has an inverse: there exists a unique function $t(s)$ such that $t(s(t)) = t$ for all $t \in [a, b]$. The composition $\boldsymbol{d}(s) = \boldsymbol{c}(t(s))$ is another parametrisation of the curve. It is easy to show that the derivative of t (with respect to s) is $1/\|\boldsymbol{c}'\|_2$ and the derivative of the parametrisation $\boldsymbol{d}$ (with respect to s) for all s is a unit vector. By changing the parameter from t to s, we obtain what is called an **arc length** or **unit speed parametrisation.**[12]

[12]Arc length parametrisations are useful in theoretical considerations and in practice, but in most practical cases (where numerical results are needed) they may only be approximated. If the parametrisation $\boldsymbol{c}$ is described by cubic polynomials, then the expression for the arc length is an elliptic integral which cannot be expressed by a closed algebraic formula.

Let $\boldsymbol{c}(t)$ be a parametrisation of class C^d of a curve located in the d-dimensional space, $\mathbb{R}^d$. Suppose that the vectors $\boldsymbol{c}'(t), \boldsymbol{c}''(t), \ldots, \boldsymbol{c}^{(d-1)}(t)$ are linearly independent for all $t \in [a, b]$. By the Gram–Schmidt orthonormalisation of the sequence of vectors $\boldsymbol{c}'(t), \boldsymbol{c}''(t), \ldots, \boldsymbol{c}^{(d-1)}(t)$ for all t, we obtain a sequence of vector functions, $\boldsymbol{e}_1(t), \ldots, \boldsymbol{e}_{d-1}(t)$, which may be extended by appending the function[13] $\boldsymbol{e}_d(t) = \boldsymbol{e}_1(t) \wedge \cdots \wedge \boldsymbol{e}_{d-1}(t)$. The functions $\boldsymbol{e}_1, \ldots, \boldsymbol{e}_d$ form the **Frenet frame** of the parametrisation $\boldsymbol{c}$. For any t, the vectors $\boldsymbol{e}_1(t), \ldots, \boldsymbol{e}_d(t)$ are mutually orthogonal, their length is 1 and the orientation of the Frenet frame is positive (i.e., $\det[\boldsymbol{e}_1(t), \ldots, \boldsymbol{e}_d(t)] = 1 > 0$).

For the parameters t such that the vectors $\boldsymbol{c}'(t), \boldsymbol{c}''(t), \ldots, \boldsymbol{c}^{(d-1)}(t)$ are linearly dependent, the Frenet frame is undefined. An example of a curve in $\mathbb{R}^3$ with an undefined Frenet frame is a line segment (but the Frenet frame is defined for any parametrised line segment in $\mathbb{R}^2$, except at the points t for which the derivative of the parametrisation is the zero vector).

The vector $\boldsymbol{e}_1(t)$ is **tangent** to the curve at the point $\boldsymbol{c}(t)$ (below we omit the argument t). In the two- and three-dimensional space it is traditionally denoted by the symbol $\boldsymbol{t}$. The vector $\boldsymbol{e}_2$ is a linear combination of $\boldsymbol{c}'$ and $\boldsymbol{c}''$ perpendicular to $\boldsymbol{e}_1$. It is called the **normal vector** and is traditionally denoted by $\boldsymbol{n}$. In $\mathbb{R}^3$ the Frenet frame also has the vector $\boldsymbol{e}_3$, called the **binormal vector** and usually denoted by $\boldsymbol{b}$.

Components of the Frenet frame in $\mathbb{R}^d$ satisfy the system of d differential equations

$$\begin{bmatrix} \boldsymbol{e}'_1 \\ \vdots \\ \vdots \\ \boldsymbol{e}'_d \end{bmatrix} = \|\boldsymbol{c}'\|_2 \begin{bmatrix} 0 & \kappa_1 & & \\ -\kappa_1 & 0 & \ddots & \\ & \ddots & \ddots & \kappa_{d-1} \\ & & -\kappa_{d-1} & 0 \end{bmatrix} \begin{bmatrix} \boldsymbol{e}_1 \\ \vdots \\ \vdots \\ \boldsymbol{e}_d \end{bmatrix}, \tag{A.47}$$

called the **Frenet–Serret equations**. The coefficients $\kappa_1, \ldots, \kappa_{d-1}$ are called the **curvatures** of the curve $\boldsymbol{c}$. They are functions of t; all curvatures with the possible exception of the last one are positive.

Planar curves have just one curvature, $\kappa = \kappa_1$. The value $\kappa(t)$ is the inverse of the radius of the **osculating circle**, i.e., the circle which passes through the point $\boldsymbol{c}(t)$ and may be parametrised so as to have the same first and second-order derivatives at t as those of the parametrisation $\boldsymbol{c}$. Curves in the three-dimensional space have two curvatures, called the **curvature**, $\kappa = \kappa_1$, and the **torsion,** $\tau = \kappa_2$, respectively. By solving the Frenet–Serret equations, one can derive formulae which express the curvatures in terms of derivatives of the parametrisation $\boldsymbol{c}$. For a planar curve, there is

$$\kappa(t) = \frac{\det[\boldsymbol{c}'(t), \boldsymbol{c}''(t)]}{\|\boldsymbol{c}'(t)\|_2^3}. \tag{A.48}$$

[13]Here we use the vector product in $\mathbb{R}^d$, which is an operation with $d-1$ arguments. In particular, in $\mathbb{R}^2$ it has one argument, and the result is a vector orthogonal to this argument and having the same length. The vector product in $\mathbb{R}^3$ is often called the cross product because of the popular symbol $\times$, used in this book to denote Cartesian products.

For curves in $\mathbb{R}^3$, we have the following formulae:

$$\kappa(t) = \frac{\|c'(t) \wedge c''(t)\|_2}{\|c'(t)\|_2^3}, \tag{A.49}$$

$$\tau(t) = \frac{\det[c'(t), c''(t), c'''(t)]}{\|c'(t) \wedge c''(t)\|_2^2}. \tag{A.50}$$

Exercise. Derive the formulae expressing the curvature and torsion of a Bézier curve p of degree n, using the points obtained by de Casteljau's algorithm (p. 164):

$$\kappa(t) = \frac{n-1}{n} \frac{\det[\Delta p_0^{(n-2)}, \Delta p_1^{(n-2)}]}{\|\Delta p_0^{(n-1)}\|_2^3} \quad \text{in } \mathbb{R}^2, n \geq 2,$$

$$\kappa(t) = \frac{n-1}{n} \frac{\|\Delta p_0^{(n-2)} \wedge \Delta p_1^{(n-2)}\|_2}{\|\Delta p_0^{(n-1)}\|_2^3} \quad \text{in } \mathbb{R}^3, n \geq 2,$$

$$\tau(t) = \frac{n-2}{n} \frac{\det[\Delta p_0^{(n-3)}, \Delta p_1^{(n-3)}, \Delta p_2^{(n-3)}]}{\|\Delta p_0^{(n-2)} \wedge \Delta p_1^{(n-2)}\|_2^2} \quad \text{in } \mathbb{R}^3, n \geq 3,$$

where $\Delta p_i^{(j)} = p_{i+1}^{(j)} - p_i^{(j)}$. Note the presence of expressions for the length of a line segment, the area of a parallelogram and the volume of a parallelpiped determined by the difference vectors in these formulae; an example for a quartic curve is in Figure A.25.

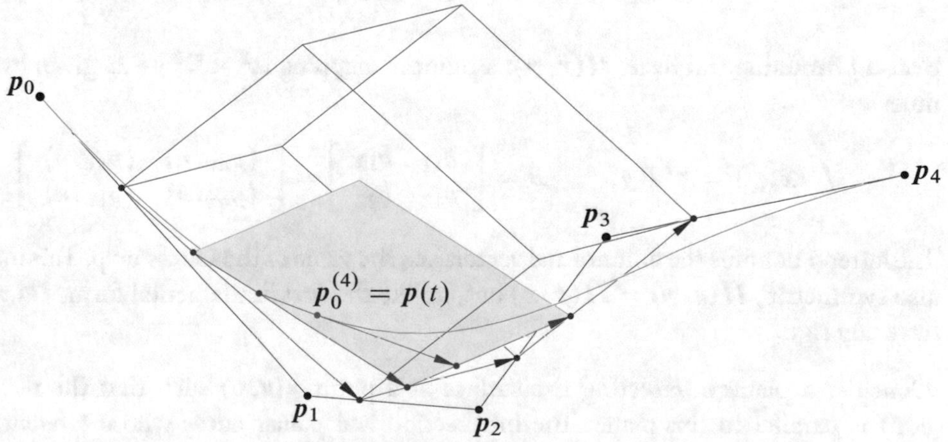

Figure A.25: Computing curvature and torsion of a Bézier curve with de Casteljau's algorithm

A.10.2 CURVATURES OF SURFACES

We consider a regular parametrisation $\boldsymbol{p}(u, v)$ of a surface $\mathcal{M}$ in $\mathbb{R}^3$ whose derivatives of the first and second order are continuous in an area A being the domain of $\boldsymbol{p}$. The **Gauss map** is a vector function $A \to \mathbb{R}^3$ defined with the formula

$$\boldsymbol{n}(u, v) = \frac{1}{\|\boldsymbol{p}_u(u, v) \wedge \boldsymbol{p}_v(u, v)\|_2} \boldsymbol{p}_u(u, v) \wedge \boldsymbol{p}_v(u, v).$$

Its value is the unit normal vector of the surface at the point $\boldsymbol{p}(u, v)$.

For each point $(u, v) \in A$ of $\boldsymbol{p}$ the following mappings are defined:

- **Differential** $\mathrm{D}\boldsymbol{p}$: a linear mapping $\mathbb{R}^2 \to T_{\mathcal{M}} \subset \mathbb{R}^3$ whose matrix is made of the partial derivatives of $\boldsymbol{p}$: $\mathrm{D}\boldsymbol{p}(\boldsymbol{x}) = [\boldsymbol{p}_u, \boldsymbol{p}_v]\boldsymbol{x}$. The linear vector subspace $T_{\mathcal{M}}$, called the **tangent space** of the surface $\mathcal{M}$, is parallel to the tangent plane of $\mathcal{M}$ at the point $\boldsymbol{p}(u, v)$. The parametrisation $\boldsymbol{p}$ is regular if and only if the mapping $\mathrm{D}\boldsymbol{p}$ is one-to-one.

- **First fundamental form** $\boldsymbol{I}(\boldsymbol{x}, \boldsymbol{y})$: a bilinear mapping $\mathbb{R}^2 \times \mathbb{R}^2 \to \mathbb{R}$:

$$\boldsymbol{I}(\boldsymbol{x}, \boldsymbol{y}) = \langle \mathrm{D}\boldsymbol{p}(\boldsymbol{x}), \mathrm{D}\boldsymbol{p}(\boldsymbol{y}) \rangle.$$

  Its arguments $\boldsymbol{x}$, $\boldsymbol{y}$ are vectors in $\mathbb{R}^2$ and the value is the scalar product of the vectors $\mathrm{D}\boldsymbol{p}(\boldsymbol{x})$ and $\mathrm{D}\boldsymbol{p}(\boldsymbol{y})$ in $\mathbb{R}^3$. Note that the first fundamental form is a (non-standard) scalar product in $\mathbb{R}^2$, i.e., a bilinear mapping such that $\boldsymbol{I}(\boldsymbol{x}, \boldsymbol{y}) = \boldsymbol{I}(\boldsymbol{y}, \boldsymbol{x})$ for all vectors $\boldsymbol{x}$, $\boldsymbol{y}$ and $\boldsymbol{I}(\boldsymbol{x}, \boldsymbol{x}) > 0$ for all non-zero vectors $\boldsymbol{x}$. The first fundamental form is represented by the matrix G:

$$\boldsymbol{I}(\boldsymbol{x}, \boldsymbol{y}) = \boldsymbol{x}^T G \boldsymbol{y}, \qquad G = \begin{bmatrix} g_{11} & g_{12} \\ g_{12} & g_{22} \end{bmatrix} = \begin{bmatrix} \langle \boldsymbol{p}_u, \boldsymbol{p}_u \rangle & \langle \boldsymbol{p}_u, \boldsymbol{p}_v \rangle \\ \langle \boldsymbol{p}_u, \boldsymbol{p}_v \rangle & \langle \boldsymbol{p}_v, \boldsymbol{p}_v \rangle \end{bmatrix}.$$

- **Second fundamental form** $\boldsymbol{II}(\boldsymbol{x}, \boldsymbol{y})$: a bilinear mapping $\mathbb{R}^2 \times \mathbb{R}^2 \to \mathbb{R}$ given by the formula

$$\boldsymbol{II}(\boldsymbol{x}, \boldsymbol{y}) = \boldsymbol{x}^T B \boldsymbol{y}, \qquad B = \begin{bmatrix} b_{11} & b_{12} \\ b_{12} & b_{22} \end{bmatrix} = \begin{bmatrix} \langle \boldsymbol{p}_{uu}, \boldsymbol{n} \rangle & \langle \boldsymbol{p}_{uv}, \boldsymbol{n} \rangle \\ \langle \boldsymbol{p}_{uv}, \boldsymbol{n} \rangle & \langle \boldsymbol{p}_{vv}, \boldsymbol{n} \rangle \end{bmatrix}.$$

  The letter $\boldsymbol{n}$ denotes the unit normal vector, i.e., the value of the Gauss map. This mapping is also symmetric, $\boldsymbol{II}(\boldsymbol{x}, \boldsymbol{y}) = \boldsymbol{II}(\boldsymbol{y}, \boldsymbol{x})$ but, unlike the first fundamental form, $\boldsymbol{II}(\boldsymbol{x}, \boldsymbol{x})$ may have any sign.

Consider a plane intersecting the surface at a point $\boldsymbol{p}(u, v)$ such that the normal vector $\boldsymbol{n}(u, v)$ is parallel to this plane. The intersection is a planar curve whose tangent vector $\boldsymbol{t}$ is orthogonal to $\boldsymbol{n}$ (we omit the arguments). The curvature of this curve is called the **normal curvature** of the surface in the direction of the vector $\boldsymbol{t}$. If $\boldsymbol{y}$ is a vector in $\mathbb{R}^2$ such that $\mathrm{D}\boldsymbol{p}(\boldsymbol{y}) = \boldsymbol{t}$, then the normal curvature[14] in the direction of $\boldsymbol{t}$ is equal to

$$\varkappa = \frac{II(\boldsymbol{y}, \boldsymbol{y})}{I(\boldsymbol{y}, \boldsymbol{y})}. \tag{A.51}$$

The length of the vector $\boldsymbol{t}$ or $\boldsymbol{y}$ does not influence the normal curvature, which takes its minimal and maximal values for the vectors $\boldsymbol{t}$ having two different directions. One can prove that the extremal curvatures satisfy the equation $(B - \varkappa G)\boldsymbol{y} = \boldsymbol{0}$, with vectors $\boldsymbol{y} \neq \boldsymbol{0}$, which determine the directions of the corresponding vectors $\boldsymbol{t}$. Such a vector exists if the matrix $B - \varkappa G$ is singular. Hence, we obtain the quadratic equation $\det(B - \varkappa G) = 0$, which may be rewritten as

$$(\varkappa - \varkappa_1)(\varkappa - \varkappa_2) = 0 \qquad \text{or} \qquad \varkappa^2 - 2H\varkappa + K = 0.$$

The solutions of this equation, $\varkappa_1$ and $\varkappa_2$, are real, and they are the extremal normal curvatures. By the Viète formulae, $H = (\varkappa_1 + \varkappa_2)/2$, $K = \varkappa_1\varkappa_2$. The average of the extremal normal curvatures, H, is called the **mean curvature**, and their product K is called the **Gaussian curvature**. They depend on the matrices of the first and second fundamental form in the following way:

$$\begin{aligned} H &= \frac{1}{2\det G}\left(\det\begin{bmatrix} b_{11} & g_{12} \\ b_{12} & g_{22} \end{bmatrix} + \det\begin{bmatrix} g_{11} & b_{12} \\ g_{12} & b_{22} \end{bmatrix}\right) \\ &= \frac{b_{11}g_{22} - 2b_{12}g_{12} + b_{22}g_{11}}{2(g_{11}g_{22} - g_{12}^2)}, \end{aligned} \tag{A.52}$$

$$K = \frac{\det B}{\det G} = \frac{b_{11}b_{22} - b_{12}^2}{g_{11}g_{22} - g_{12}^2}. \tag{A.53}$$

Let $\mathcal{C}$ be a curve obtained as a planar section of a surface $\mathcal{M}$ that has a regular parametrisation $\boldsymbol{p}$ of class C^2 (see Figure A.26). If the intersecting plane is not tangent to the surface, then this curve has a regular parametrisation $\boldsymbol{c}$ of class C^2. Consider the Frenet frame $(\boldsymbol{e}_1(t), \boldsymbol{e}_2(t), \boldsymbol{e}_3(t))$ of the curve $\mathcal{C}$ at a point $\boldsymbol{c}(t) = \boldsymbol{p}(u, v)$. By Meusnier's theorem, the normal curvature $\varkappa$ of the surface $\mathcal{M}$ in the direction of the vector $\boldsymbol{e}_1$ and the curvature κ of the curve $\mathcal{C}$ at that point are related as follows:

$$\kappa|\langle \boldsymbol{e}_2, \boldsymbol{n}\rangle| = |\varkappa|,$$

where $\boldsymbol{n}$ is the unit normal vector of the surface; the scalar product $\langle \boldsymbol{e}_2, \boldsymbol{n}\rangle$ is the cosine of the angle between the normal vector of the curve $\mathcal{C}$ and the normal vector of the surface. Due to Meusnier's theorem, we can define the **curvature continuity of a surface** as the continuity of curvature of *all* curves being planar sections of that surface, obtained with planes that are not tangent to the surface. It is equivalent to the existence of a regular parametrisation of class C^2 of the surface. In other words, it is geometric continuity of the second order of the surface.

[14]Note that the normal curvature may have any sign, as the intersection curve is planar. The sign of the normal curvature may change if the surface is reparametrised, e.g. if the parameters u, v are exchanged.

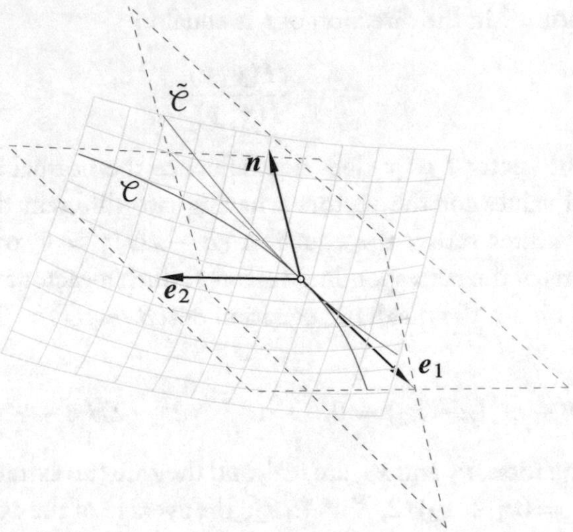

Figure A.26: Two planar sections of a surface patch. The curvature of the curve $\tilde{\mathcal{C}}$ at the point where the two curves are tangent is equal to the normal curvature in the direction of $\boldsymbol{e}_1$

A.11 FÀA DI BRUNO'S FORMULA

In [1855], Fàa di Bruno published a formula to express derivatives of arbitrary order of a composition of two functions in terms of their derivatives. Here it is: let $f(u)$ and $g(t)$ be functions of class C^n and let $h = g \circ f$, i.e., $h(u) = g\big(f(u)\big)$. Then,

$$\frac{\mathrm{d}^n h}{\mathrm{d}u^n} = \sum_{k=1}^{n} a_{nk} \frac{\mathrm{d}^k g}{\mathrm{d}t^k}, \tag{A.54}$$

$$\text{where} \quad a_{nk} = \sum_{\substack{m_1+\cdots+m_k=n \\ m_1,\ldots,m_k>0}} \frac{n!}{k!m_1!\ldots m_k!} \frac{\mathrm{d}^{m_1} f}{\mathrm{d}u^{m_1}} \cdots \frac{\mathrm{d}^{m_k} f}{\mathrm{d}u^{m_k}}.$$

Fàa di Bruno's formula has a direct application in constructions of smooth junctions of curve arcs; note that g and h may be vector functions, i.e., parametrisations. If two parametrisations, $\boldsymbol{p}(u)$ and $\boldsymbol{g}(t)$, are regular and have a common point, say, $\boldsymbol{p}(u_1) = \boldsymbol{g}(t_0)$, then their junction is of class G^n when one of the curves, say, $\boldsymbol{g}$, may be reparametrised so as to obtain a parametrisation $\boldsymbol{h}(u) = \boldsymbol{g}(f(u))$ whose derivatives up to the order n match the derivatives of $\boldsymbol{p}$ at the junction point. We explore this possibility in Chapter 2. However, Formula (A.54) is too complicated to be convenient. Usually we need its particular cases for rather small n. We can derive the needed formulae recursively, using the formula for the first-order derivative of a composition and the formula for the derivative of the product of functions. This calculation gives us

$$\begin{aligned}
\boldsymbol{h}' &= f'\boldsymbol{g}',\\
\boldsymbol{h}'' &= f''\boldsymbol{g}' + f'^2\boldsymbol{g}'',\\
\boldsymbol{h}''' &= f'''\boldsymbol{g}' + 3f'f''\boldsymbol{g}'' + f'^3\boldsymbol{g}''',\\
\boldsymbol{h}'''' &= f''''\boldsymbol{g}' + (3f''^2 + 4f'f''')\boldsymbol{g}'' + 4f'^2f''\boldsymbol{g}''' + f'^4\boldsymbol{g}'''',
\end{aligned}$$

and it may be carried on as long as necessary.

A generalisation of Fàa di Bruno's formula for functions of more than one variable was found by Constantine and Savits [1996]. Let $\boldsymbol{f}\colon \mathbb{R}^k \to \mathbb{R}^l$ and $\boldsymbol{g}\colon \mathbb{R}^l \to \mathbb{R}^m$ be functions of class C^n and let $\boldsymbol{h} = \boldsymbol{g} \circ \boldsymbol{f}$. The formula is

$$\frac{\partial^{|\boldsymbol{\alpha}|}}{\partial u_1^{\alpha_1} \dots \partial u_k^{\alpha_k}} \boldsymbol{h}(u_1, \dots, u_k) = \sum_{\boldsymbol{\beta}:1\leq|\boldsymbol{\beta}|\leq|\boldsymbol{\alpha}|} a_{\boldsymbol{\alpha},\boldsymbol{\beta}} \frac{\partial^{|\boldsymbol{\beta}|}}{\partial t_1^{\beta_1} \dots \partial t_l^{\beta_l}} \boldsymbol{g}(t_1, \dots, t_l). \tag{A.55}$$

The symbols $\boldsymbol{\alpha} = (\alpha_1, \dots, \alpha_k)$ and $\boldsymbol{\beta} = (\beta_1, \dots, \beta_l)$ denote multi-indices, i.e., vectors made of nonnegative integers, and $|\boldsymbol{\alpha}| = \alpha_1 + \dots + \alpha_k \leq n$, $|\boldsymbol{\beta}| = \beta_1 + \dots + \beta_l$. The general formula which describes the coefficients $a_{\boldsymbol{\alpha},\boldsymbol{\beta}}$ in terms of derivatives of $\boldsymbol{f}$ is even more impractical than the one for functions of one variable, and so it is not reproduced here; an interested reader may find it in the original paper. Instead of using this general formula, one can recursively derive the formulae for derivatives needed in a particular application. In this way, for $k = l = 2$, assuming that the functions $s(u, v)$ and $t(u, v)$ describe the coordinates of $\boldsymbol{f}$, we obtain

$$\begin{aligned}
\boldsymbol{h}_u &= s_u\boldsymbol{g}_s + t_u\boldsymbol{g}_t,\\
\boldsymbol{h}_v &= s_v\boldsymbol{g}_s + t_v\boldsymbol{g}_t,\\
\boldsymbol{h}_{uu} &= s_{uu}\boldsymbol{g}_s + t_{uu}\boldsymbol{g}_t + s_u^2\boldsymbol{p}_{ss} + 2s_ut_u\boldsymbol{p}_{st} + t_u^2\boldsymbol{p}_{tt},\\
\boldsymbol{h}_{uv} &= s_{uv}\boldsymbol{g}_s + t_{uv}\boldsymbol{g}_t + s_us_v\boldsymbol{p}_{ss} + (s_ut_v + s_vt_u)\boldsymbol{p}_{st} + t_ut_v\boldsymbol{p}_{tt},\\
\boldsymbol{h}_{vv} &= s_{vv}\boldsymbol{g}_s + t_{vv}\boldsymbol{g}_t + s_v^2\boldsymbol{p}_{ss} + 2s_vt_v\boldsymbol{p}_{st} + t_v^2\boldsymbol{p}_{tt},
\end{aligned}$$

and so on.

The mapping which associates a function $\boldsymbol{g}$ with the composition $\boldsymbol{g} \circ \boldsymbol{f}$ is linear with respect to $\boldsymbol{g}$; the sum $\boldsymbol{g}_1 + \boldsymbol{g}_2$ is mapped to $\boldsymbol{g}_1 \circ \boldsymbol{f} + \boldsymbol{g}_2 \circ \boldsymbol{f}$, and the product of $\boldsymbol{g}$ with a scalar a is mapped to the product $a(\boldsymbol{g} \circ \boldsymbol{f})$. As a consequence, the derivatives of order n of the function $\boldsymbol{h}$ are linear combinations of the derivatives of $\boldsymbol{g}$ up to the order n. We can write these relations in matrix form, e.g.:

$$\begin{bmatrix} \boldsymbol{h}_u \\ \boldsymbol{h}_v \end{bmatrix} = \begin{bmatrix} s_u & t_u \\ s_v & t_v \end{bmatrix} \begin{bmatrix} \boldsymbol{g}_s \\ \boldsymbol{g}_t \end{bmatrix},$$

$$\begin{bmatrix} \boldsymbol{h}_{uu} \\ \boldsymbol{h}_{uv} \\ \boldsymbol{h}_{vv} \end{bmatrix} = \begin{bmatrix} s_{uu} & t_{uu} \\ s_{uv} & t_{uv} \\ s_{vv} & t_{vv} \end{bmatrix} \begin{bmatrix} \boldsymbol{g}_s \\ \boldsymbol{g}_t \end{bmatrix} + \begin{bmatrix} s_u^2 & 2s_ut_u & t_u^2 \\ s_us_v & s_ut_v + s_vt_u & s_vt_v \\ s_v^2 & 2s_vt_v & t_v^2 \end{bmatrix} \begin{bmatrix} \boldsymbol{g}_{ss} \\ \boldsymbol{g}_{st} \\ \boldsymbol{g}_{tt} \end{bmatrix}$$

etc. In general, for functions of class C^n, we can gather the partial derivatives of order $p \leq n$ of the functions $\boldsymbol{g}$ and $\boldsymbol{h}$ in matrices as follows:

$$G_p = \begin{bmatrix} \frac{\partial^p}{\partial s^p}\boldsymbol{g} \\ \frac{\partial^p}{\partial s^{p-1}\partial t}\boldsymbol{g} \\ \vdots \\ \frac{\partial^p}{\partial t^p}\boldsymbol{g} \end{bmatrix}, \quad H_p = \begin{bmatrix} \frac{\partial^p}{\partial u^p}\boldsymbol{h} \\ \frac{\partial^p}{\partial u^{p-1}\partial v}\boldsymbol{h} \\ \vdots \\ \frac{\partial^p}{\partial v^p}\boldsymbol{h} \end{bmatrix}.$$

The coefficients $a_{\boldsymbol{\alpha},\boldsymbol{\beta}}$ from (A.55) may be gathered in matrices F_{pq}, where $p = |\boldsymbol{\alpha}|$, $q = |\boldsymbol{\beta}|$ are orders of derivatives in G_p and F_q, and $q = 1, \ldots, p$. Then, we can write

$$H_p = \sum_{q=1}^{p} F_{pq} G_q.$$

If the function $\boldsymbol{f}$ is regular, i.e., if the matrix

$$F_{11} = \begin{bmatrix} s_u & s_v \\ t_u & t_v \end{bmatrix}$$

is non-singular, then the matrices F_{pp} for $p = 2, \ldots, n$ are also non-singular. In that case for $p = 1, \ldots, n$ we can set up and solve the systems of linear equations

$$F_{pp} G_p = H_p - \sum_{q=1}^{p-1} F_{pq} G_q$$

in order to evaluate the partial derivatives of the function $\boldsymbol{g} = \boldsymbol{h} \circ \boldsymbol{f}^{-1}$. In particular, we can evaluate the derivatives of $\boldsymbol{f}^{-1}$ of arbitrary order by substituting the identity mapping for $\boldsymbol{h}$.

Exercise. Derive the formulae for the derivatives of the third and fourth order of the composition of functions of two variables and find the matrices F_{pq} for $1 \leq q \leq p = 4$.

Bibliography

Barsky, B. A. "*The Beta-spline: A Local Representation Based on Shape Parameters and Fundamental Geometric Measures*", Ph.D. thesis, University of Utah, Salt Lake City, UT, December 1981. 16

Barsky, B. A. and DeRose, T. D. "*Geometric Continuity of Parametric Curves*", Technical Report No. UCB/CSD 84/205, Computer Science Division, Electrical Engineering and Computer Science Department, University of Califormina, Berkeley, CA, October 1984. 16

Bartels, R. H., Beatty, J. C. and Barsky, B. A. "*An Introduction to the Use of Splines in Computer Graphics*", Technical Report, University of Waterloo TR CS-83-09, University of California, Berkeley TR UCB/CSD 83/136, Revised May 1985. 20

Beier, K.-P. and Chen, Y. "Highlight-line algorithm for realtime surface-quality assessment", *Computer-Aided Design*, 1994, 26, pp. 269–277. DOI: 10.1016/0010-4485(94)90073-6. 154

Bernstein, S. "Démonstration du théorème de Weierstrass fondée sur le calcul des probabilités", *Proc. Kharkov Math. Soc.*, ser. 2, 13(1912), pp. 1–2. 163

Bloor, M. I. G. and Wilson, M. J. "Generating blend surfaces using partial differential equations", *Computer-Aided Design*, 1989, 21(3), pp. 165–171. DOI: 10.1016/0010-4485(89)90071-7. 109

Boehm, W. "Inserting new knots into B-spline curves", *Computer-Aided Design*, 1980, 12, pp. 199–201. DOI: 10.1016/0010-4485(80)90154-2. 181

Boehm, W. "Curvature continuous curves and surfaces", *Computer Aided Geometric Design*, 1985, 2, pp. 313–323. DOI: 10.1016/s0167-8396(85)80006-6. 36

Boehm, W. "Rational geometric splines", *Computer Aided Geometric Design*, 1987, 4, pp. 67–77. DOI: 10.1016/0167-8396(87)90025-2. 36

Catmull, E. and Clark, J. "Recursively generated B-spline surfaces on arbitrary topological meshes", *Computer-Aided Design*, 1978, 10(6), pp. 350–355. DOI: 10.1016/0010-4485(78)90110-0. 191

Ciarlet, P. G. "*The Finite Element Method for Elliptic Problems*", North Holland, Amsterdam, 1978. DOI: 10.1137/1.9780898719208. 109, 133

Cohen, E. and Schumaker, L. L. "Rates of convergence of control polygons", *Computer Aided Geometric Design*, 1985, 2, pp. 229–235. DOI: 10.1016/0167-8396(85)90029-9. 183

Constantine, G. M. and Savits, T. H. "A multivariate Faa di Bruno formula with applications", *Transactions of the American Mathematical Society*, 1996, 348(2). DOI: 10.1090/S0002-9947-96-01501-2. 219

Cox, M. "The numerical evaluation of B-splines", *J. Inst. Maths. Applics.*, 1972, 10, pp. 134–149. DOI: 10.1093/imamat/10.2.134. 173

de Boor, C. "On calculating with B-splines", *J. Approx. Theory*, 1972, 6(1), pp. 50–62. DOI: 10.1016/0021-9045(72)90080-9. 173

de Boor, C. "*A Practical Guide to Splines*", Springer, 1978. DOI: 10.1007/978-1-4612-6333-3. 163

Degen, W. L. F. "Explicit continuity conditions for adjacent Bézier surface patches", *Computer Aided Geometric Design*, 1990, 7, pp. 181–189. DOI: 10.1016/0167-8396(90)90029-q. 44

do Carmo, M. "*Differential Geometry of Curves and Surfaces*", Prentice Hall, Inc., Englewood Cliffs, NJ, 1976. DOI: 10.1007/978-3-642-57951-6_5. 213

Doo, D. W. H. and Sabin, M. A. "Behaviour of recursive division surfaces near extraordinary points", *Computer-Aided Design*, 1978, 10(6), pp. 356–360. DOI: 10.1016/0010-4485(78)90111-2. 191, 196

Dyn, N. and Micchelli, C. A. "Piecewise Polynomial Spaces and Geometric Continuity of Curves", *Numerische Mathematik*, Vol. 54, 1988, pp. 319–333. Also Research Report No. 11390, IBM Thomas J. Watson Research Center, Yorktown Height, NY, 1985. DOI: 10.1007/bf01396765. 18

Evans, L. C. "*Partial Differential Equations*", American Mathematical Society, 1998. DOI: 10.1090/gsm/019. 133

Fàa di Bruno, F. "Sullo sviluppo delle funzioni", *Annali di Scienze Matematiche e Fisiche*, 1855, Tomo 6, Tipografia delle Belle Arti, Roma, pp. 479–480. 218

Farin, G. "Visually C^2 cubic splines", *Computer-Aided Design*, 1982, 14, pp. 137–139. DOI: 10.1016/0010-4485(82)90326-8. 36

Farin, G. "*NURBS: from Projective Geometry to Practical Use*", A. K. Peters, Natick, MA, 1999. 197

Farin, G. "*Curves and Surfaces for Computer Aided Geometric Design. A Practical Guide*", 5th ed., Academic Press, 2002. 202, 208

Geise, G. and Jüttler B. "A geometrical approach to curvature continuous joints of rational curves", *Computer Aided Geometric Design*, 1993, 10, pp. 109–122. DOI: 10.1016/0167-8396(93)90014-t. 14

Giaquinta, M. and Hildebrandt, S. "*Calculus of Variations I*", Springer, 1996. DOI: 10.1007/978-3-662-03278-7. 129

Greiner, G. "Surface construction based on variational principles", In Laurent, P. J., Le Méhauté, A. and Schumaker, L. L. *Wavelets, Images and Surface Fitting*, pp. 277–286, A. K. Peters, 1994. 137

Habib, A. W. and Goldman, R. N. "Theories of contact specified by connection matrices", *Computer Aided Geometric Design*, 1996, 13, pp. 905–929. DOI: 10.1016/s0167-8396(96)00015-5. 9

Hahn, J. "Filling polygonal holes with rectangular patches", in "*Theory and Practice of Geometric Modeling*", Blaubeueren, 1988, Springer, pp. 81–91. DOI: 10.1007/978-3-642-61542-9. 69

Hermann, T., Peters, J. and Strotman, T. "Curve networks compatible with G^2 surfacing", *Computer Aided Geometric Design*, 2012, 29, pp. 219–230. DOI: 10.1016/j.cagd.2011.10.003. 80

Hungerford, T. W. "*Algebra*", Springer, 1974. DOI: 10.1007/978-1-4612-6101-8. 47

Joe, B. "Knot insertion for Beta-spline curves and surfaces", *ACM Transactions on Graphics*, 1990, 9(1), pp. 41–65. DOI: 10.1145/77635.77638. 32

Kahmann, J. "Continuity of Curvature Between Adjacent Bézier Patches" in Barnhill, R. E. and Boehm, W. (eds.), "*Surfaces in Computer Aided Geometric Design*", North Holland, Amsterdam, 1983, pp. 65–75. 48

Karčiauskas, K. and Peters, J. "Biquintic G^2 surfaces via functionals", *Computer Aided Geometric Design*, 2015, 33, pp. 17–29. DOI: 10.1016/j.cagd.2014.11.003. 109

Kiciak, P. "Constructions of G^1 continuous joins of rational Bézier patches", *Computer Aided Geometric Design*, 1995, 12, pp. 283–303. DOI: 10.1016/0167-8396(94)00014-j. 46

Kiciak, P. "Conditions for geometric continuity between polynomial and rational surface patches", *Computer Aided Geometric Design*, 1996, 13, pp. 709–741. DOI: 10.1016/0167-8396(96)00006-4. 46, 48, 50

Kiciak, P. "Trigonometric splines and geometric continuity of surfaces", *WSCG'99 Conference Proceedings*, Plzen, 1999, pp. 502–509. 80

Kiciak, P. "Spline surfaces of arbitrary topology with continuous curvature and optimized shape", *Computer-Aided Design*, 2013, 45, pp. 154–167. DOI: 10.1016/j.cad.2012.09.003. 139

Kincaid, D. and Cheney, W. "*Numerical Analysis. Mathematics of Scientific Computing*", The Wadsworth Group, 2002. 135

Klass, R. "Correction of local surface irregularities using reflection lines", *Computer-Aided Design*, 1980, 12(2), pp. 73–77. DOI: 10.1016/0010-4485(80)90447-9. 151

Klingenberg, W. "*A Course in Differential Geometry*", Springer, NY, 1978. DOI: 10.1007/978-1-4612-9923-3. 213

Lane, J. M. and Riesenfeld, R. F. "A theoretical development for the computer generation and display of piecewise polynomial surfaces", *IEEE Transactions on Pattern Analysis and Machine Intelligence*, 1980, 2(1) pp. 35–46. DOI: 10.1109/tpami.1980.4766968. 184

Loop, C. and DeRose, T. "Generalized B-spline surfaces of arbitrary topology", *Computer Graphics*, 1990, 24(4), pp. 347–356. DOI: 10.1145/97880.97917. 194

Nielson, G. "Some Piecewise Polynomial Alternatives to Splines Under Tension", In Barnhill, R. E. and Riesenfeld, R. (eds.), *Computer Aided Geometric Design*, Academic Press, NY, 1974, pp. 209–235. DOI: 10.1016/c2013-0-10333-9. 15

Peters, J. "C^2 free-form surfaces of degree $(3, 5)$", *Computer Aided Geometric Design*, 2002, 19, pp. 113–126. DOI: 10.1016/s0167-8396(01)00081-4. 109

Prautzsch, H. "Smoothness of subdivision surfaces at extraordinary points", *Advances in Computational Mathematics*, 1998 (9), pp. 377–389. 196

Prautzsch, H., Boehm, W. and Paluszny, M. "*Bézier and B-spline Techniques*", Springer, 2002. DOI: 10.1007/978-3-662-04919-8. 190, 196

Putz, B. "On curvature analysis of free-form surfaces", *Second International CAD/CAM Conference CAMP'92*, Budapest, 1992, pp. 126–133. 156

Putz, B. "On normal curvature discontinuity between tangent plane continuous patches", *Computer Aided Geometric Design*, 1996, 13, pp. 95–99. DOI: 10.1016/0167-8396(95)00045-3. 156

Reif, U. "A unified approach to subdivision algorithms near extraordinary vertices", *Computer Aided Geometric Design*, 1995, 12, pp. 153–174. DOI: 10.1016/0167-8396(94)00007-f. 196

Sarraga, R. "Recent methods for surface shape optimization", *Computer Aided Geometric Design*, 1998, 15, pp. 417–436. DOI: 10.1016/s0167-8396(97)00030-7. 137

Seidel, H. P. "*Geometric Constructions and Knot Insertion for Geometrically Continuous Spline Curves of Arbitrary Degree*", Research Report No. CS-90-24, University of Waterloo, Waterloo, Ontario, June 1990. 32

Seroussi, G. and Barsky, B. A. "*A Symbolic Derivation of Beta-splines of Arbitrary Order*", Computer Science Division, Technical Report No. UCB/CSD-91-633, University of California, Berkeley, CA, June 1991. 17, 25, 26

Spivak, M. "*Calculus on Manifolds*", Westview Press, 1965. 132, 135

Veltkamp, R. C. "Survey of continuities of curves and surfaces", *Computer Graphics Forum*, 1992, II, pp. 93–112. DOI: 10.1111/1467-8659.1120093. 2

Ye, X. "The Gaussian and mean curvature criteria for curvature continuity between surfaces", *Computer Aided Geometric Design*, 1996, 13, pp. 549–567. DOI: 10.1016/0167-8396(95)00044-5. 156

Ye, X. and Nowacki, H. "Ensuring compatibility of G^2-continuous surface patches around a nodepoint", *Computer Aided Geometric Design*, 1996, 13, pp. 931–949. DOI: 10.1016/s0167-8396(96)00016-7. 103

Zheng, J., Wang, G. and Liang, Y. "GC^n continuity conditions for adjacent rational parametric surfaces", *Computer Aided Geometric Design*, 1995, 12, pp. 111–129. DOI: 10.1016/0167-8396(94)00005-d. 48

Author's Biography

PRZEMYSŁAW KICIAK

Przemysław Kiciak was born in 1961 in Warsaw. He studied at the Warsaw University of Technology (Faculty SiMR—Cars and Working Machines) and then at the University of Warsaw (Faculty MIM—Mathematics, Informatics and Mechanics). Since 1992 he has been working at the University of Warsaw (Faculty MIM). His scientific interests are in the area of computer graphics, geometric design and numerical analysis.

Przemysław Kiciak is the author of an open source software package called BSTools, which consists of a number of procedures (written in C) processing spline curves and surfaces. All figures in this book (except 6.8 and the author's photo) have been drawn or rendered using this package.

Index